ANNOTATED INSTRUCTOR'S EDITION

9th Edition

English for Careers

Business, Professional, and Technical

Leila R. Smith

with contributions from
Barbara Moran, Ph.D.

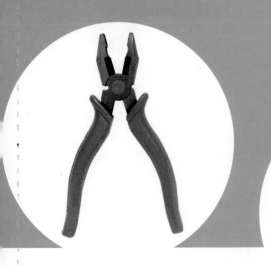

PEARSON

Prentice
Hall

Upper Saddle River, New Jersey 07458

RECOGNIZING LANGUAGE PARTS

The nuts and bolts highlight the easy ways to identify the language elements from individual words to groups of words.

WORD TO THE WISE

This mortarboard will alert you to take special care with a certain aspect of English.

Word Power

This key will point out interesting facts about words or the roots of the English language.

Writing for Your Career

The person at the computer indicates important advice about effective writing styles for business careers.

THE GLOBAL MARKETPLACE

The globe will precede interesting facts for communicating internationally and in a diverse workplace.

> **When you see a sentence in a screened box like this one, you will find a rule or definition that will help you gain control of English for your career.**

Dear Instructor

An English text for workplace communication should enable you to teach language skills nobody notices. When an employee or job applicant makes grammar or usage errors or otherwise communicates poorly, better educated co-workers notice the gaffes and may be distracted from the intended message. However, if English is correctly and effectively used, hardly anyone notices—since the listener or reader is thinking about the content. A negative objective for students could be: Upon completing this business English course, the student will have eliminated grammar, vocabulary, and usage errors that distract from the message—and will use English that nobody notices!

Eight editions of *English for Careers* have helped instructors teach language skills nobody notices. In the Ninth Edition, we retain the instructional innovations that have worked well, eliminate what is no longer needed, and add what today's students do need. Students, instructors, and businesspeople in the United States, Canada, and other English-speaking countries have let us know how and why *English for Careers* enables their students to improve English for their careers:

USER-FRIENDLY STYLE + PERSONALITY AND WARMTH

The warm, friendly approach to learning what for many had been a dreaded subject made the first *English for Careers* (1976) a happy change for students and instructors. The Ninth Edition continues the user-friendly, but updated, style to balance the pervasiveness of technology in our lives.

FOCUS ON REAL-WORLD USAGE

The *English for Careers* course encourages students to communicate in Standard English for their careers as well as personal lives. They will avoid the noticeable English errors by learning and practicing correct and effective English language style of successful business and professional people. *English for Careers* students will learn and then practice:

- Correct sentence structure, grammar, punctuation, spelling, word choice, vocabulary, pronunciation
- Correct and current format and style for letters, short reports, faxes, e-mail, memos, multimedia presentations, and oral presentations
- Proofreading, proofreading, and proofreading

With sample sentences and other devices, students learn about conventional courtesies in the workplace as well as attitudes that help in career growth. In addition, they develop an awareness of the diversity in today's workplace and the importance of today's international marketplace to most careers. There, but de-emphasized and made interesting, are the structure and grammar of the English language.

The focus, however, is clearly on real-world English skills that contribute to good workplace communication as well as social communication.

WHAT'S NEW IN THE NINTH EDITION?

- The 14 chapters are arranged in a different order so that students can start with what probably interests them most—vocabulary improvement.

- Grammar rules, parts of speech, etc.—which discourage many students—begin in Chapter 4 after students have had some good experiences with vocabulary and dictionary practice.

- Chapter 13 starts with reviews of English basics and then introduces more sophisticated concepts of sentence construction.

- Added to the updated Chapter 14 is an introduction to techniques of giving oral presentations at work.

- Each chapter opens with a famous person's quotation relevant to the chapter's topic. Instructors might conduct a brief class discussion about the quotation or the author or simply have students write a short paragraph about it.

- Updated international boxes ending each chapter now also include how to say "I don't understand" in the language of that box. This addition can be quite useful to international business travelers.

- A new dictionary is recommended: *Microsoft Encarta College Dictionary*, which has been billed as "The First Dictionary for the Internet Age."

- For Chinese-speaking students, a new version of *English for Careers* is available with explanations in Chinese for business English instruction.

Finally, I quote from the Foreword of the Fourth Edition of *Webster's New World College Dictionary*:

> As society changes, adapting to technological innovation and cultural shifts, language changes along with it. New words are coined, existing words take on new meanings, old grammar rules gently fade away and are replaced with new language styles, pronunciations change, words shift in tone—all part of the continuing process by which a language maintains its vigor and usefulness.

As business English instructors, we should be prepared to accept the new and not try to hang on to yesterday's English styles. Fortunately, it's a gradual process.

The Ninth Edition adapts the tried and true features of *English for Careers* to current workplace needs, to our changing student population including young men and women from all over the world, to the potential for online courses, and to the early 21st century culture in general.

Workplace ethics, etiquette, human relations, technology, and the global marketplace are still essential topics for all students. These subjects are included and updated within sample sentences, practice exercises, memo and letter contents, proofreading exercises, and other instructional exercises. Students absorb a wide range of the latest information and vocabulary and acquire desirable workplace attitudes while improving their English skills. Crowded curricula often don't allow extra time for these peripheral, but essential, topics.

SUPPLEMENTS FOR THE INSTRUCTOR

- **Annotated Instructor's Edition, 9th edition** (ISBN 0–13–192130–4)
  Instructor's copy of the Student Textbook with all the answers in place and margin notes right where you can use them as teaching tips.

- **Instructor's Resource Kit with Test Item File and PowerPoint Slides, 9th edition** (ISBN 0–13–119980–3)
  Instructor's Manual and a CD-ROM of essential information you need to manage a successful program in Business English.

General Information About *English for Careers*
Classroom, lab, or individualized instruction management: teaching suggestions; testing; grading methods; motivational activities; sample syllabi; and more—with detailed Contents page to save you time.

Complete Testing Program
Updated and extended test item file includes copy of all questions now available on computerized testing program, Test Gen.

Supplementary Practice
Assignments for each chapter; these may be used instead of objective tests if desired. They are interesting and fun to do.

Answers and Solutions
Available for all textbook and *Instructor's Resource Kit* assignments, tests, other handouts, and Proofreading Supplement in the *Student Prep Book.*

Instructor's CD-ROM features include:
Fully automated **PowerPoint Slide Show** that can be offered in the classroom or over the Internet for Distance Learning. More than 140 slides will run independently or can be adjusted to suit your specific needs. All slides are coordinated with the text so that you can accompany any teaching situations with lively visuals.

The contents of the *Instructor's Resource Kit* in **word processing files** to allow easy access and quick adjustment to suit your specific audience.

- **Test Gen** (ISBN 0–13–192131–2)
  Software program for designing, managing, and generating tests or practice exercises, based on printed tests in the *Instructor's Resource Kit.*

- **Ask the Author**
  If you have questions or comments about *English for Careers* or about English for careers, get in touch with the author. She will be pleased to hear from you and to respond.

ASK THE AUTHOR

Voice 310.377.5293   **Fax** 310.533.2739   **E-mail EngLRSmith@aol.com**
Postal 30711 Ganado Drive, Rancho Palos Verdes, CA 90275

Call your Prentice Hall representative or 1.800.526.0485 to request the supplements you wish to use. If you don't know who your representative is or how to reach him or her, call 1.800.526.0485.

SUPPLEMENTS FOR THE STUDENT

- **Student Prep Book, 9th edition** (ISBN 0–13–171115–6)
  This free, 100+ page, extra-practice book is an option for every *English for Careers* student when you order the text and Prep Book package through your bookstore. The book provides additional practice in the form of quizzes and proofreading exercises coordinated with the text. Answers are in the back of the Student Prep Book.

- **Companion Website with PowerPoint Tutorials on CD-ROM, 9th edition**
  Packaged free with every textbook, the Companion Website offers additional questions for reinforcement and practice. Questions are available in a variety of formats, including multiple choice, true/false, and more. Students receive

automatic feedback for incorrect answers. The PowerPoint tutorials for each chapter provide an additional resource for students.

- **Companion Website, 9th edition** (ISBN 0–13–119979–X)
  http://www.prenhall.com/business_studies
  Online course support includes expanded multiple choice, true/false, and essay questions with immediate feedback for each chapter; links to a multitude of writing and research resources; and instructor-only access to downloadable files for all instructor resources. *English for Careers,* 9th edition online resources provide all you need when offering a distance learning course.

- **Distance Learning Courses, 9th edition**
  Contact your Prentice Hall representative for information regarding distance learning courses.

LEARNING, RETENTION, AND CARRYOVER

The warm, relaxed, and amusing style focusing on real-world English skills results in high-level learning and retention. Our objective is for students to apply these principles to on-the-job communications. Carryover more readily results with a text that encourages students to interact with it frequently. If *English for Careers* is new to you, glance through the Contents; then thumb through a chapter to see how the format leads to learning, retention, and carryover.

Contributing to learning, retention, and carryover are color coding, layout, graphics, and art. Design features synchronized with current media and technology help put students at ease about learning this subject. Studies show improved student progress with lively, colorful, and creatively designed texts and other media.

Organization of each of the 14 chapters is explained in the "Dear Student" preface on page xi. I suggest you encourage your students to read it. Of course, the magenta answers are only in your *Annotated Instructor's Edition*—for your convenience.

CHALLENGE AND SUPPORT FOR ALL STUDENTS

Challenging, interesting, and supportive assignments are provided for students at all levels of ability from those with serious language skill deficiencies or English-as-a-second language problems, to businesspeople with advanced degrees. The wide variety of instructional materials in the text and supplements enables you to teach to where your students are now, not where others think they should be.

Many entering students have less knowledge of good English than in previous years. Moreover, too many students are unwilling or unable to spend enough time on study or homework. In other cases, fewer hours are available for the class. To deal with these realities, select what will be most useful to your students from the wide variety of projects in each chapter of the text, your *Instructor's Resource Kit,* and the various other supplements in book or electronic format.

A TEACHER'S ROLE

"The role of the teacher is to provide victories for students," wrote Roman educator Quintilian. I view this text as my way of supporting your role. I appreciate and enjoy hearing from you with questions or comments and am pleased to respond. E-mail is the most efficient way for us to communicate.

HANDY PROOFREADING REFERENCE GUIDE

Proofreader's Marks

Proofreader's Mark	What It Means	How to Use It	Corrected Version
ᵍ	Delete or omit	beginn	begin
∧	Insert	occurᵣence	occurrence
∽	Transpose	revelant, decied	relevant, decide
STET ∘∘∘	Retain crossed-out characters with a dot underneath	if you, Harry, and I go	if you and I go
#	Insert space	fountainpen	fountain pen
⌒	Close up space	stock holder	stockholder
ℙ	Start a new paragraph	days. We are ready	days. We are ready
⌐	Move left	⌐ Dear Ms. Adams:	Dear Ms. Adams:
⌐┐	Move right	Sincerely,	Sincerely,
/	Change capital letter to lower case	the Advertising Budget	the advertising budget
≡	Change lower case letter to capital	new year's eve	New Year's Eve
SP	Spell out	5 days in NYC	five days in New York City
SS	Single space	This plan is under consideration now.	This plan is under consideration now.
DS	Double space	This plan is under consideration now.	This plan is under consideration now.
↻	Move as shown	This is $25 only.	This is only $25.
~	Run in; no new line	four years. We'll be	four years. We'll be
⌐ ┐	Center	⌐memo┐	memo

For examination copy requests, fill out the following form and send to Melissa Orsborn via fax or to the mailing address below.

Professor: _____ School: _____

Address: _____

City/State/Zip: _____

Phone: (____) _____ Fax: (___) _____ E-mail: _____

Courses You Teach	Book You're Using	Yearly Enrollment	Next Decision Date for Text

Please send this completed form to:

Melissa Orsborn
Prentice Hall
445 Hutchinson Ave., Fourth Floor
Columbus, OH 43235

Or Fax to : (614) 841-3701
E-mail: melissa_orsborn@prenhall.com
Phone: 1-800-228-7854 ext. 3622

English for Careers

Business, Professional, and Technical

Leila R. Smith

with contributions from
Barbara Moran, Ph.D.

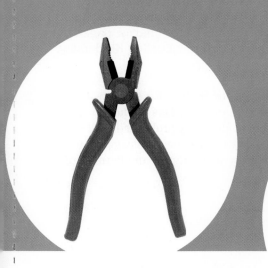

PEARSON

Prentice
Hall

Upper Saddle River, New Jersey 07458

Library of Congress Cataloging-in-Publication Data

Smith, Leila R., 1928–
    English for careers : business, professional, and technical / Leila R. Smith.— 9th ed.
        p. cm.
    Includes index.
    ISBN 0-13-118386-9
  1. English language—Business English—Problems, exercises, etc. 2. English language—
Technical English—Problems, exercises, etc. I. Title.
    PE1115.S62 2004
    428.2'024' 65—dc22   2004028761

Director of Production & Manufacturing: Bruce Johnson
Editorial Assistant: Cyrenne Bolt de Freitas
Marketing Manager: Leigh Ann Sims
Managing Editor—Production: Mary Carnis
Manufacturing Buyer: Ilene Sanford
Production Liaison: Denise Brown
Full-Service Production: Lori Dalberg/Carlisle Publishers Services
Composition: Carlisle Communications, Ltd.
Design Director: Cheryl Asherman
Design Coordinator and Interior Design: Mary E. Siener
Cover Design: Wanda España
Director, Image Resource Center: Melinda Reo
Manager, Rights and Permissions: Zina Arabia
Interior Image Specialist: Beth Brenzel
Cover Image Specialist: Karen Sanatar
Image Permission Coordinator: Robert Farrell
Cover Printer: Phoenix Color
Printer/Binder: Banta/Menasha

Interior Images: page 1, Ron Chapple, Getty Images, Inc.; page 17, © Dorling Kindersley; page 38, Freudenthal Verhagen/Stone/Getty Images; page 60, Andy Crawford, © Dorling Kindersley; page 86, Getty Images, Inc.; page 110, Getty Images, Inc.; page 137, Mike Powell/Stone/Getty Images; page 204, © Dorling Kindersley; page 228, Photodisc/Getty Images; page 300, Steve Gorton, ©Dorling Kindersley.

Pearson Education Ltd.   Pearson Education Australia PTY, Limited
Pearson Education Singapore, Pte. Ltd.   Pearson Education North Asia Ltd.
Pearson Education Canada, Ltd.   Pearson Educacíon de Mexico, S.A. de C.V.
Pearson Education—Japan   Pearson Education Malaysia, Pte. Ltd.

10 9 8 7 6 5 4 3 2
ISBN 0-13-118386-9
(Annotated Instructor's Edition) ISBN 0-13-192130-4

To Seymour, Eric, Alice, Roberta, Nina Beth, Sheela Danielle, Sarala Rose, Sean Suresh, Jonathan

"A word fitly spoken is like apples of gold in settings of silver."
—The Bible

Contents

Chapter 1

Word Power
Enlarge Your Business Vocabulary   1
Read, Recap, Replay

Chapter 2

Weather or Knot
Choose the Right Words   17
Read, Recap, Replay

Chapter 6

Be Kind to Substitutes
Pronouns Substitute for Nouns 110

Read, Recap, and Replay

Chapter 7

Looking for the Action?
Then Find the Verbs! 137

Read, Recap, and Replay

Chapter 8

Words That Describe
Describe with Adjectives and Adverbs 164

Read, Recap, and Replay

Chapter 9

The Taming of the Apostrophe

Avoid Apostrophe Catastrophe 187

Read, Recap, and Replay

Chapter 10

Secret Life of a Sentence Revealed

Say No to Blunders and Gaffes 204

Read, Recap, and Replay

Chapter 11

The Pauses That Refresh

Use Commas, Exclamation Marks, Periods, Question Marks 228

Read, Recap, and Replay

Chapter 12

Punctuation Potpourri
Use ! . ? , as well as (; : "- –') 253

Read, Recap, and Replay

Chapter 13

Get Your Act Together
Write and Speak Clearly, Correctly, Logically, and Concisely 275

Read, Recap, and Replay

Dear Student

Despite learning the subject matter and advanced technology of your chosen field, you may not get the job you want if your communication ability is inadequate. Someone who does get the job may lose it when it's discovered that the employee can't spell or write a clear, correct sentence. Others may keep their jobs but be stuck at a dead end, unable to advance to meaningful careers because of poor or mediocre oral or written communication.

THE EMPEROR'S GRAMMAR

In the year 1414, Sigismund, Emperor of the Holy Roman Empire, said (in Latin) to an important church official who had objected to His Majesty's grammar: "Ego sum rex Romanus et supra grammaticam." (I am the Roman king and am above grammar.) If you are a Roman king, don't bother to read on. For the rest of us, the language we use, both spoken and written, significantly affects our ability to earn a good living, advance in a career, and even enjoy good social contacts.

THE FRESH START

Most of us enjoy the feeling of making a fresh start when we begin a new course. The new textbook, perhaps a new CD-ROM and a new notebook, fresh pencils, and a pen that writes as though it will never run out—these all contribute to the enthusiasm and the resolve to do well. You can turn this fresh start into a successful experience that you enjoy and that will help your career.

THE LANGUAGE OF CAREERS

What kind of language does a business, professional, or technical career require? "Career English" is not a special or separate language. It is the language of network television newscasters and is often called Standard English. It includes English principles you already know, those you learned in the past and forgot, and those you wish you had learned.

DIFFERENT STROKES FOR DIFFERENT FOLKS

> We all use several language styles to help communicate successfully with different people in various situations.

- **To a Child:** Imagine talking with a group of adults at a party; now picture yourself warning a young child away from a hot stove. Think about how your communication style would differ.

- **To Friends and Family:** Perhaps you use slang or a regional or ethnic dialect in everyday conversation with certain friends and family. You might use a different communication style with other friends or acquaintances. We all vary our communication style with the circumstances.

- **At Work:** The style essential for success in business, professional, and technical careers as well as in many personal relationships is called Standard English.
- **Standard English:** With a good command of *English for Careers,* you will communicate confidently and correctly in Standard English with business and professional colleagues.

THE SYSTEM

> You learn only the Standard English usage principles needed by adults to communicate successfully and confidently in the workplace.

- I've excluded or simplified certain grammar terms and rules. In *English for Careers,* you focus only on principles needed for oral and written communication in today's Standard English, as used by well-informed and well-educated people. Most adults need instruction in Standard English for careers to be sure of being right.
- The information you need is presented in an interesting and amusing way that makes learning efficient.

ENGLISH FOR CAREERS IS DIFFERENT

> This book is different. You don't browse through it. You don't read it like other books. What you DO is learn your way through it!

CHAPTER ORGANIZATION

STARTING PAGES
Each of the 14 chapters has starting pages that include a painting of people at work, the chapter's objectives, and a quotation from a famous person about that chapter's topic. Then the chapter introduction tells you exactly what skills and knowledge you should expect to acquire by the time you complete the chapter.

READ, RECAP, AND REPLAY
Next come unique learning steps called *Read, Recap, and Replay.* When you **Read,** you get information in small portions. These short learning units are more efficient than longer ones, and you enjoy a feeling of accomplishment as you complete each portion. After each short portion, you apply what you just read by doing a **Recap.** After another recap or two, you verify that you've learned by responding to the **Replay** questions.

CHECK ANSWERS
As soon as you complete a Recap or a Replay, check your answers in Appendix E, beginning on page 385. Write answers with a blue or black pen; then use a different color pen to show corrections. When ready to review, you'll know which ones, if any, you originally had wrong. Then if you have a question, you can reread that text portion or ask your instructor.

CLOSING PAGES
After several Reads, Recaps, and Replays, each chapter concludes with the **Checkpoint** (which usually summarizes the chapter), **Special Assignment, Proofreading for Careers,** and a **Practice Quiz.** A Global Marketplace box ends the chapter.

Studying the Checkpoint and taking the Practice Quiz are minimum essentials for the closing pages of each chapter. Depending on time and students' needs, your instructor might also assign the Proofreading, one or more Special Assignments, and/or other chapter-related practice on CD-ROM, online, or on paper.

THE STEPS WORK

> The recommended learning steps result in student success; skipping steps results in lower achievement. So please play the game according to the rules: Read before you Recap and before you Replay. Check your answers carefully, and ask about anything not clear to you.

Most students are enthusiastic about this way of learning. However, because doing the Replays is interesting and challenging, some students are tempted to pretest their English knowledge by responding to the questions without reading the explanations and studying the examples. Please resist such shortcuts. By following the recommended steps, you learn more, do better on tests, and end up saving time.

Because of interacting with the textbook so often, you immediately apply what you learn, enabling you to understand it better and remember it. Immediate feedback (with answers in back of the book) is satisfying and encourages you to continue with enthusiasm.

WHAT'S IN IT FOR YOU?

A PROVEN METHOD THAT WORKS

What's in *English for Careers* for you? After successfully completing this textbook, you will enjoy confidence in the correctness and effectiveness of your speech and writing. Good communication skills, more than any other single factor, determine who gets the good job, who keeps it, and who gets the promotion.

SIDE-BY-SIDE LEARNING

While learning *English for Careers,* you also learn more about today's workplace, and you increase or develop a success-oriented attitude. Side-by-side learning happens because many of the sentences illustrating English points deal with business practices, workplace cultural diversity, successful behavior for today's international marketplace, workplace etiquette, and helpful attitudes for self-development.

RESPECT OF CO-WORKERS

You'll find that co-workers and even supervisors will come to you for business English help. They will soon sense that you are the company expert in grammar, punctuation, spelling, and communication style.

FUN AND GAMES

Athough learning isn't all fun and games, people don't learn very much unless they enjoy the experience at least some of the time. You'll find bits of humor hidden in the various exercises; smiling helps us feel better and puts our minds in a learning mode. Enjoy *English for Careers.* With a positive attitude, you'll have some fun along the way. Give it a chance; you'll find your command of English will be a lifelong asset to your career (and personal life too)!

Leila R. Smith

Replay
i

Write your answers in the blanks of this sample Replay. If you don't know an answer, look for the information in the preceding "Dear Student" letter on page xi. Fill in the blank for Question 1 and answer **T** (true) or **F** (false) for the rest of the questions.

1. The language style appropriate for most business, professional, and technical careers is called <u>Standard English</u> .

_F_ 2. It's wrong to use slang when writing a letter to a friend.

_F_ 3. You should be sure to use the same style of language in all communications.

_F_ 4. Standard English is always superior to other types of English.

_F_ 5. To succeed in this course, you must memorize a long list of traditional grammar rules and terms.

_T_ 6. This course includes learning a little about international business.

_T_ 7. The single most important ability required to get a good job or a promotion is communication ability.

_F_ 8. "Side-by-side" learning means it's best to lie on your side while doing a Read and Replay.

_F_ 9. If you look at the answers in the back of the book, you are cheating.

_F_ 10. The most successful students start by first answering the Replay questions and then reading the explanations and examples.

_F_ 11. By completing *English for Careers* carefully, you achieve thorough mastery of traditional grammar terms and rules.

_F_ 12. Look up the answers in the back of the book the day after you complete the Recaps and Replays.

_T_ 13. The Checkpoint near the end of each chapter usually summarizes the chapter.

_F_ 14. The last item in nearly every chapter is Proofreading for Careers.

_T_ 15. After completing this course successfully, you will be confident of your ability to speak and write English in a style that leads to a successful career.

Check your answers below.

How did you do? _____

Answers to Replay i
1. Standard English   2. F   3. F   4. F   5. F   6. T   7. T   8. F   9. F   10. F
11. F   12. F   13. T   14. F   15. T

PRETEST

Write the letter of the correct answer in the blank.

__b__   **1.** The carton of books and papers (a) have been (b) has been (c) were lost.

__d__   **2.** Etymology is the study of (a) insects (b) synonyms and antonyms (c) grammar and word usage (d) the history of words.

__d__   **3.** If the first line of the inside address of a business letter is "Mr. Samuel E. Smith," the preferred salutation is (a) Ladies and Gentlemen: (b) Dear Sir: (c) Dear Sirs: (d) Dear Mr. Smith: (e) Dear Mr. Samuel E. Smith:

__c__   **4.** **Outsource** means to (a) import needed supplies for manufacturing (b) export high technology products (c) use services of workers who are not employees of your organization (d) wastefully use valuable natural resources (e) help laid-off employees find new jobs.

__c__   **5.** Such a (a) phenomena (b) phenomenae (c) phenomenon (d) phenomeni (e) phenomenae has never before occurred.

__a__   **6.** George's wife is the (a) president of the company (b) President of the Company (c) President of the company (d) president of the Company.

__F__   **7.** When preparing slide presentations, the words are relatively unimportant. (True or False)

__c__   **8.** After three (a) year's (b) years' (c) years of being on this merry-go-round, George decided to get off and change his ways.

__T__   **9.** When you're laying out material for a Web site, place the most interesting and important information at or near the beginning. (True or False)

__b__   **10.** Please give the reports to Frank Hitt and (a) I (b) me (c) myself.

__a__   **11.** Ms. Denova is the one (a) who (b) whom we believe danced the hoochie koochie.

__a__   **12.** If you had (a) gone (b) went to work today, you would have seen the sunset.

__c__   **13.** Which would be (a) easyer (b) easiest (c) easier (d) easyest for you to prepare, a letter or a short report?

__b__   **14.** We hope to receive (a) a (b) an 18 percent discount.

__d__   **15.** After working at the computer all day, his eyes were tired.
The preceding sentence has a (a) comma splice (b) lack of parallel construction (c) misplaced part (d) dangling verbal (e) lack of subject/verb agreement.

For items 16–20, write **C** in the blank if all the punctuation is correct; otherwise, correct the punctuation.

__C__   **16.** "Telecommuting," he said, "is good if you are disciplined."

_____   **17.** "Money, beauty, intelligence, and charm—she has them all," said George's friend Jesse.

_____   **18.** Mr. Crane is not here; however, I can help you.

__C__   **19.** We've mailed you a copy of the new book we told you about and hope it will reach you before the end of the month.

_____   **20.** Greet your clients by name; then welcome them with a smile.

After you find out how many you answered correctly, read your "fortune."

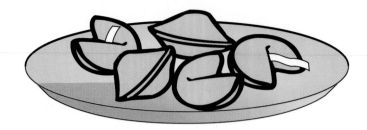

Number Right _____

18–20 You have a good command of English for your career. This course will serve as a brushup, and you will become an expert.

15–17 Your English skill is fair. After the practice provided in this course, you'll have excellent skill in the English required for careers.

12–14 Your English for your career needs improvement. Because your basic language skill is all right, you'll enjoy the rapid learning that will result from completing all assignments.

0–11 You came to the right place! Taking this course is a wise decision. Because you are now motivated to learn English for your career, you'll capture those principles that escaped you in the past.

About the Author

Leila R. Smith, Los Angeles Harbor College Professor Emeritus of Office Administration, has a New York University Bachelor of Science degree in business education and a University of San Francisco Master's degree in education. In addition to Harbor College, Professor Smith taught at Bay Path College in Massachusetts, in California's Pierce and Valley Colleges, and in New York City proprietary schools. Among her many professional activities, she has been a Fulbright exchange instructor, teaching English and communication in the business department of City and East London College in London, England, for an academic year.

A federal grant enabled her to study methods of applying brain research to business English instruction, as reflected in the unique teaching/learning styles of this text. This study also culminated in the writing of the text *RSVP—Relaxation, Spelling, Vocabulary, Pronunciation.* Other publications include the texts *Communication and English for Careers* and *Basic English for Business and Technical Careers,* as well as professional newsletters and articles in professional journals.

She has served as Communication Editor for the *Business Education Forum,* the journal of the National Business Education Association. Professor Smith, a recipient of the Pimentel Award for Excellence in Education, has conducted workshops and seminars on business English and communication and on teaching and learning methods for educators, corporate groups, and government agencies and has worked in various business capacities.

Acknowledgments

The author and editors of *English for Careers,* 9th edition, thank you for your assistance:

Professor Elaine Johnson—New Mexico State University
Professor Connie Jo Clark—Lane Community College

A finely crafted book can result only with the assistance of a talented and dedicated publishing staff. Thank you, Prentice Hall, especially Elizabeth Sugg, Deborah Hoffman, Cyrenne Bolt de Freitas, Denise Brown, and Mary Siener. A special thanks is also extended to Lori Dalberg and the staff at Carlisle Communications.

Word Power

Enlarge Your
Business Vocabulary

> "Words have weight, sound, and appearance; it is only by considering these that you can write a sentence that is good to look at and good to listen to."
>
> —*W. Somerset Maugham*

After Completing Chapter 1, You Will

> Have enlarged your business vocabulary—spelling, pronunciation, and meaning.

Everyone involved in or planning a business, professional, or technical career needs a good current vocabulary. In fact, even if you are independently wealthy and don't want a career, you need to be familiar with Chapter 1 words. An extensive vocabulary pays off in job and social success, as well as in managing personal affairs.

Regardless of the job we have, we're involved in business, professional, and technical matters all our lives: We make decisions about banking, home ownership, insurance, attorneys' services, credit, voting, investing, and electronic devices. It is difficult to obtain information and advice about these matters without appropriate vocabulary.

You probably already know many of these Chapter 1 words. In that case, make sure you spell and pronounce them correctly. If some words are new to you, become sufficiently familiar with them so that you'll understand them in print or conversation. Obviously one chapter will not provide you with a complete vocabulary. Continue to increase your workplace vocabulary through reading business and technical news in the newspaper, a weekly news magazine, and information from the Internet.

If some of the following words are unfamiliar, prepare index cards for ease of study. Write the word on one side of a card and the definition on the other. Carry the cards with you and test yourself frequently. Look at the definition side and try to think of the word and vice versa. For some of the words, not all definitions are given here. For example, look up *audit,* Word 12. You'll find several definitions in your dictionary besides the one given in this chapter.

See *Instructor's Resource Kit* for Chapter 1 Pretest.

You've come to the right place—if you want to eliminate the most embarrassing mistakes from your speech and writing. Mastery of the English language and the ability to communicate effectively opens many doors to you. One's future is often determined by skill in communicating.

MEMO FROM THE WORDSMITH

In this actual printed invitation, I've changed the location names to keep the misspellers anonymous. Possibly it was a belly dancing party?? Circle the error.

United States Navy League

Cordially Invites You to Participate in the

Anchors Aweigh Ceremony

Mayville Navel Station

Read 1

ACCOUNT EXECUTIVE THROUGH BROKERAGE

1. **account executive**   Person who manages a customer's account in a service business such as an advertising agency or a financial services organization.
2. **affidavit**   Written statement sworn to before a notary public—a person authorized by law to administer oaths.
3. **amalgamation**   Joining of two or more businesses into a single body; also called a **merger.**
4. **AMEX (American Stock Exchange)**   Second largest stock exchange in the United States; located in New York City.
5. **annual report**   Printed annual message to stockholders providing information about the progress of the corporation.
6. **antitrust legislation**   Laws against monopoly-type business practices that result in a business making unfair profits.

7. **appreciate**   To increase in value.

8. **APR**   Annual percentage rate—what a borrower must pay in interest.

9. **arbitration**   A third party appointed to settle a dispute between two groups.

10. **arrears**   An amount overdue and still unpaid is "in arrears."

11. **attachment**   Court order authorizing seizure of property for failure to meet obligations.

12. **audit**   Examination of financial records of a business to determine correctness; such an examination is made by an **auditor.**

13. **balance sheet**   Statement of assets, liabilities, and net worth as of a certain date.

14. **bar code**   A code, imprinted on consumer goods and mail, that can be read by a computerized scanner designed for that purpose.

15. (a) **bear market**   The stock market when prices are declining, and many stockholders are selling their stock because they think prices will continue to decrease. These sellers are called *bears.*
    (b) **bull market**   The stock market when prices are increasing, and many stockholders are buying stock because they expect prices to continue to increase. These buyers are called *bulls.*

16. **beneficiary**   A person designated to receive benefits from an insurance policy, a will, or a trust fund.

17. **bid**   An offer to buy or sell services or goods at a certain price.

18. **bill of lading**   A form made out by a transportation company and issued to the shipper as a receipt listing goods to be shipped.

19. **blogger**   Somebody who creates or runs a Web log (called a **blog**).

20. **brokerage**   A business that buys and sells stocks and bonds for its clients.

Blue Chips: stocks of the largest and most profitable corporations. This term originated from the most valuable chips (the blue ones) in poker games.

Word Power

Sign on a hospital door in a West Coast city: NO BEAR FEET ALLOWED.

Replay 1

Insert the appropriate number in the blank.

1. account executive	2. affidavit	3. amalgamation
4. AMEX	5. annual report	6. antitrust legislation
7. appreciate	8. APR	9. arbitration
10. arrears	11. attachment	12. audit
13. balance sheet	14. bar code	15. bear market
16. beneficiary	17. bid	18. bill of lading
19. blogger	20. brokerage	

a. ___18___ The form made out by a shipping company listing goods to be shipped.

b. ___14___ This code is imprinted on consumer goods and mail; it is read by a computerized scanner.

c. ___15___ A pessimistic stock market. Stock prices are declining, and stockholders are selling.

d. ___8___ Percentage of interest to be paid on one's debts.

e. ___11___ A court order authorizing seizure of property for failure to meet obligations.

f. ___4___ Second largest stock market in the United States.

g. ___7___ To increase in value.

h. ___2___ A written statement sworn to before a person authorized to administer oaths.

i. ___10___ A debt that is overdue and still unpaid.

j. ___6___ Laws against monopoly-type business practices.

k. ___9___ A third party appointed to settle a dispute.

l. ___16___ One who receives benefits from an insurance policy, a will, or a trust fund.

m. ___12___ Examination of business records to determine their correctness.

n. ___13___ Statement of assets, liabilities, and net worth as of a certain date.

o. ___3___ Joining two or more organizations into a single group.

p. ___19___ Slang expression for someone who creates Web sites.

q. ___1___ Person who manages a customer's account in an advertising agency or a financial services organization.

r. ___20___ Business that buys and sell stocks and bonds for its clients.

s. ___5___ Annual message to stockholders about the corporation's progress.

t. ___17___ An offer to buy or sell services at a certain price.

Answers are on page 385.

Read 2

CERTIFICATE OF DEPOSIT THROUGH EXCHANGE RATE

Money trivia: The only American coin not honoring a past U.S. president is the $1 coin honoring women-suffragist Susan B. Anthony. The average life of a $1 bill is 13 to 18 months before it wears out.

21. **(CD) certificate of deposit** A written message from a bank to a depositor showing the percentage of interest to be paid during a specified time period.

22. **certified check** Personal or company check drawn on funds made available by the depositor and therefore guaranteed by the bank to be valid.

23. **COLA** Cost-of-living adjustment for wages or Social Security payments.

24. **collateral** Borrower's property held by a lender as security for payment of a loan.

25. **concierge** (pronounced *con.SYERJH*) A person employed at a hotel or an apartment building to help the residents in various ways.

26. **consumer price index (CPI)** A monthly survey of changes in consumer prices; it is used to measure inflation.

27. **copyright** Legal protection of documents, computer software, films, and other creative works of authors, composers, artists, etc.

28. **corporation** Artificial "person" created by law, operating under a charter granted by a state and authorized to do business under its own name.

29. **cyberlaw** Laws relating to computers, information systems, and networks.

30. **deficit**   Shortage of money; the opposite of *surplus.*
31. **depreciation**   Decline in value of property because of wear and age; the opposite of appreciation.
32. **direct mail**   Advertisements mailed directly to homes or businesses.
33. **diversification**   To expand a commercial organization and make it more varied by engaging in additional or different areas of business.
34. **dividend**   A payment to a stockholder of a portion of the corporation's profits.
35. **dotgov**   An informal term for a government official.
36. **Dow Jones Industrial Average**   The daily average of the closing prices of specific stocks on the New York Stock Exchange. This figure is widely publicized and indicates current stock market trends.
37. **down time**   Time during which equipment is unable to be used until it has been adjusted or repaired.
38. **e-blocker**   Employer who uses special software to prevent employees from visiting certain Web sites while at work.
39. **e-signature**   Electronic signature.
40. **ethics**   Moral principles—such as the loyalty, honesty, and integrity required of employees in business and professions.

Replay 2

Insert the appropriate number in the blank.

21. CD—certificate of deposit	22. certified check
23. COLA	24. collateral
25. concierge	26. CPI
27. copyright	28. corporation
29. cyberlaw	30. deficit
31. depreciation	32. direct mail
33. diversification	34. dividend
35. dotgov	36. Dow Jones
37. down time	38. e-blocker
39. e-signature	40. ethics

a. ___21___ Written acknowledgment to a depositor showing the interest the bank will pay.

b. ___22___ Check guaranteed by a bank to be worth the amount for which the depositor wrote it.

c. ___27___ Legal protection of creative work produced by authors, composers, artists, etc.

d. ___30___ Shortage of money; opposite of surplus.

e. ___38___ Special software preventing employees from visiting certain Web sites while at work.

f. ___23___ Cost-of-living adjustment for salaries and Social Security payments.

g. ___39___ Electronic signature.

h. ___34___ Money paid by a company to a stockholder.

i. ___40___ Moral principles required of business and professional employees.

j. ___31___ Decline in value of property.

k. ___37___ Time during which equipment cannot be used because it needs repair.

l. ___36___ Daily average of closing prices of 30 stocks traded on the New York Stock Exchange.

m. ___26___ consumer price index.

n. ___32___ Advertisements mailed directly to homes or businesses.

o. ___29___ Laws relating to computers and information systems.

p. ___24___ Borrower's property held by a lender until debt is paid.

q. ___28___ "Artificial person" created by law and acting under a state-granted charter.

r. ___25___ A hotel employee who helps guests in a variety of ways.

s. ___33___ A business organization that engages in various areas of business.

Check answers on page 385.

WORD TO THE WISE

Like it or not, using *like* in every other sentence makes young women sound uneducated and unbusinesslike. The *like* habit has become known as "Valley Girl" slang (for the San Fernando Valley, a Los Angeles suburb; however, it's used all over the United States by some high school and college-age women). Avoid this expression when interviewing for a job, scholarship, or college acceptance.

Read 3

Global Marketplace: Other than in the United States, bills of various denominations usually have different colors and/or sizes, making it easy to quickly distinguish one from another. Tourists and businesspeople doing business in the United States often are confused because all currency is the same size and color; they wonder whether Americans mistake one bill for another.

EURO THROUGH LIABILITIES

41. **euro** The currency unit used in many of the European Union countries.
42. **exchange rate** Price of one country's money in relation to the price of another's; for example, the rate at which pesos, euros, francs, yen, rupees, etc., can be exchanged for dollars.
43. **exemption** A specified amount of money not subject to taxation, such as money used to support a dependent.
44. **FAQ** Frequently asked questions.
45. **fiber media** Media that use paper—rather than online publishing.
46. **financial planner** A specially trained and licensed stockbroker who advises clients how to handle their financial assets in the most effective manner.
47. **fiscal year** A period of 12 months between one annual balancing of accounts and another. (An annual report is prepared at the end of a fiscal year.)
48. **foreclosure** A lender taking over property when a debtor is not making payments on the mortgage.
49. **401k** A retirement plan for employees.
50. **fringe benefits or fringes** Paid vacations, insurance coverage, pension plans, part-time college tuition fees, etc., given to employees in addition to salary or wages.
51. **Ginnie Mae** Bond sold by the Government National Mortgage Association.
52. **glass ceiling** Perceived or actual invisible barrier that prevents career advancement beyond a certain level for reasons of gender, age, race, ethnicity, or sexual preference.
53. **graphics** Visual display of data such as graphs, charts, and diagrams.

54. **gross**   The total, such as gross income, gross profit, gross weight; 12 dozen.
55. **hacker**   A person who disturbs other people's computer data.
56. **infomercial**   A long TV commercial made to seem like an informative talk show (information + commercials).
57. **insolvent**   Unable to pay one's debts (broke—slang).
58. **Internet hotel**   A place where a great many computers are made available to businesses.
59. **itinerary**   A written travel plan showing arrival and departure dates and times, as well as hotels where reservations have been made.
60. **liabilities**   The obligations or debts of a business or an individual.

Replay
3

Insert the appropriate number in the blank.

41. euro	**48.** foreclosure	**55.** hacker
42. exchange rate	**49.** 401k	**56.** infomercial
43. exemption	**50.** fringe benefits	**57.** insolvent
44. FAQ	**51.** Ginnie Mae	**58.** Internet hotel
45. fiber media	**52.** glass ceiling	**59.** itinerary
46. financial planner	**53.** graphics	**60.** liabilities
47. fiscal year	**54.** gross	

a. __41__ Currency used in the European Union.

b. __44__ Frequently asked questions.

c. __43__ Sum of money not subject to taxation.

d. __56__ A long TV commercial.

e. __53__ Visual display of data, including graphs and charts.

f. __57__ Unable to pay one's debts.

g. __60__ Debts.

h. __52__ Invisible barrier that usually prevents job promotions.

i. __42__ Price of a country's money in relation to the price of another country's money.

j. __59__ Written travel plan.

k. __48__ Debtor losing property because of inability to pay debts.

l. __50__ Benefits given to employees in addition to salaries.

m. __54__ The total amount.

n. __49__ A retirement plan for employees.

o. __45__ Media that use paper for publishing.

p. __58__ Place where many computers are available for business use.

q. __51__ Bonds sold by federal government.

r. __55__ Person who disturbs other people's electronic data.

s. __47__ Annual report is prepared at the end of this.

t. __46__ Specially trained stockbroker who advises clients on handling financial matters.

Check answers on page 385.

Read
4

LIBEL THROUGH POWER OF ATTORNEY

To distinguish *libel* from *slander*, remember that both *slander* and *speech* begin with *s*.

English uses more words than any other language; German is probably second.

The New York Stock Exchange (NYSE) lists thousands of stocks. Thousands more are listed on the American Stock Exchange (AMEX), NASDAQ, and other regional and foreign exchanges. Unlisted securities are traded on the over-the-counter (OTC) market. London's stock exchange, the oldest in the world, opened in 1773.

61. **libel**   An untrue statement, usually in writing, that injures another's reputation.
62. **lien**   Claim against property preventing the owner from selling it until a debt (such as taxes) is paid.
63. **liquidate**   To close the affairs of a business and sell the assets; to turn assets into cash.
64. **markup**   The difference between the cost price and the selling price in a retail business.
65. **merit rating**   Rating employees by their job performance so that fair decisions may be made about raises, promotions, demotions, etc.
66. **monopoly**   Exclusive control of the supply of a commodity or service.
67. **mortgage**   Pledge of property (real estate) as security for a loan.
68. **multinational corporation**   A company with subsidiaries or branches in many nations.
69. **mutual fund**   An investment in a fund that invests shareholders' money in various stocks or bonds.
70. **NASDAQ**   **N**ational **A**ssociation of **S**ecurities **D**ealers **A**utomated **Q**uotations System—average computed from prices of a number of stocks; this average is announced weekdays along with the Dow Jones average, which is for a different group of stocks.
71. **negotiable instruments**   Documents in which ownership is easily transferred to another person; for example, stock certificates or checks made out to "cash."
72. **net**   Amount remaining after all deductions are made; for example, net profit is the balance after all expenses have been deducted from gross profit.
73. **New York Stock Exchange (NYSE)**   Largest stock exchange in the United States; it's on Wall Street in New York City. If you live in New York or visit, be sure to go to the NYSE and take the tour.
74. **outplacement**   A company's assistance in finding jobs for its employees who have been terminated.
75. **outsource**   Using outside sources for labor, parts, and various services instead of personnel and services within the company.
76. **overhead**   General costs of running a business, such as taxes, rent, heating, lighting, and depreciation of equipment.
77. **per annum**   By the year.
78. **per capita**   By the person.
79. **per diem**   By the day.
80. **power of attorney**   The right granted by one person to another to act for him or her legally; the legal document granting this right.

Replay
4

Insert the appropriate number in the blanks.

61. libel

62. lien

63. liquidate

64. markup

65. merit rating		**73.** New York Stock Exchange
66. monopoly		**74.** outplacement
67. mortgage		**75.** outsource
68. multinational corporation		**76.** overhead
69. mutual fund		**77.** per annum
70. NASDAQ		**78.** per capita
71. negotiable instruments		**79.** per diem
72. net		**80.** power of attorney

a. __61__ An untrue statement, usually published, that injures another's reputation.

b. __62__ A claim against property that prevents the owner from selling it until a debt (such as taxes) is paid.

c. __73__ Largest stock exchange in the United States.

d. __64__ Difference between a retailer's cost price and the selling price.

e. __63__ Turning assets into cash.

f. __79__ By the day.

g. __70__ National Association of Securities Dealers Automated Quotations System

h. __65__ Rating employees by their job performance.

i. __67__ Using property as security for a loan.

j. __72__ Amount remaining after deductions.

k. __74__ Finding jobs for terminated employees.

l. __77__ By the year.

m. __76__ General costs of running a business.

n. __71__ Documents easily transferred to another person or organization.

o. __69__ Fund that invests shareholders' money in various stocks or bonds.

p. __66__ Exclusive control of a commodity or a service.

q. __68__ A company with branches in many nations.

r. __75__ Use of outside sources for labor instead of the company's own employees.

From a **community college instructor**'s proposal for funding sex-equity workshops in colleges: "Staff can be exposed to the sex equity problems through a hands-on involvement."

Check your answers on page 385. Take the Pop Quiz on page 334.

Writing for Your Career

- Do use occasional contractions such as *they're* and *you're* in business writing. An informal, conversational tone is desirable in American business communication. Avoid contractions and use a more formal tone, however, in international communciation.

- Better writers try to avoid beginning sentences with *there*.

  There's a box of disks in the top drawer. IS REWRITTEN A box of disks is in the top drawer.

PROXY THROUGH WYSIWYG

81. **proxy**   A written authorization by a stockholder for someone to vote in his or her place at a stockholders' meeting.
82. **quorum**   The number of members of an organization required to be present to have a formal meeting at which business is transacted.
83. **ream**   500 sheets of paper.
84. **reconciliation**   Bringing into agreement a bank's records with the depositor's records of checks and deposits.
85. **requisition**   A written request, made within an organization, for supplies or equipment.
86. **retainer**   A fee paid to a lawyer or other professional for services to be rendered in the future.
87. **shareholder**   Shareholder and stockholder have the same meaning—one who owns stock (or shares) in a corporation.
88. **slander**   Untrue *spoken* remarks about someone that harm the person's reputation. (Libel refers to untrue *written* remarks.)
89. **solvent**   Having the funds necessary to pay all debts.
90. **spyware**   Software installed on a hard disk without the user's knowledge; it can be used to relay encoded information.
91. **start-up**   A new company that is just beginning to operate as a business.
92. **stock**   Investment in a corporation; stockholders are entitled to vote on various corporate matters and to share in the company's profits—based on the amount of stock they possess.
93. **tariff**   A tax, called a *duty,* on imported items.
94. **turnaround time**   Time elapsed between starting a task and completing it.
95. **turnover**   Change in employees because of people being terminated or quitting; also refers to sale of products that are then replaced by additional merchandise.
96. **viral marketing**   A form of marketing in which customers tend to act as advertisers for a company's products by talking about them and by spreading the word over the Internet (like a virus that spreads rapidly).
97. **voice recognition**   Ability of a computer to accept speech and translate it into characters on the screen.
98. **voxel**   The smallest unit of space in a computer image.
99. **Weblish**   The form of English used online around the world—very informal English characterized by omission of punctuation and capital letters and by use of abbreviations and new words (such as Weblish). Avoid Weblish in business letters, formal meetings, interviews, etc.
100. **Webmaster**   One who organizes or updates information on a Web site.
101. **WYSIWYG**   An acronym for **W**hat **Y**ou **S**ee **I**s **W**hat **Y**ou **G**et.

From college students' papers: pier pressure, designer genes.

Replay

5

Insert the appropriate number in the blank.

81. proxy

82. quorum

83. ream

84. reconciliation

85. requisition

86. retainer

87. shareholder

88. slander

89. solvent

90. spywear	**94.** turnaround time	**98.** voxel
91. start-up	**95.** turnover	**99.** Weblish
92. stock	**96.** viral marketing	**100.** Webmaster
93. tariff	**97.** voice recognition	**101.** WYSIWYG

a. __98__ The smallest unit of space in a computer image.

b. __88__ Untrue spoken remarks about someone that harm the person's reputation.

c. __81__ Written authorization by a stockholder for someone else to vote in his or her place.

d. __87__ One who owns shares in a corporation.

e. __83__ 500 sheets of paper.

f. __86__ Fee paid to a lawyer or other professional for services to be rendered in the future.

g. __82__ The number of members who must be present in order to have a formal meeting.

h. __99__ Very informal English often used online that shouldn't be used for business letters or formal documents.

i. __94__ Time elapsed between starting and completing a task.

j. __101__ What You See Is What You Get.

k. __89__ Having funds needed to pay all debts.

l. __100__ One who organizes or updates Web site information.

m. __92__ Investment in a corporation entitling the investor to vote and to share in profits.

n. __91__ A new company.

o. __95__ Replacement of employees or sale of merchandise to be replaced by additional employees or items.

p. __97__ Computer program enabling the user to translate speech into screen characters.

q. __93__ Tax, often called a *duty,* on imported items.

r. __85__ Written request for supplies made within an organization.

s. __84__ Determining that a bank's records agree with the depositor's records.

t. __90__ Software that can be used illegally and without the user's knowledge.

Check answers on page 385. Take the Pop Quiz on page 334.

Word Power

Workplace Jargon

Bean Counter   Someone who makes business judgments based mostly on numerical calculations; also slang for an accountant

Comp Time   Compensating employees with time off instead of overtime pay

Green   Concerned with ecology such as protecting the environment

SPELLING AND PRONUNCIATION FOR MAVENS

Avoid nonstandard expressions if you want to sound businesslike and educated: lotsa, coulda, shoulda, woulda—when *of* or *have* is required; looking to (use *expecting* or *planning*); had ought to (use *should*); wait on (unless you're a salesperson); try and (use *try to*); irregardless (regardless is correct); the reason is because (*that* instead of *because*); enthused (should be enthusiastic).

A. According to the *Microsoft Encarta College Dictionary,* a maven is an "expert or a knowledgeable enthusiast of something." Be a word maven by reading the sentences below to increase your familiarity with the meaning, spelling, and pronunciation of Chapter 1 highlighted words. If you're unsure of a pronunciation, look it up in your dictionary.

1. Beginning in 2000, 13 **stocks** on the **NYSE** and **AMEX** began to be traded in decimals instead of fractions; that is, if a stock was listed in the newspaper as selling for 141/4, it would now appear as 14.25, meaning $14.25 a share.

2. Our bank's FDIC insured account is ideal as an income **CD.** (See abbreviations in Appendix D.)

3. **WYSIWYG** and **NASDAQ** are acronyms.

4. The **annual report** provides details about our **amalgamation** with the June Company Department Stores.

5. We bought the stock because we expect it to **appreciate,** not **depreciate**.

6. We have a **deficit** because of all the **down time** on the new equipment.

7. The **balance sheet** had already been **audited** before the **exchange rate** changed.

8. His grandsons are the **beneficiaries** of the **mutual funds.**

9. This **corporation** is very sensitive to **cyberlaw**.

10. The **exchange rate** is affected by the value of the **euro.**

11. The advice of a **financial planner** would have prevented the **foreclosure** of his home.

12. The **balance sheet** shows the **liabilities** as well as the assets.

13. The **concierge** is grateful for the **COLA.**

14. He used a **certified check** and a **CD** as **collateral.**

15. One of his **fringes** is a **401k,** and another is a low-interest **mortgage.**

16. **Outplacement** is often helpful when a business is about to be **liquidated.**

17. A **multinational corporation** usually has many **shareholders,** and some may be guilty of **slander** and even **libel.**

18. A **bill of lading** and an **affidavit** are attached to the **beneficiary's itinerary**.

19. **Antitrust legislation** prohibits most **monopolies,** but **amalgamation** is often legal.

20. This **start-up** will not be successful because of the rapid **turnover** of employees, slow **turnaround time** for production, **spyware** installed by a dishonest competitor, and failure of its **viral marketing** plan.

21. Pepsi Cola and Nissan Motors are **multinational corporations** doing business within the **European Union** and are always watchful of the **euro's exchange rate.**

During the 1870s, NYSE members, wanting to appear more respectable, developed a system for fining one another—$10 for standing on a chair, $5 for smoking a cigar, and 50 cents for knocking off another member's hat.

22. The **hacker** was showing off by communicating in **Weblish** at an **Internet hotel.**

23. Since we don't have a **quorum,** the **ream** of costly paper we **requisitioned** isn't needed.

24. The lawyer was given a **retainer** and will file suit for **slander** against the **Webmaster** who holds the **mortgage** on the property.

25. He was also guilty of **libel** for claiming the **financial planner** is a **hacker.**

B. Avoid "behind your back" criticism. These words are frequently mispronounced. To increase confidence in your communication ability, practice saying them correctly.

accessories—ak.SES.uh.rees	mischievous—MIS.chiv.us
affluence—AF.loo.ens	naive—NI.EEV
affluent—AF.loo.ent	nuclear—NOO.cle.er
asked—askt (never: ast)	pageant—PĂJ.nt
debris—de.BREE	picnic—PIK.nik
Des Moines—de.MOYN	picture—pik.cher
genuine—JEN.u.in	preface—PREH.fis
grievous—GREEV.us	preferable—PREF.er.abl
height—hIt	probably—PROB.ub.lee
Illinois—ILL.in.oy	realtor—REE.l.tor
incomparable—in.COM.per.a.ble	realty—REE.l.tee
irrelevant—ir.EL.e.vint	recognize—REC.og.nize
irrevocable—ir.EV.uh.cub.l	similar—SIM.i.ler
Italian—i.TAL.yin	statistics—sta.TIS.tiks
jewelry—JEW.el.ree	subtle—sut.l
library—LI.brer.ee	superfluous—su.PERF.lu.us

Replay
6

A. In the blanks write the correct spelling of the following incorrectly spelled words.

1. amalgemation ___amalgamation___

2. dipreciate ___depreciate___

3. dificet ___deficit___

4. balence sheet ___balance sheet___

5. benificiarys ___beneficiaries___

6. uro ___euro___

7. forclosure ___foreclosure___

8. slandar ___slander___

9. lible ___libel___

10. morgage ___mortgage___

B. Fill in the blanks with a Chapter 1 word that makes sense in the sentence.

11. A _____concierge_____ is usually a hotel employee.

12. A corporation with branches in several countries is known as
_____multinational_____.

13. Payment to a stockholder of a portion of the company's profits is
called a _____dividend_____.

14. Software preventing employees from visiting certain Web sites is
called _____spyware_____.

15. The second largest stock exchange in the United States is the
_____American Stock Exchange (or AMEX)_____.

Check your answers on page 385.

MEMO FROM THE WORDSMITH

CYBER is a combining form meaning *computer*—as in *cyberspace, cybercrime,* or *cybernetics.* If your dictionary is old, you won't find these words and many others that are now in the ever-growing, ever-changing English language.

CHECKPOINT

A. The student I assigned to key this short list of business terms for your review misspelled every one of them. Would you please rewrite them correctly.

1. afidavid ____affidavit____   **2.** lein ____lien____
3. defecit ____deficit____   **4.** exemtion ____exemption____
5. bite ____byte____   **6.** itenirarie ____itinerary____
7. dough jones ____Dow Jones____   **8.** morgaje ____mortgage____
9. tarrif ____tariff____   **10.** kworim ____quorum____
11. Nasdak ____NASDAQ____   **12.** liabilitys ____liabilities____

B. Pronounce correctly all the words in Read 6, Part B.

SPECIAL ASSIGNMENT

A. Please spell out the following abbreviations and acronyms: Use a dictionary when needed.

1. APR — annual percentage rate

2. NASDAQ — National Association of Securities Dealers Automated Quotations

3. e-signature — electronic signature

4. CD — certificate of deposit

5. COLA — cost-of-living adjustment

6. AMEX — American Stock Exchange

7. dotgov — Government official who communicates via the Internet

8. FAQ — Frequently asked questions

9. NYSE — New York Stock Exchange

10. WYSIWYG — What you see is what you get

11. CPI — consumer price index

12. WWW — World Wide Web

13. Ginnie Mae — Government National Mortgage Association

14. URL — Identifies a file located on the Internet—(uniform resource locator)

PROOFREADING FOR CAREERS

Special symbols are often used to mark changes needed or errors found by proofreading. These symbols, called "proofreaders' marks," are shown in college dictionaries, keyboarding textbooks, and reference manuals—and the inside back cover of this book.

Proofreading tests alertness and knowledge.

Proofreading is slower than other types of reading.

Proofreading requires more than one reading.

Proofreading is more than checking for spelling and typographical errors—often called "typos." Don't let anything get past your desk that doesn't make sense to you or is incorrect. Proofread and make corrections in the following:

A. From a classified ad in a Tennessee newspaper:

Telephone receptionist needed for doctors' office. Duties are to relay massages between patients and doctors.

B. Under a supermarket advertisement in a Los Angeles newspaper:

NOT RESPONSIBLE FOR TYPORGRAPHICAL ERRORS

C. An answer from a science student's test that made biology history:

An example of animal breeding is the farmer who mated a bull that provided good meat with a bull cow that gave a great deal of milk.

D. Another newspaper typographical error:

Downtown has its fair share of buried treasurers.

Your computer's spell checker will find only two of the preceding errors. Which ones? typographical, receptionist

Practice Quiz
www.prenhall.com/smith

A. Mark these statements **T** (true) or **F** (false).

EXAMPLE ____F____ AMEX is the stock exchange that is in Chicago.

1. ____T____ A statement of assets, liabilities, and net worth is called a **balance sheet.**
2. ____T____ An overdue amount that is unpaid is in **arrears.**
3. ____F____ An offer to buy or sell goods at a certain price is a **bar code.**
4. ____F____ To increase in value is to **appreciate.**
5. ____F____ Because many polar bears roam Alaska, it has a **bear market.**
6. ____T____ The printed message sent to stockholders annually is called an **annual report.**
7. ____T____ Joining two or more businesses into a single organization is called a **merger.**
8. ____T____ The term **arrears** refers to an overdue, unpaid bill.
9. ____F____ **COLA** is the favorite drink at the NYSE.
10. ____F____ **Fiber media** refers to the visual display of data via charts and diagrams.
11. ____F____ A **glass ceiling** is an untrue statement that injures someone's reputation.
12. ____F____ A pledge of property as security for a loan is called **monopoly.**
13. ____F____ **Per annum** is the Latin expression for **by the day.**
14. ____T____ A stockholder's written authorization for someone to vote in his or her place at a stockholders' meeting is a **proxy.**
15. ____T____ A written lie about someone that harms the person's reputation is called **libel.**
16. ____F____ The smallest unit of space in a computer image is called a **Weblish.**
17. ____T____ Time between starting a task and completing it is called **turnaround time.**
18. ____T____ A new company just beginning to operate as a business is called a **start-up.**
19. ____F____ One who organizes information for a Web site is often called a **WYSIWYG.**
20. ____T____ The number of members required for a formal meeting is a **quorum.**
21. ____T____ A pledge of property as security for a loan is a **mortgage.**
22. ____T____ The **markup** is the difference between the cost price and the selling price.
23. ____T____ **Stockholders** share in a company's profits.
24. ____F____ **Ginnie Mae** is the receptionist at an **Internet hotel.**
25. ____T____ One who disturbs other people's computer data is a criminal called a **hacker.**

B. Are you sure of your pronunciation of all Chapter 1 words? If in doubt, this is a good time to check with your dictionary or instructor.

THE GLOBAL MARKETPLACE—NORWAY

Norwegian	Spell	Say	Currency
GREETING	God morgen—good morning	Good morning	Krone
DEPARTURE	Adjo	Adjo	
PLEASE	Var so snill	Varso snill	
THANK YOU	Takk	Tăk	
YES	Yah	Yah	
NO	Nay	Nay	
I DON'T UNDERSTAND	Jeg forstaar ikke	Yeg forstaar igguh	

Courtesy of Alice Smith

Weather or Knot

Choose the Right Words

"Words are, of course, the most powerful drug used by mankind."

—*Rudyard Kipling*

After Completing Chapter 2, You Will

> Have further enlarged your business vocabulary—spelling, definitions, pronunciation, and usage.

Chapter 2 Pretest is in your *Instructor's Resource Kit*.

The words we use tell a listener or reader much about us. They make us seem foggy (weather or knot) or clear (whether or not) thinkers; they make us appear well educated and informed or ignorant. They *affect** (or is it *effect*) how people react to us or to the organization we represent.

Choosing the right word enables us to communicate clearly and confidently. Someone may say, "I know what I mean, but I just can't explain it." Those who want to express themselves clearly can develop this ability if it is important to them. The methods are not secret:

- Read books, magazines, and newspapers on various subjects
- Listen to articulate and well-informed speakers (radio, TV, lectures, etc.)
- Converse with well-informed, educated people
- Develop a large and precise vocabulary

MEMO FROM THE WORDSMITH

In *Alice's Adventures in Wonderland*, Humpty Dumpty says scornfully, "When I use a word, it means exactly what I choose it to mean—neither more nor less." Instead of a Humpty Dumpty philosophy, choose accuracy and precision in word choices.

Read
7

AN APPLE HAS A PEEL

Chapter 2 words meet the double criteria of (a) being used in the workplace and (b) language authorities generally agree on their usage. Many words listed in other books—e.g., lend/loan, like/as, raise/rear, anxious/eager, ensure/insure—are excluded because language authorities now accept those pairs as virtually synonymous, although their meanings differed in the past.

Computer checkers are simply not smart enough to inform us with certainty when we mean *appeal*, not *a peel*. For accurate documents, start your proofreading with the spelling checker—but remember that's not good enough by itself. With closed pen or other pointer, "underline" each word as you proofread it on the screen. Remember that only careful human proofreading catches most errors that result from soundalikes. English has hundreds of homophones (soundalikes, such as *right, write, rite*). The following are just some of them—those most often causing embarrassing errors.

ACCEPT	agree to receive
EXCEPT	excluding, as in everyone *except* me
ACCESS	ability to enter, communicate, or use something
EXCESS	more than is needed or wanted
AD	short for advertisement; one *d* like **ad**vertisement
ADD	to join; two *d*'s as in **add**ition
AFFECT	verb meaning to change, to influence, to pretend (His limited education will *affect* his ability to do the job.)
EFFECT	verb meaning to bring about or to result in (We can *effect* no changes without your approval.)
EFFECT	a noun that means *result* (We all know the *effect* would be disastrous.)

In Replay 7, No. 20, *affect* is a verb meaning *influence,* and in No. 21, *effect* is a noun.

*affect

ALOT—no such word	if you don't believe me, try to look it up; you won't find it in your dictionary. Do not write this "nonword!" Your spelling checker *will* alert you to this error.
ALLOT	with two *l*'s, the word is correct when it means to apportion or distribute
A LOT	correctly written as two words; use this expression only in conversation or informal writing, such as memos; instead use *a great deal, very much,* or *many.* Of course, *a lot* also refers to a small piece of land.
ALRIGHT	avoid this spelling
ALL RIGHT	this spelling is preferred: two *l*'s; two words
ALTER	*change* (The groom said, "Don't try to **alter** me after we leave the **altar.**")
ALTAR	a place for sacred rituals
BAZAAR	a sale or marketplace (note three *a*'s)
BIZARRE	odd; grotesque; strange (note two *r*'s)
CAPITAL	wealth; a city that is the capital of a state; an upper case letter; execution, as in capital punishment
CAPITOL	a building where legislators make laws
CITE	summon to court; to honor; to quote
SIGHT	ability to see; a view
SITE	noun meaning a location
CITE	to mention, name, or quote
COARSE	rough; of poor quality; crude
COURSE	school subject; portion of meal; place where golf is played; a direction taken
COUNSEL	lawyer or advice or to give advice
COUNCIL	group that meets to discuss, plan, or decide action
DIE, DYING	to pass from life
DYE, DYEING	to change a color, such as fabric or hair
DISSENT	disagreement
DESCENT	a downward movement; the verb is **descend**
HERE	at this place
HEAR	*ear* **is in hear;** proofread carefully to avoid careless here/hear errors
HEIR	**person** who inherits (Female who inherits is an heiress.)
AIR	referring to atmosphere; pronounced like *heir*

ILLICIT   not legal; prohibited; improper

ELICIT   to draw forth or to bring out

PRINCIPLE   a rule; a fundamental truth; a law

PRINCIPAL   main or most important; the chief administrator of a school

Replay

7

Write the correct word in the blank.

accept	**1.** We'll ___accept___ deliveries every
except	**2.** day ___except___ Sunday.
ad	**3.** We'll ___add___ the figures before
add	**4.** we place the ___ad___ in the newspaper.
dyeing	**5.** I'm ___dying___ to hear what he'll say after finding out
dying	**6.** that I've been ___dyeing___ my hair.
descent	**7.** Peaceful ___dissent___ should be encouraged.
dissent	**8.** The ___descent___ from Mt. Baldy will be difficult.
bazaar	**9.** His behavior was so ___bizarre___
bizarre	**10.** that she left him at the ___bazaar___.
coarse	**11.** Her ___coarse___ manners were distasteful to him, but the
course	**12.** professor, of ___course___, couldn't exclude her from the course.
site	**13.** A magnificent ___site___ was selected for the new theater.
sight	**14.** We could ___cite___ several examples of unfair taxation.
cite	**15.** What a ___sight___ he was with his torn clothing!
council	**16.** Richard is the ___counsel___ for the defense.
counsel	**17.** The City ___Council___ meets every Friday morning.
access	**18.** Do you have ___access___
excess	**19.** to the ___excess___ funds?
affect	**20.** How does the hot weather ___affect___ you?
effect	**21.** What ___effect___ does the heat have on you?
capital	**22.** Is the ___capitol___ in
capitol	**23.** the downtown section of the ___capital___?
a lot	**24.** We cannot ___allot___ any funds for the company
allot	**25.** to purchase ___a lot___ for the new parking structure.
principle	**26.** An important ___principle___ for students to remember is to be
principal	**27.** respectful to the ___principal___.

elicit   **28.** We are trying to _____*elicit*_____ the full details concerning

illicit   **29.** the _____*illicit*_____ affair.

Answers are on page 385.

Read 8

LETTUCE DEVISE A DEVICE

Words about words: An **eponym** is a word named for a person; for example Napoleon, once emperor of France, is now a fattening, flaky pastry. Nicotine is from the name of another French government official—Jean Nicot, who cultivated tobacco plants sent from America. Perhaps students can think of other eponyms.

My spell checker doesn't find anything wrong with the preceding title—proving it isn't as smart as you and I are.

Although the pairs of words below on the left in bold print are similar in appearance, spelling, and pronunciation, they differ greatly just like *lettuce* and *let us.* Be sure to observe the distinctions in your reading, writing, and speech.

BESIDE	by the side of; near; next to
BESIDES	in addition to
CHOSE	past tense of choose; pronounce the *o* like the alphabet sound of *o.*
CHOOSE	present tense; pronounce the *oo* as in *pool.*
COMPLIMENT	to praise
COMPLEMENT	to complete
CONSCIENCE	the part of us that hurts when we do wrong
CONSCIOUS	alert; awake; aware
DEFER	to put off or to postpone (accent on the second syllable)
DIFFER	disagree

Notice what happens when *desserts* is spelled backwards?

DESERT	accent on first syllable—where camels hang out
DESERT	accent on second syllable—to leave behind or to abandon
DESSERT	last course of a meal (Taking **seconds** on dessert is the memory hook for spelling this word with two *s*'s.)
DEVICE	a machine, tool, or method to achieve or do something (rhymes with rice)
DEVISE	to plan or figure out (rhymes with rise)
ELIGIBLE	have the qualifications to participate
ILLEGIBLE	not readable or difficult to read
EMINENT	well known for accomplishments; outstanding; famous
IMMINENT	about to happen
ENVELOPE	paper container for a letter to be mailed
ENVELOP	to wrap or surround

IRREGARDLESS NOT Standard English. Do NOT say it or write it! Instead use *regardless.*

REGARDLESS no matter what else happens

FISCAL pertaining to financial affairs

PHYSICAL pertaining to the body

GUISE a false outward appearance

GUYS informal word for *men* or *boys*

IT'S with the apostrophe—contraction for *it is*

ITS no apostrophe—shows possession; for example, *Its WINGS were flapping.*

LED past tense of lead, as in "He *led* the parade."

LEAD He'll lead it again tomorrow.

MINOR under 18; unimportant

MINER worker in a mine

Write the correct word in the blank. Use your dictionary for help with those Replay 8 words that are not defined in Read 8.

eminent
imminent

1. A storm is ___imminent___.

2. The ___eminent___ statesman Winston Churchill used the English language more effectively than anyone else in the 20th century.

regardless
irregardless

3. He plans to attend ___regardless___ of the weather. (Do not say or write *irregardless;* don't even *think* about using it.)

eligible
illegible

4. He is not ___eligible___ to play soccer this semester

5. because his handwriting is ___illegible___.

devise
device

6. Can you ___devise___

7. a ___device___ that is less expensive?

desert
dessert

8. Will you join me for ___dessert___ when we meet

9. in the ___desert___,

10. but don't ___desert___ me.

choose
chose

11. Did you ___choose___ the same books

12. that I ___chose___?

biographical
bibliography

13. I read a ___biographical___ sketch of Julio Iglesias.

14. On the last page of this biography is a ___bibliography___.

defer	15. I will always ___defer___ to your wishes.
differ	16. I do not ___differ___ with you on any subject.
conscience	17. My ___conscience___ is clear.
conscious	18. He was ___conscious___ but not in pain.
beside	19. No one ___besides___ Ms. Muffet sits on tuffets.
besides	20. However, a spider did come along and sit down ___beside___ her.
envelop	21. When I see him, I'll ___envelop___ him in my arms.
envelope	22. He'll carry an ___envelope___ with a great deal of money in it.
fiscal	23. During this ___fiscal___ year, we won't have the funds to
physical	24. construct a ___physical___ education building at the high school.
compliment	25. She ___complimented___ me on the quality of my work.
complement	26. The fabric and color of the curtains ___complemented___ the new sofa.
guys	27. She hid her true self under the ___guise___ of friendliness.
guise	28. That girl likes ___guys___ (informal word) who are intelligent.
realty	29. In ___reality___
reality	30. that ___realty___ office is in an undesirable neighborhood.
weather	31. Do you know ___whether___ the
whether	32. ___weather___ will change within the next few days?
wrapped	33. She listened to the instructor with ___rapt___ attention
rapt	34. while she ___wrapped___ the package.

You'll find the answers on page 386. After checking your answers, take the Pop Quiz on page 335.

Read 9

BEE QUITE QUIET

Carefully check your writing for be/bee and quite/quiet mistakes—as well as for other errors that may occur as a result of hasty proofreading.

LOSE to misplace or be unable to locate something

LOOSE not tight or not fastened

MORAL a concept of right behavior

MORALE spirit; sense of common purpose

PERQUISITE a privilege, a benefit, a payment, or a profit in addition to salary; this word is usually used in the abbreviated form *perk* or the plural *perks*

PREREQUISITE something required beforehand, such as taking a beginning class before being permitted to enroll in the advanced class

PERSECUTE	to mistreat or injure, often because of a belief or a way of life
PROSECUTE	to take legal action against someone accused of a crime
PERSONNEL	employees of a particular company or others who make up a group
PERSONAL	private
PERSPECTIVE	ability to see objects in terms of their relative distance from one another or to consider ideas in terms of their relative importance to one another
PROSPECTIVE	expected; likely to happen in the future
PROCEED	to go ahead, advance, or continue
PRECEDE	to go before; to be earlier
PROCEEDS	the money or profits derived from a business transaction; pronounced PRO.ceeds
QUITE	positively; completely
QUIET	without noise
REALITY	what is real or true
REALTY	real estate; property
REASON IS BECAUSE	avoid this combination
REASON IS THAT	use **reason is that** or simply **because** without the word *reason*
RESPECTFULLY	with respect; however, do not use *respectfully* as a closing for typical business letters—unless a letter is to someone warranting an unusual degree of respect, such as a high-ranking official or religious leader
RESPECTIVELY	in the order named
RYE	a grain or seeds used for making flour or whiskey
WRY	twisted; perverse; ironic—read the dictionary entry for *wry*
SUIT	clothing consisting of a matched outfit; a legal action
SUITE	group of items forming a unit, such as matched furniture, or a group of adjoining rooms or offices
THEN	at that time; next
THAN	use in comparisons such as *better than, rather than, more than,* and so on
THROUGH	across or from one side to another
THOROUGH	with attention to detail; complete
WERE	When writing, be careful (avoid carelessness)
WE'RE	to make appropriate distinctions
WHERE	among these three words.

WHETHER indicates a choice

WEATHER climate condition

Replay 9

From the Read 9 words, select the appropriate word for each blank.

personal, personnel   **1.** Mail clerks shouldn't open envelopes marked ___Personal___.

lose, loose   **2.** Robin Hood would never ___lose___ his arrow.

than, then   **3.** His work is usually better ___than___ anyone else's.

thorough, through   **4.** She is doing a ___thorough___ job of redesigning the executive
suit, suite   ___suite___.

weather, whether   **5.** ___Whether___ you like it or not, men are required to wear a
suit, suite   ___suit___ to work every day except Friday.

prosecute, persecute   **6.** Mr. Chandra would be the best attorney to ___prosecute___ this case.

that, because   **7.** The reason Sarala was promoted is ___that___ she now has an MBA.

weather, whether   **8.** Do you know ___whether___ the ___weather___ will change tonight?

proceeds, procedes   **9.** With the ___proceeds___ from the sale, we'll buy a dozen ___rye___
rye, wry   breads.

we're, were   **10.** ___We're___ ___quite___ sure he receives many perks.
quite, quiet

proceed, procede/moral,   **11.** Are you going to ___proceed___ with evaluation of the ___morale___
morale/personal, personnel   of our ___personnel___?

prospective, perspective/   **12.** The ___prospective___ instructor wanted the ___perks___ that only full
perquisite, prerequisite/   professors receive, but the ___wry___ grin of the University's
wry, rye/personal,   ___Personnel___ Director indicated refusal.
personnel

moral, morale   **13.** The ___morale___ is good in this company because the executives
make ___moral___ decisions.

where, wear/we're, were   **14.** ___Where___ are you when ___we're___ doing all the work?

perspective, prospective   **15.** Seen in ___perspective___, the incident was not too serious.

respectfully, respectively   **16.** I ___respectfully___ request a month's paternity leave.

We're, were/weather,   **17.** ___We're___ wondering ___whether___ we'll ever get ___through___
whether/thorough, through   with this job.

persecuted, prosecuted   **18.** They were ___persecuted___ because of their beliefs.

moral, morale/personal,   **19.** ___Morale___ was good in this department because the
personnel   ___personnel___ were all well trained.

proceed, precede/suit, soot   **20.** The attorney was asked to ___proceed___ with the ___suit___.

wry, rye/lose, loose   **21.** He expressed himself with ___wry___ humor when told he would
___lose___ his job.

complements, compliments   **22.** He appreciated the ___compliments___ about the quality of his work.

We're, were/where, ware   **23.** ___We're___ going ___where___ the jobs are.

| then, than | **24.** _____Then_____ he said that I look better _____than_____ ever. |
| that, because/realty, reality | **25.** The reason is _____that_____ he works in the _____realty_____ business. |

See answers on page 386.

CONFUSING PEARS

Confusing *pears* with *pairs* or *pares* is what happens if you misspell homonyms and homophones (soundalikes). You'll also find useful words for vocabulary growth that will help you avoid embarrassing errors. Keep your dictionary handy for further information.

APPRAISE   to estimate the value of an item

APPRISE   to inform

BLOC   a group of persons or countries combined to achieve a purpose

BLOCK   a large solid piece of a heavy material; see dictionary for multiple meanings

CANVAS   coarse cloth

CANVASS   to ask for votes, opinions, etc.

EVERYDAY   ordinary

EVERY DAY   each day

FOREWORD   an introduction to a book or an article

FORWARD   toward the front

HALVE   verb meaning to reduce to half

HALF   one of the two equal parts of something

HAVE   possess

KEY   a device used for unlocking

QUAY   (pronounced the same as KEY) concrete or stone waterfront structure

LESSON   something to learn

LESSEN   decrease

MARQUEE   a rooflike projecting structure over an entrance

MARQUIS   (quis pronounced *key*) royalty ranking above a count

MARQUISE   (pronounced *eez*) wife or widow of a marquis

Mnemonic for navEL—as in bELly button; the sillier a mnemonic device, the easier it is to remember.

NAVAL   referring to a navy

NAVEL   small scar in the abdomen—and a kind of orange

ODE	a dignified poem
OWED	responsibility to repay
PEAK	the top
PEEK	to look without anyone knowing you're looking
PIQUE	to be annoyed
REIGN	royal power
RAIN	water from the sky
REIN	means of controlling an animal
SERGE	a strong fabric
SURGE	a sudden, strong increase, as in power or water
STATIONERY	writing paper
STATIONARY	unmovable
TAUGHT	past tense for teach
TAUT	tightly pulled or stretched
THEIR	belonging to them
THERE	at that place
THEY'RE	contraction of *they are*
THROES	spasm or pangs of pain
THROWS	tosses
VISE	a device for holding an object so that it can be worked on
VICE	an evil action or habit
WAIVE	to give up or postpone
WAVE	to signal by moving a hand or an arm
WARY	cautious
WEAR AND TEAR	business term for loss and damage as a result of using a product

Provide example; e.g., the throes of passion

Replay 10

Write the correct word in the blank.

appraise
apprise

1. After we _____apprise_____ you of the cost, you can decide whether you want Mr. Gold to _____appraise_____ the ring.

bloc
block

2. The European _____bloc_____ wants to _____block_____ further action. (See No. 3 for additional use of *block*.)

block
canvass
canvas

3. We hope to _____canvass_____ the neighborhood to get the majority opinion regarding the _____canvas_____ tent for the _____block_____ party.

every day
everyday

4. I wear my _____everyday_____ clothes _____every day_____.

foreword
forward

5. Read the _____foreword_____, and then tell the members about going _____forward_____ with the project.

halve
half
have

6. To _____have_____ a smaller cake, simply _____halve_____ the ingredients. Then serve the cake with coffee and _____half_____ and _____half_____. (milk and cream)

key
quay

7. You don't need a _____key_____ until you arrive at the _____quay_____.

lesson
lessen

8. A _____lesson_____ is something we sometimes learn the hard way. If we try to do the right thing, perhaps we'll _____lessen_____ the consequences.

marquis
marquise
marquee

9. The _____Marquis_____ and the _____Marquise_____ stood under the _____marquee_____ before entering the theater.

navel
naval

10. It isn't advisable to wear a diamond in one's _____navel_____ when attending a party at the _____Naval_____ Academy.

ode
owed

11. Did you know that an _____ode_____ is a lovely poem expessing romantic emotion? After her bad behavior at the prom, she _____owed_____ him one.

proceed
precede

12. His speech will _____precede_____ our arrival. When he finishes speaking, we will _____proceed_____ to the refreshment tables.

rain
reign
rein

13. He needs to _____rein_____ in his horse and return to the stable because of the _____rain_____. Long may he _____reign_____.

peek
pique
peak

14. The mountain _____peak_____ will _____pique_____ your attention but you must not _____peek_____ until you reach the top.

surge
serge

15. I don't know anyone who still wears _____serge_____ clothing. When the water surges, you may experience a _____surge_____ of anger.

stationery
stationary

16. We have a large supply of _____stationery_____ in the desk that is _____stationary_____; that is, fastened to the floor.

taut
taught

17. I was _____taught_____ that one of the meanings of _____taut_____ is emotionally tense.

vise
vice

18. A _____vise_____ is closed with a screw or a lever and holds the object being worked on. It is totally different from a _____vice_____, which is a negative activity.

wary
wear

19. Be _____wary_____ of the

20. _____wear_____ and tear on your new tools when you lend them to friends with limited experience in using such tools.

waive
wave

21. If you _____waive_____ your right to be first at the buffet, you might as well _____wave_____ goodbye to the best appetizers.

Check your answers on page 386.

Read 11

WORDS OFTEN CONFUSED AND ABUSED

The following pairs of words require special care because they seem similar, but their meanings differ considerably.

ANXIOUS	worry or fear
EAGER	looking forward to something
BESIDE	by the side of; near
BESIDES	in addition to
DISINTERESTED	impartial; one who listens to all sides of an issue
UNINTERESTED	not interested or lacking enthusiasm
ENTHUSE/ENTHUSED	Avoid using these words; they are not "Standard" English.
ENTHUSIASTIC/ENTHUSIASM	These words are "Standard" English and are appropriate to use.
EMIGRATE	to move out of a country
IMMIGRATE	to move into a country
EXPLICIT	clearly expressed
IMPLICIT	not stated, but understood—"between the lines"
FLAMMABLE	these two words mean the same: can burn.
INFLAMMABLE	these two words mean the same: can burn.
NONFLAMMABLE	cannot burn; the opposite of *flammable* and *inflammable*
INDIGENOUS	people, wildlife, plants, etc., native to a particular area
INDIGENT	poor or needy
INGENIOUS	original, clever; ESL tip: pronounce *gen* with long *e* as in *jeans*
INGENUOUS	seeming to be innocent, simple, natural; pronounce *gen* as in *gen*tle
IRREGARDLESS	**not** Standard English; do NOT use this nonword
REGARDLESS	Standard English word meaning *without regard*
LESS	refers to bulk that can't be counted
FEWER	means a smaller number that can be counted

LESSON	something to learn
LESSEN	decrease
PER ANNUM	by the year, annually (Latin)
PER DIEM	by the day (Latin)

These terms refer to wages or salaries

PERQUISITE	a privilege or extra benefit; usually shortened to *perk*
PREREQUISITE	something required in advance
RSVP	French abbreviation for *Respond if you please* or Please respond.
PLEASE RSVP	**Avoid** beginning with *please:* it's redundant (saying the same thing twice).

English has about 550,000 words. About 2,000 are used in speech, and about 600 to 700 words are used in most books.—Gordon Dryden

SIMPLE/ SIMPLISTIC	These words do not have the same meaning; *simple* means easy-to-understand; *simplistic* means using poor judgment by making complex ideas sound deceptively easy.

Replay
11

Insert the appropriate Read 11 word in each blank. Use your dictionary when needed. Have fun with this "romp" through a number of interesting words.

anxious

eager

1. We are _____eager_____ to see you but are
2. _____anxious_____ about your health.

beside

besides

3. No one _____besides_____ Ms. Muffet would sit
4. _____beside_____ a tuffet.

disinterested

uninterested

5. Although he is a _____disinterested_____ observer,
6. he is not _____uninterested_____.

enthuse/enthused

enthusiastic, enthusiasm

7. We are _____enthusiastic_____ about this new project, and our
8. _____enthusiasm_____ shows.

implicit

explicit

9. _____Explicit_____ orders were not given; however, there was
10. _____implicit_____ acceptance of the orders by the entire staff.

per annum

per diem

11. His _____per annum_____ is enormous, but her
12. _____per diem_____ is low.

flammable

inflammable

nonflammable

13. _____Inflammable_____ blankets are illegal in this city's hospitals.
14. Only _____nonflammable_____ blankets may be used.

regardless

irregardless

15. Excellent methods have been developed and will be used
16. _____regardless_____ of the cost. When should you use *irregardless*? NEVER!

indigenous	17. ___Indigenous___ food is available for the
indigent	18. ___indigent___ workers and their families.
emigrate	19. To ___emigrate___ is to leave a country, while
immigrate	20. to ___immigrate___ means to move to another country.
less	21. We bought ___fewer___ apples than planned and made
fewer	22. ___less___ applesauce.
perquisite	23. A ___prerequisite___ is necessary before you can take the
prerequisite	advanced course. Upon completion of the advanced course, you will be eligible for many perks—also known as ___perquisites___.
thorough	24. We ___thought___ we did a ___thorough___ job of going
through	25. ___through___ the test items.
thought	
simple	26. The problem is ___simple___ to solve. However, his
simplistic	27. generally ___simplistic___ attitude toward serious problems is dangerous.
please rsvp	28. ___RSVP___ to let us know whether you can attend.
rsvp	29. Don't use PLEASE before RSVP; it's redundant. (Look it up if you're not sure about *redundant*.)
proceed	30. Let's ___proceed___ with the rehearsal.
proceeds	31. The ___proceeds___ from admission to the performance will be divided equally.
ingenuous	32. The ___ingenious (cleverly inventive)___ politicians of that era
ingenious	33. did an ___ingenuous (pretending to be direct, honest, or innocent)___ job of thought control.

Verify your answers on pages 386–387. Take the Pop Quiz on page 336.

Read
12

LET'S TALK BUSINESS

Correct pronunciation increases your confidence when speaking with intelligent and well-educated personal friends—but even more so with clients, colleagues, or supervisors in the workplace. Some mispronounced words may result in individuals being thought of as undereducated, uneducated, or even ignorant. A colleague may even laughingly tell a superior or other colleague about how a co-worker mispronounced some word. However, if English is your second language, friends and colleagues are more forgiving and understanding.

In the frequently mispronounced words below, the correct pronunciation follows the spelling; the syllable receiving the most emphasis is capitalized.

The common mispronunciations are in parentheses AFTER the sample sentence.

ACCESSORIES	**ak SES a rees** The interior designer recommended that the accessories be selected last. (uhsessories)
AFFLUENCE	**AF loo ens AF loo ent** Palm Beach, Florida, and Beverly Hills,
AFFLUENT	California, are examples of affluent communities. (a FLOO ent, a FLOO ens)
APPLICABLE	**AP lik uhble** This information is not applicable to our dilemma. (uh PLIK uh bl)
ASKED	**askt** He asked the four questions. (ast or axt)
ATHLETICS	**ath LET iks** Participation in athletics contributes to good health. (ath uh LET iks)
DEBRIS	**deBREE** After the storm, debris was everywhere. (DEBris)
DEBUT	**daybYOO** She made her debut in a British film. (DEB yoot)
DES MOINES	**de MOYN** The site for our new factory is in Des Moines. (des MOYNS)
ENTRÉE	**ON tray** The interviewer ordered the cheapest entrée on the menu. (EN tree)
ETCETERA	**et SET e ra** or **et SET ra** (usually abbreviated as etc.) The king of Siam was fond of saying "etcetera, etcetera, etcetera." (ek SET era)
FEBRUARY	**FEB ru er ee** Valentine's Day is February 14. (FEB u e ree)
GENUINE	**JEN u in** The stock certificates are genuine. (JEN u wine)
GOURMET	**goor MAY** A diet of burgers and fries doesn't qualify one as a gourmet. (goorMET)
GRIEVOUS	**GREEVus** A grievous crime has been committed. (greev e us)
HEIGHT	**HITE** The height of the new building has not been decided. (hithe)
HOSTILE	**HOS til** His hostile attitude made us uncomfortable. (HOS tile)
ILLINOIS	**ILL i noy** The salesman's territory is the entire state of Illinois. (ILL i noys)
INCOMPARABLE	**in COMP er able** Our widgets are incomparable. (in com PAR able)
IRRELEVANT	**ir REL a vint** The course is irrelevant to my major but is related to my hobby. (ir REV a lint)
IRREVOCABLE	**i REV uh kuh bl** An irrevocable decision cannot be revoked. (i rev OK able)
ITALIAN	**i TAL yin** If you move to Rome, you'll need to learn Italian. (**I** tal yin)
JEWELRY	**JOO el ree** Many jewelry manufacturers are still on 45th Street. (jool e ree)
LACKADASICAL	**LAK uh DAY zuh kul** Employers don't hire applicants who appear lackadasical. (lax uh DAY zuh kul)
LIBRARY	**LI brer ee** Be sure to pronounce both *r*'s in library. (LI ber ee)
LIEU	**LOO** *In lieu of* means *instead of* or *in place of*. (leu)

MISCHIEVOUS	**MIS chiv us** Some children are mischievous on Halloween. (mis CHEEV e us)
NAÏVE	**NI eev** He is naïve to think he will get a raise without asking for it. (NAV)
PAGEANT	**PAJ nt** The fishermen's pageant takes place in San Pedro, California, every year at Christmastime. (PAGE ant)
PICNIC	**PIK nik** I look forward to meeting your husband at the company picnic. (PIT nik)
PICTURE	**PIK cher** Pictures of past presidents hang in the gallery. (pitch er)
PREFACE	**PRE fis** Have you read the preface to this book? (PREE face)
PREFERABLE	**PREF er able** or **PREF rable** I think the old equipment would be preferable to the new. (pre FER able)
PROBABLY	**PROB ub lee** Two baristas will probably serve coffee at the meeting. (PROB lee)
PRONUNCIATION	**pro NUN see A shun** Pronounce the second syllable *nun*. (pro NOUN see A shun)
REALTY	**REE ul tee** Several new realty offices opened last year in Fairbanks, Alaska. (REEL uh tee)
RELEVANT	**REL uh vint** Mr. Goldfinger included only relevant statistics in his report. (REV uh lint)
STATISTICS	**sta TIS tks** The sales statistics are included in Ms. Gomez' report. (sis TIS tiks)
SUBPOENA	**suh PEE nuh** A subpoena was issued for the murder witness. (sub PEE na)
SUBTLE	**SUT l** George was so subtle that Jesse didn't understand he had been fired. (sub tl)
SUPERFLUOUS	**soo PER floo us** Those items are superfluous and should be returned. (su perf e lus and sooper FLU us)
VEHICLE	**VEE i kil** Do not pronounce the *h* in vehicle. (vee HICK el)
VERSATILE	**VER suh tl** A versatile object can be used for various purposes. A versatile person can do many different things. (VER suh tile)
VISA	**VEE zuh** The word **VISA** came into the English language from French in the early 1800s. (VEE suh)

Replay
12

Say each word aloud before responding to these questions.

A. How many syllables does each word have?

1. grievous _____2_____

2. superfluous _____4_____

3. mischievous _____3_____

4. probably _____3_____

5. naïve _____2_____

B. Which letters are silent in these words?

    6. Des Moines ___s es___

    7. Illinois ___s___

    8. debris ___s___

    9. subtle ___b e___

    10. vehicle ___h e___

Challenge for would-be word mavens: fix the incorrectly used words: 1. When my piers signed the packed, the belles peeled. 2. I'll pear a pare to have with my serial. 3. He maybe studying the pole results too find out the will of the populous. 4. "Aisle altar hymn," said the bride on the knight of the wedding. 5. The law student studied his forth lessen on leans in a careless manor.

C. Write **T** (true) or **F** (false) in the blank.

___T___ 11. Pronounce *affluent* with the accent on the first syllable.

___F___ 12. The *i* in *versatile* is pronounced the same as the alphabet sound of *i*.

___T___ 13. When saying *preferable*, the accent is on the first syllable.

___F___ 14. The capital *i* in *Italian* sounds like the alphabet sound of *I*.

___F___ 15. *Height* ends with a *th* sound.

___F___ 16. Pronounce *jewelry JOOL e ree*.

___F___ 17. *Realty* should be pronounced *reel i tee*.

___T___ 18. The second syllable of *pronunciation* sounds different from the second syllable of *pronounce*.

___F___ 19. The first *a* in *pageant* sounds like the alphabet sound of *a*.

___F___ 20. The first syllable of *statistics* sounds like *sis*.

___T___ 21. To remember how to pronounce and spell *irrelevant* and *relevant*, notice that the *l* goes before the *v*.

___F___ 22. *Picture* is pronounced the same as the word for a baseball player who pitches.

___F___ 23. *Irrevocable* has the primary accent on the third syllable.

___F___ 24. The second syllable of *preface* is pronounced like the word *face*.

___F___ 25. When identifying the nationality of a native of Italy, the first syllable should sound like the word *eye*.

Check your answers on page 387.

CHECKPOINT

Review all words in this chapter now so that you'll know the appropriate use of each one. Take advantage of this opportunity to improve your vocabulary and to use words correctly.

QUICK SELF-CHECK

Which does the sick man hope is right? No. 1 __x__ or No. 2 _____

1. A glass of brandy will effect his recovery.
2. A glass of brandy will affect his recovery.

Which requires more cleverness and originality? No. 1 __x__ or No. 2 _____

1. He adapted the plan.
2. He adopted the plan.

Which theater was almost deserted? No. 1 _____ or No. 2 __x__

1. In the theater there were five people beside me.
2. In the theater there were five people besides me.

USEFUL DEFINITIONS

Complete the word in each blank.

Words with similar meanings such as *satisfied* and *contented* are __synonyms__.
Words with opposite meanings such as *rich* and *poor* are __antonyms__.
Words pronounced alike but different in meaning such as *to, too,* and *two* are __homonyms__.

SPECIAL ASSIGNMENT

This article wasn't really in the wedding-news section of a local newspaper. However, if it had been published and you were the editor, you would probably fire the reporter. Draw a line through the incorrectly spelled word, and write the correct word in the space above.

The bride ~~dissented~~ [descended] the ~~isle~~ [aisle] in an off-white satin gown crowned with a ~~waste~~ [waist]-length ~~tool vale~~ [tulle veil] and carrying a lavender ~~orchard~~ [orchid].

PROOFREADING FOR CAREERS

Below is an article about the profession called "Certified Financial Planner." While this article was written by an expert, I inserted numerous English mistakes. Why did I do that? You can probably guess! I want you to proofread carefully while you absorb information about financial planning and additional business vocabulary. Correct misspelled words, repeated words, use of a symbol instead of a word, or one word written as two separate words.

Keyboard the corrected article. Start about an inch and a half from the top of the page, double-space, indent the first line of each paragraph, and use side margins of about one inch. For additional formatting details, see the Mini Reference Manual beginning on page 370.

WHAT IS A CERTIFIED FINANCIAL PLANNER?

by Eric A. Smith, CFP

A Certified financial Planner is a member of a distinguished proffession dedicated to serving financial needs of individuals, familys, and businesses.

"Certified Financial Planner" is far more than a tittle. It is a precice definition of a Planner's compitance, experience, and intelligence in the complex proffession of financial planning. Not every person described as a financial planner is is a "CFP" licensee. To earn this title, the financial planner must demonstrate competence in analyzing and developing personal and business financial plans through successful completion of rigerous financial planning examinations.

The CFP professional's area of expertise is analyzing needs & prudentley arranging over all financial plans, rather than promoting individual financial products. In carrying out this task, one CFP licencee may be more knowledgeable in a particular feild than in another and may consult with qualified individuals in other specialties. The talent of the licensed CFP is seeing that all the elements of the of the plan are implemented and that the advice of specialists, such as attorney or an accountant, are coordinated in the best interests of the client.

Call 206-632-3337 for an appointment with Eric A. Smith, CFP

Verify your corrections with your instructor, and select a word to evaluate your business communication and proofreading skills: Excellent ___ Good ___ Not bad ___ Or ? ___

PRACTICE QUIZ

www.prenhall.com/smith

Fill in the blank with the correct word. These nonsensical sentences will make you smile or groan while you study.

EXAMPLE   Your ___advice___ is needed at this time. (advice advise)

1. We cannot ___allot___ any more funds to this project. (alot, allot, a lot)
2. The ___eminent___ professor prepared the ___bibliography___. (eminent, imminent/ bibliography, biographical)

3. Yesterday we _____chose_____ to _____defer_____ the legal action. (choose, chose/differ, defer)
4. Even though his writing is _____illegible_____, he is _____eligible_____ to win the prize. (eligible, illegible)
5. I _____respectfully_____ report that I don't know _____whether_____ the _____weather_____ will be any better today than it was yesterday. (respectfully, respectively/whether, weather)
6. _____We're_____ going to _____proceed_____ to _____their_____ home before the _____rain_____ begins. (We're, Were, Where/proceed, precede/there, their/reign, rein, rain)
7. She wrote an _____ode_____ to a _____canvas_____ umbrella, which had been _____appraised_____ for over $1,000. (ode, owed/canvas, canvass/appraised, apprised)
8. I will _____halve_____ the ice cream portions because the _____Marquise_____ is on a diet, and this will _____lessen_____ her calorie intake. (have, halve, half/Marquis, Marquise, marquee/lesson, lessen)
9. _____Regardless_____ of your position in the company, you will get more _____compliments_____ if you don't behave in a _____bizarre_____ manner when you tour the _____bazaar_____. (Regardless, Irregardless/complements, compliments/bizarre, bazaar)
10. _____It's_____ a fundamental _____principle_____ that when you tour the _____Capitol_____ building, you don't try to _____elicit_____ confidential information from our _____eminent_____ governor. (It's, Its/principle, principal/capital, capitol/elicit, illicit/eminent, imminent)
11. _____We're_____ doing a _____thorough_____ job of refurbishing the _____suite_____. (Were, We're, Where/through, though, thorough/suit, suite)
12. With a _____wry_____ grin, he asked what the _____perks_____ would be if he had concealed his true character with a _____guise_____ of friendliness. (rye, wry/perks or perquisites, prerequisites/guys, guise)
13. We met the _____Marquise_____ at the _____naval base_____ near the _____quay_____. (Marquise, Marquis, Marquee/naval, navel/key, quay)
14. He _____led_____ the coal _____miner_____ to the building where the _____lead_____ is stored in _____canvas_____ bags. (led, lead/minor, miner/canvas, canvass)
15. _____Regardless_____ of the company's _____fiscal_____ situation we cannot in good _____conscience_____ encourage _____fiscal_____ irresponsibility. (Regardless, Irregardless/fiscal, physical/conscious, conscience)
16. You don't need a _____visa_____ to use the _____proceeds_____ from the sale for an _____incomparable_____ _____gourmet_____ dinner in Phoenix. (visa, viva/precedes, proceeds/incomparable, uncomparable/gourmay, gourmet)
17. During the next _____fiscal_____ year, we cannot _____allot_____ funds to rebuild the historic _____altar_____ in the _____Capitol_____. (fiscle, fiscal/allot, alot/alter, altar/Capitol, capital)

THE GLOBAL MARKETPLACE—CHINA

Cantonese	Spell	Say	Currency
GREETING	lihoma	lay.HO.mah	yuan (Y)
DEPARTURE	doigen or bye bye	doi.jen or bye bye	
PLEASE	ng goi	mm goy	
THANK YOU	toutge	thaw jay	
I DON'T UNDERSTAND	Cantonese: Ngoh m-ming	Mandarin: Wo bu dong	

Courtesy of Hi Lin Yau and Yet Yee

3

Ain't Is in the Dictionary

Learn Dictionary Smarts

"When I feel inclined to read poetry, I take down my dictionary. The poetry of words is quite as beautiful as that of sentences. The author may arrange the gems effectively, but their shape and lustre have been given by the attrition of the ages."

—*Oliver Wendell Holmes*

After Completing Chapter 3, You Will

> Use dictionaries efficiently to improve vocabulary, word choices, spelling, and pronunciation.

> Know how to locate and interpret the various kinds of information in dictionaries.

> Choose the dictionaries that best meet *your* needs.

> Have the "dictionary habit."

How do words get into the dictionary to start with? Who makes the decisions? When Moses received the Ten Commandments, did he also receive a list of spelling rules? Then why can't I spell *enough* with just four letters—*enuf*? You'll discover answers to these questions as you read on.

See pretest in *Instructor's Resource Kit.*

Lexicographer Noah Webster spent 20 years handwriting *An American Dictionary of the English Language.* This first comprehensive American dictionary had 70,000 entries and an extensive preface and appendix. Many dictionary publishers include "Webster" in the title because of his fame.

LEXICOGRAPHERS AT WORK

Lexicographers (dictionary compilers) review tremendous numbers of words used throughout the English-speaking world. Who makes the decisions as to which words get into the dictionary? The lexicographers record meanings, spellings, and pronunciations of people using English, both formally and informally, and in various activities—from a̲rmed robbery to z̲oology. They discover new words, new spellings, new pronunciations, new meanings for old words, and old words used in new ways or discarded. The information is gleaned from numerous books, magazines, trade journals, brochures, radio and TV programs, speeches, conversations, and all other forms of spoken and written communication. The lexicographers record each word of interest, along with when, where, and how it was used and by whom.

Thousands of entries called *citations* may be added daily to files insured for millions of dollars. When compiling a new edition of a dictionary, lexicographers sort the entries to decide what will be added, deleted, or changed.

DICTIONARIES—DESCRIPTIONS, NOT PRESCRIPTIONS

If you have a dictionary originating in Great Britain or a "British"-speaking country, be aware that there are differences—in spelling, pronunciation, and meanings—between British and American English. One of **many** examples is: In American English, the word is spelled *practice* while in British English, it is *practise*. Another: In British English, pronounce the first *y* in *dynasty* like the *i* in *if,* but in American English it's pronounced like *die.* A couple of new British English words are *zydeco* and *bling bling*. Americans will probably adopt them, just as Brits do with new American words.

Dictionaries **describe** how English is used. Slang, street language, and other non-Standard English words are used, whether we like the words or not. Contrary to the slogan you may have heard in your childhood, "Ain't ain't in the dictionary," *ain't* has been in some dictionaries since 1951. Modern dictionaries don't prescribe—or tell you what to do—as when a physician writes a prescription. Instead the dictionary informs you of how people worldwide are using English at the time of compiling that particular dictionary.

If the dictionary doesn't tell how to speak and write good English, what good does it do to look up a word? The dictionary does tell you which pronunciations, spellings, and vocabulary are Standard English—that is, widely used and respected by well-educated people—and which are non-Standard, informal, slang, vulgar, or at some other level. When you understand dictionary "code," you can make intelligent choices depending on your purpose. For example, knowing dictionary information about *ain't*, you won't use it during job interviews. However, you might write, "He ain't here," as dialogue for an uneducated character in a film. If you look up *hospitable,* you'll find two acceptable pronunciations, meaning both ways to say this word are used by educated English speakers; however, the one shown first is heard more often.

Many Americans have emigrated (or are students) from a country where British English is used. If you are one of these, you might have a British English dictionary, such as the eminent *Oxford English Dictionary* or the *Longman's Dictionary.* British and American spelling, vocabulary, and pronunciation differ for many words. Canadian English is more like American than British, but includes some British spellings and pronunciations.

The British equivalent to "Standard English" is generally known as either the *Queen's English, BBC (British Broadcasting Company) English,* or *Received English.* Don't assume British English is "correct" and that American English is somehow inferior. Although neither one is superior to the other, they do differ somewhat. Use the style appropriate to where you're living or working. The focus in *English for Careers* is Standard American English.

Following are examples in which Americans and British use the same word, but with a different meaning: for example, *homely* in American English means *ugly* or *unattractive,* while in British English, it means *comfortable* or *cozy.*

Word Power

Here are examples of differences in the education field between British and American English:

BRITISH ENGLISH	AMERICAN ENGLISH
a re-sit	a make-up test
supply teacher	substitute teacher
headmaster/headmistress	principal
principal	college president
public school	private school
lollipop lady	female crossing guard

New editions of dictionaries add thousands of words, as well as new meanings to old words, that didn't exist at the time of the previous edition or that were too new to be accepted. Most new words are related to business, professions, technology, and popular culture—making them important to workplace communication. Here are some of the latest: *micropower, spyware, tweenager, romvelope, cybrary, beltway, cybernate.* Yesterday's grammar errors are often today's idioms. (That's one reason business English textbooks need to be updated frequently.)

School in American English refers to any educational institution regardless of level, while in British English the word is restricted to elementary level.

Many words are spelled differently, such as the American word *practice*—spelled *practise* in British English. If you brought an English dictionary to the United States from another country, check to see whether it is British or American. While speaking British English is usually understood and always respected in the United States, if you spell in British English, many Americans will simply think you're spelling incorrectly.

However, English is now used all over the world—not just in the United States, Great Britain, and Canada. Businesspeople (and others) need to be aware that English has many varieties of accents, usage, vocabulary, and spelling from countries such as South Africa, India, Australia, New Zealand, and others.

THESAURUSES—THE RIGHT WORD

A thesaurus is another useful reference for people interested in words. It's what you reach for (the hard copy book or a computerized version) when you're writing and just can't think of the right word. Thesauruses provide synonyms (words with similar meanings), antonyms (words with opposite meanings), and other related words. Do not use your thesaurus for definitions, pronunciation, or other information that requires dictionary reference. However, like dictionaries, thesauruses are available in hardcover or paperback as well as on the Internet or on CD or other word processing software. In addition, some dictionaries are combined with thesauruses; this combination is part of the book title. Roget, pronounced *ro.ZHA,* is the famous name in thesauruses, like Webster is in dictionaries, though other thesauruses and dictionaries may be as good or even better.

Read
13

ALL SHAPES AND SIZES

Since the dictionary is the single most valuable reference tool for acquiring good language skills, think of it as a friend you wish to know better as you work your way through this chapter. Just as friends come in all shapes, sizes, and formats, so also do comprehensive dictionaries: abridged, unabridged, paperback, hardcover, and electronic.

ABRIDGED DICTIONARIES

Abridged dictionaries are available in pocket, college, electronic, and encyclopedic formats. They are updated frequently, some as often as annually. It's important for your career as well as home use to own the newest pocket-size and college dictionaries.

POCKET DICTIONARIES

Usually paperback, these have far fewer entries than a college dictionary. In addition, explanations, introductory material, and appendixes are minimized. However, since they are convenient to carry and use without power or batteries, students should carry one to classes and elsewhere as needed.

COLLEGE DICTIONARIES

Usually hardcover, these average at least 100,000 entries and include thousands of new words, making it important to use a comprehensive, up-to-date dictionary. Ample front matter (introductory pages) and back matter (appendixes) are included. Keep a current college dictionary both at home and at work—if your job or classes require good English skills.

ENCYCLOPEDIC DICTIONARIES

Larger than most college dictionaries, they include an encyclopedia of general information. I don't recommend them for career purposes because they supply less information about words than many smaller dictionaries do. However, these are useful for school-age children who don't have easy access to a multivolume encyclopedia.

DICTIONARY-THESAURUS

This, as the name implies, is a combination book. Though a convenient combination, the dictionary portion might not be as complete as a college dictionary by itself. Compare the number of dictionary entries with other dictionaries.

A good college dictionary is useful for several years for an entire family or for an office, and it costs less than many textbooks. Students with a college dictionary tend to use it more often, more intelligently, and more enthusiastically than those with only a pocket dictionary.

When you shop for a new dictionary, look for the publication year and latest year updated. You'll find dictionaries on bookstore shelves with copyright dates of several years earlier, and they are likely to be less expensive than the newest ones. I suggest you **get the newest one available.** English changes rapidly along with our constantly changing lifestyles: new words are added to the language, old words disappear, and meanings of old words change. The following college and pocket dictionaries are good for workplace communication and for home and family use. When you select one of these, **be sure it is the latest edition of that title:**

- *American Heritage Dictionary of the American Language*
- *American Heritage College Dictionary*
- *Merriam-Webster's Collegiate Dictionary*
- *Microsoft Encarta College Dictionary*
- *Random House Webster's College Dictionary*
- *Webster's New World College Dictionary*
- *Oxford **American** College Dictionary*

UNABRIDGED DICTIONARIES

An unabridged dictionary typically has at least 250,000 entries, or all the words the lexicographers believe belong in an unabridged dictionary.

Information for each entry in an unabridged dictionary is extensive, as are the front matter and appendixes. Unabridged dictionaries are in libraries, schools, newspaper offices, and similar places. However, they are not updated as often as smaller dictionaries, making them less helpful than college dictionaries for career purposes. Unless you have a professional or a strong personal interest in language, you probably don't need an unabridged dictionary. When needed, use the one in your local or college library.

Even unabridged dictionaries do not list all words in the language. Millions of different kinds of insects, animals, and plants have been named. General-use dictionaries simply have no room for all these words. The largest and most prestigious unabridged American dictionary as of this writing is *Merriam-Webster's New International Dictionary of the English Language* (latest edition).

ELECTRONIC DICTIONARIES

Electronic dictionaries are on CDs, the Internet, and word processing software. Use electronic dictionaries as needed and when convenient. An example of a good electronic dictionary is *Microsoft Encarta Dictionary on CD-ROM*.

When proofreading on screen, use the spelling dictionary in your word processing software. However, do not assume that electronic dictionaries, whether portable or part of your software, replace the need for a college dictionary in book form. Dictionaries in book form are likely to provide more information about language and are more portable.

WHAT'S IN IT FOR ME?

As you proceed through this chapter, learn about the kinds of information in your college dictionary and where to find each. Your college dictionary is divided into front matter (including the inside front cover), A–Z entries with guide words on each page, and appendixes (including the inside back cover).

> *Front matter* is everything before the first *A* word. *Appendixes* are everything after the last *Z* word.

Information about how to interpret dictionary code is found mostly in the front matter, including the table of contents, or the index in the books, or both. Exactly what you find and where to find it varies from one dictionary to another and from one edition to the next—somewhat like comparing features of various kinds of cars.

Entries are all the A–Z words (including word parts like prefixes and suffixes) with information about each. Entries include some or all of the following: spelling, syllables, pronunciation, definitions, parts of speech, etymology (history of the word), usage label (in what context the word is used), picture of the item, synonyms, run-ins (which are words derived from the entry word, such as *typographical* near the end of the *typography* entry), plurals, capitalization, year the word was first seen in print, and other helpful notes.

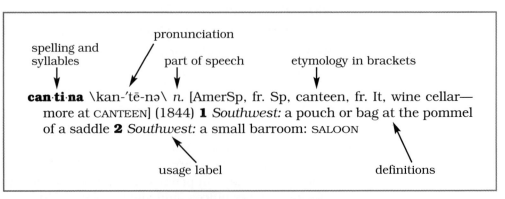

By permission from *Merriam-Webster's Collegiate Dictionary*, Tenth Edition.

> *Guide words* at the top of each page show the alphabetic range of that page so that words can be located quickly.

For example, in *Webster's New World College Dictionary,* Fourth Edition, the guide words on page 928 look like this:

<u>moldy/mom and pop store 928</u>

This means the first entry on page 928 is *moldy* and the last entry on that page is *mom and pop store.* When you look something up, you save time by referring to the guide words for the alphabetic range of that page.

Replay
13

Fill in the blanks based on **answers obtained from your dictionary and Read 13.**

EXAMPLE Name a dictionary recommended in Read 13.

<u>Oxford American College Dictionary</u>

Many answers are shown in the Recaps and Replays as "see your dictionary" or "answers will vary" because symbols and information differ from one dictionary to another. Think of your dictionary like an "owner's manual" for a car.

1. What is the full name and copyright date of your newest dictionary? <u>Answers will vary</u>

2. In what continent is Afghanistan? <u>Asia</u>

3. What is the purpose of the *guide words*? <u>They show the first and last words</u> <u>of that page</u>.

4. What guide words are in your dictionary for the page that defines *bungee jump*? <u>Answers will vary</u>.

5. The English usually spoken and written in London is the same as that usually used in Chicago. True _____ False <u>X</u>

6. What basic information follows a typical entry word? <u>Spelling, syllables, pronunciation, definitions, part of speech</u>

7. A comprehensive dictionary containing at least 250,000 entries is known as a/an <u>unabridged</u> dictionary.

8. The four types of comprehensive abridged dictionaries are called <u>pocket</u>, <u>college</u>, <u>encyclopedic</u>, <u>online</u>.

9. If you like to use good English, you should have a <u>college</u> dictionary at home and at work. If you're a student or for some other reason need to carry a dictionary, you should also have a <u>pocket or small electronic</u> dictionary.

10. To find another word to express your idea—perhaps a synonym or an antonym—consult a <u>thesaurus</u>.

11. The history of a word is called the <u>etymology</u>, which in most dictionaries is enclosed in <u>brackets</u>.

12. Explanations about using the dictionary most effectively are usually in the <u>appendixes</u>, the <u>index</u>, or the <u>front matter</u>.

13. On what pages (in small Roman numerals) does the front matter begin and end in your dictionary? <u>Answers will vary</u>.

14. Look at the list of appendix items in the table of contents or index of your college dictionary. Then look through the appendix. What are the titles of each part? <u>Answers will vary but should include everything after the main section Z words.</u>

15. Which dictionary might you purchase as a gift for an American friend whose work is taking her to London for a yearlong assignment? <u>*Longman's*</u>

Answers to this Replay are on page 387.

Read
14

MISSPELLERS ANONYMOUS

George Bernard Shaw bequeathed money for spelling reform—which never happened. He joked about spelling inconsistencies by spelling *fish* as "ghoti"— *gh* as in tou**gh**, *o* as in w**o**men, *ti* as in addi**ti**on.

If you've ever wished you could join Misspellers Anonymous, here's your chance to improve your spelling and remain anonymous; no one has to know about that "spelleton" in your closet.

> **When writing at your computer, use the spelling checker, even though it won't always find miss steaks—an example of what it's likely to overlook.**

Although the spelling checker is important, it doesn't replace the computer between your ears for intelligent human proofreading. Errors result when writers hastily accept the spelling checker's "word" that there are no mistakes. If you substitute *lead* for *led* or *wired* for *weird*, the computer's brain might not highlight the error. (When in doubt about *lead/led*, see your college dictionary.)

> **Use a good current dictionary in book form as well as your word processing dictionary.**

If you often consult a dictionary for spelling help, you're probably a good speller; remember, no one needs to know your secret. The reader sees the words correctly spelled.

> **Some words have more than one correct spelling. The dictionary entry will show them joined by *or*.**

>   the.ater or the.atre   mea.ger or mea.gre   pi.zazz or piz.zazz

Although both spellings are correct, the form seen more often than the other is first. It is called the "preferred" spelling; choose it for workplace writing so that an unusual spelling does not distract the reader.

> **However, when *also* precedes a second spelling, the second one is less acceptable; avoid it when writing for your career—unless you have a special reason to use it.**

>   lovable *also* loveable   lasagna *also* lasagne

> **Some words are spelled differently in British English.**

>   misdemeanor . . . chiefly British: misdemeanour
>   colour, chiefly British variant of *color*

British English dictionaries show spellings like honour (honor), centre (center), and both practi<u>c</u>e and practi<u>s</u>e, dependent on the meaning. If you live or work in Canada or another country using British English, or if your work requires communication with clients in countries using British English, it's useful to know the differences. However, bear in mind that British spellings are sometimes seen in American advertising to create an aura of quality.

> **Some second spellings are shown as separate entries.**

If a second spelling is a separate entry, it is not the preferred form. A note like *non-Standard* or *disputed sp.* may appear in the separate entry. For the definition, look up the preferred spelling. For example, the second spelling for *all right* is shown in *Webster's New World College Dictionary*, Fourth Edition, as follows:

al.right (ol rit') adj., adv., interj. disputed sp. of ALL RIGHT

This means *alright* isn't all right and is a disputed spelling, not generally recognized as correct. In my *Microsoft Encarta College Dictionary*, a "usage note" on page 36 states " . . . *Alright* has never gained wide acceptance even though it is to be seen in the prose of many well known writers such as Langston Hughes, Gertrude Stein, ... "

Open your dictionary now to see how this word appears in *your* dictionary.

A compound expression means two or more words expressing one concept.

apple pie   nationwide   hand-me-down

My dictionary shows a space between *apple* and *pie*, no hyphen or space between *nation* and *wide* (the dot or accent mark shows it's one solid word), and hyphens in *hand-me-down*. If a college dictionary has no entry for a compound expression, assume it's two separate words—such as *apple tree*.

A hyphen might also be required between a prefix and a main word.

de-escalate   re-cover (when it means to cover again)

In some dictionaries, a hyphen follows the first *e* in de-escalate. In other dictionaries a second spelling with a dot between the *e*'s shows the word is correct either hyphenated or written as one solid word. Use the "preferred form"—the one that's first in your newest college dictionary.

Believe you can spell well. Most important is to want to spell correctly and to believe you can. *What you can believe, you can achieve.* Spend five minutes a day consciously looking at some words in a newspaper, book, or magazine. That's all it takes to store a tremendous number of words in the magnificent computer between your ears.

Recap

Write **T** for true or **F** for false beside each statement.

   F   **1.** To be a good speller requires a special, inborn talent.

   F   **2.** Reading newspapers and magazines is a waste of time for spelling improvement.

   F   **3.** If *also* separates two spellings in the dictionary, both spellings are equally acceptable.

Check your answers on page 387.

Relax and trust your first impulse.

You won't always have access to your computer or dictionary when you need to spell a word. When neither an electronic nor paper dictionary is available, retrieve the word from your "personal computer"—your brain—by relaxing and writing what you think of first. You might recall a test when you erased and replaced a correct word with an error. Your first hunch is often right.

Use memory devices for personal spelling demons. Then you'll be correct even if your spell checker isn't available. The sillier the memory device, the more effective it is.

Here are 14 commonly misspelled words, each with its own memory device to correct the troublesome part. Note the memory device and spell the word several times. You'll know the correct spelling for life.

accommodate: 2 c's, 2 m's

recommend: re+commend

privilege: has a leg in it

bachelor: has an ache because he's sad to be single

superintendent: He fixed the dent.

conscience: science is in it

pursue: pursue the purse snatcher

persistent: Your sister has a tent.

congratulate: Rat is in the middle.

dilemma: Emma has a dilemma.

embarrassed: railroad (RR) and steamship (SS)

villain: The villain lives in a villa.

separate: Pa is in the middle.

weird: We are weird.

"Blow up" hard-to-remember words, and stare for a few seconds at the "blown-up" letter(s). I'll bet you'll never again forget that spelling.

See pages 362–365 to increase the number of commonly used words you can spell correctly and define. Accurate spelling and proofreading are important to success in many careers. Employers and co-workers often view inaccurate spellers as uneducated, careless, or incompetent. Support your spell checker usage with your own strong spelling skills.

Writing for Your Career

What if you can't find a word in the dictionary because you don't know how to spell it?

1. Guess at the spelling and use your computer's spell checker; it might help.
2. Consult a misspellers' dictionary; they're available in bookstores.
3. If 1 and 2 don't help, phone a friend and ask for help; if your friend isn't in, then . . .
4. Think of another word to use and get back to work!

Replay
14

A. Write **T** or **F** in the blank.

EXAMPLE ___T___ "Privilege has a leg in it" is an example of a spelling memory device.

___T___ **1.** Prospective employers tend to assume a poor speller is uneducated or unintelligent.

___T___ **2.** Almost anyone who wants to be a good speller can be a good speller.

___F___ **3.** Being a good speller is unimportant because your spell checker corrects any mistakes you might make.

___F___ **4.** Poor spellers are always looking up words in the dictionary.

___F___ **5.** If a comma separates two spellings of the same word in the dictionary, it means the first spelling is non-Standard.

___T___ **6.** If *also* appears between two spellings of the same word, it means the first spelling is preferable to the second.

___F___ **7.** A computer's spell checker enables you to avoid homonym errors— for example, writing *you're* when the spelling should be *your.*

___T___ **8.** If *or* is between two spellings of the same word in the dictionary, it means either spelling is correct.

___F___ **9.** In a dictionary entry, if a dot or accent mark appears between two syllables it means you should spell it as two separate words.

10. Circle the misspelled words: pronounciation, weird, seperate, reccomend, congradulate, persue, villain, persistent, conscience, bachlor

B. Write the letter of the answer in the blank. Use your dictionary.

___b___ **1.** Which is spelled correctly? (a) counter-sign (b) countersign (c)counter sign

___c___ **2.** Which is correct? (a) epilog (b) epilogue (c) both a and b

___a___ **3.** Which is correct? (a) antitrust (b) anti-trust (c) anti trust

___a___ **4.** The adjective *buttondown* is (a) one solid word (b) hyphenated (c) two words

___c___ **5.** Which is correct? (a) ocurred (b) occured (c) occurred

C. Fill in the missing letters; then rewrite the word in the blank.

1. w__ei__rd ___weird___   **4.** congra__tu__late ___congratulate___

2. ac__commo__date ___accommodate___   **5.** pro__nun__ciation ___pronunciation___

3. bach__el__ or ___bachelor___   **6.** re__comm__end ___recommend___

7. priv___ilege___ ____privilege____   9. emba___rrass___ed ___embarrassed___

8. persist___ent___ ____persistent____   10. p___ur___sued ___pursued___

D.   Check the spelling lists in Appendix B to correct these frequently misspelled words. Then write the word correctly in the blank.

1. accomodations ___accommodations___   4. concensus ___consensus___

2. indispensible ___indispensable___   5. reccurence ___recurrence___

3. judgement ___judgment___   6. acknowledgement ___acknowledgment___

See pages 387–388 to check your responses.

CRACKING THE CODE

To save space, dictionaries use a code made up of symbols, special words, and abbreviations. The code varies somewhat from one dictionary to another. Cracking the code in at least your own dictionary is essential for intelligent dictionary use. Extremely important for career progress is knowing how and where to find information. Read 14 shows the code explaining dictionary spellings. Next you'll crack the dictionary code for syllables, pronunciation, definitions, and parts of speech.

SYLLABLES

Syllables are word parts that are sounded separately—such as di.vi.sion or in.ter.pret. Syllables help you with spelling, pronunciation, and word division at the end of a line. (See word division principles in Appendix D.) To determine syllables, look at the dots, hyphens, vertical lines, and accent marks of a few entry words in your dictionary. For example, look up *multiplication* to see how your dictionary shows syllables. Next look up *full-fashioned:* Notice a hyphen divides the first and second syllables, but an accent mark divides the second and third syllables.

Look up the syllables of the following four words. Write them in the blanks with a line between the syllables.

1. multiplication ___mul/ti/pli/ca/tion___

2. mumbo jumbo ___mum/bo jum/bo___

3. lexicographer ___lex/i/cog/ra/pher___

4. colloquial ___col/lo/qui/al___

See your dictionary for answers.

PRONUNCIATION

Pronunciation follows the entry word and is in italics (slanted letters), within parentheses ( ), or within backslashes \\. The syllables of words are spoken with either more or less force, commonly called *stress*. Stress marks show the relative amount of force for each syllable.

Diacritical Marks

Diacritical marks are a code showing how a word sounds if pronounced in Standard American English. To interpret the "code," see the pronunciation key in your dictionary. The pronunciation key is at the bottom of every page or alternate page, in the front matter, or in the appendix of your dictionary. Although the codes may differ from one dictionary to another, they result in the same pronunciation.

Using the pronunciation key in your own dictionary, show the preferred and the second pronunciations of the *a* in *apricot*.

Preferred: <u>Answers will vary</u> Second: <u>Answers will vary</u>

Answers are in your dictionary.

Schwa

The *schwa* symbol (ə) usually occurs in an unaccented syllable. Regardless of whether the vowel is *a, e, i, o,* or *u,* a schwa means the vowel sounds something like *uh,* as in the underlined vowels in the following words: <u>a</u>go, san<u>i</u>ty, c<u>o</u>mply.

Accent Marks

Diacritical marks and accent marks enable you to pronounce any word in the dictionary. The relative degree of loudness with which syllables are spoken is shown by accent marks ('). The three degrees of loudness (or stress) are:

Weak—no accent   Strong—primary accent   Medium—secondary accent

A two-or-more-syllable word always has a *primary* accent (or stress) mark and may have one or more *secondary* accent marks. In some dictionaries the darker accent mark means primary stress and the lighter one, secondary. Both **follow** the syllable to be stressed. In other dictionaries, the raised accent mark is for primary stress and the lower one for secondary. They both **precede** the syllable to be stressed. Open to any page (with word entries) of your own dictionary, and interpret the code for accenting. If necessary, see the explanation in the front matter of your dictionary.

1. Underline the schwa sound in these words: b<u>a</u>nana c<u>o</u>llect easil<u>y</u> gall<u>o</u>p circ<u>u</u>s

2. With help from the pronunciation key and the accent marks, say these words aloud correctly: *decadence, either, hors d'oeuvre, incognito, nausea, sadism, secretive*

See page 388 to check your answers.

DEFINITIONS IN THE DICTIONARY

Many words are used in more than one way; that is, they have more than one definition and may also be used as more than one part of speech. For such words, two or more numbered definitions may follow the entry word. If a definition has several shades of meaning, letters are used in addition to numbers. For words often used as more than one part of speech—for example, *fence* means one thing as a verb and another as a noun—separate definitions may be shown for each and sometimes even separate entries.

Recap

See if your dictionary can pass the NEW WORDS TEST. If yes, write the definition in the blank. If no, perhaps you need a new dictionary.

1. micropower ___electrical power generated in relatively small quantities___
2. ROMvelope ___a protective cover for a CD___
3. Cyberwar ___war in which computer systems damage or destroy enemy systems___
4. digital forensics ___obtaining legal evidence by examining computer networks and data___
5. Web log ___personal journal showing links to a Web site___

See answers on page 388.

PARTS OF SPEECH

Abbreviations show parts of speech for each entry—usually: n = noun, adj. = adjective, conj. = conjunction, vt or vi = transitive verb and intransitive verb. For now just be aware of the *v* for verb; ignore the *i* and the *t*. To interpret the abbreviations of parts of speech in your dictionary, check front matter or appendixes. Now look up *brown*. Notice that *brown* may be a noun, an adjective, or a verb.

Replay
15

To answer the following questions, use your own dictionary so that you'll become accustomed to its "code"—that is, the various symbols and abbreviations.

EXAMPLE   What part of speech is *faux pas* and what does it mean?
noun: a social blunder

1. What two letters are silent in *faux pas*? ___x and s___
2. How many syllables does *quarterback* have? ___three___
3. Which syllable in *quarterback* has the primary accent: first, second, or third? ___first___
4. Which syllable in *quarterback* has the secondary accent: first, second, or third? ___third___
5. Which syllable in *quarterback* has no accent? ___second___

6. What part of speech is *quarterback* when it means directing or leading? _____verb_____

7. Which syllable has the primary accent in the preferred pronunciation of *incomparable:* first, second, third, fourth, fifth? ____second____

8. Using a college dictionary or a thesaurus, find a synonym (word with a similar meaning) for *incomparable.* matchless, unsurpassed, etc.

9. Based on the pronunciation key in your dictionary, the first *e* in *wrecker* is pronounced the same as the *e* in what word? ____Answers will vary____ depending on the dictionary you're using

10. *Disinterested* and *uninterested* mean the same. (a) true _____(b) false ___X___ (c) maybe _____

11. What words in the pronunciation key of your dictionary illustrate the schwa (∂) sound? Answers will vary depending on the dictionary

12. What does *gymnasium* mean in Germany? A secondary school that prepares students for attending a university

13. According to the "preferred" pronunciation, which syllable has the primary stress in *affluent* and *affluence*? ____first____

14. Divide these words into syllables, and mark them for stress according to your dictionary's code: subtle ___subt' le___ rationale ___ra'. tion al'___ infrastructure ___in'fra.struc'ture (answers vary depending on dictionary)___

15. What is the other correct spelling of *catalog*? ___catalogue___ Which does your dictionary show first? ___catalog___ What two parts of speech are most common for this word? ___noun and verb___

Check your answers on page 388 and in your dictionary. The Pop Quiz for Reads 13–15 is on page 337.

AN OWNER'S MANUAL

This chapter, especially Read 16, is like an owner's manual for using your dictionary: To make good use of your college dictionary, look through the front matter, the inside covers, and the appendixes. A good dictionary is an essential reference for developing good language skills for your career. It's vital, however, to know what's in it, where to find the data, and how to use it, along with this text, to improve your English usage.

ETYMOLOGY

The origin or historical development of a word is the *etymology.*

Etymology is sometimes important to business communication and is often interesting. Most important now, however, is to recognize and interpret etymology so that you avoid the common error of confusing it with a definition.

Etymology is not provided for every word. When it is included, you usually find it within brackets [ ], double brackets [[ ]], or double lines / /. In some

Etymology of the word "sandwich": Montagu, the Earl of Sandwich, spent most of his time gambling. When hungry, he asked his servant for meat between two pieces of bread so that he wouldn't have to leave the gambling table to go to a dining table—and that's how *sandwich* got its name.

dictionaries the etymology follows the definitions. In others, however, it precedes the definitions. See the examples that follow: In the *American Heritage Dictionary*, you'll find the etymology within brackets following the definitions. However, in *Merriam-Webster's*, the etymology within brackets follows the pronunciation.

---

chau·vin·ism (shō-ˈvə-nĭzˊəm) *n.* **1.** Militant devotion to and glorification of one's country: fanatical patriotism. **2.** Prejudiced belief in the superiority of one's own gender, group, or kind: *"the chauvinism . . . of making extraterrestrial life in our own image"* (Henry, S.F. Cooper, Jr.). [French *chauvinisme*, after Nicolas *Chauvin*, a legendary French soldier famous for his devotion to Napoleon.] —**chauˊvin·ist** *n.* —**chauˊvin·isˊtic** *adj.* —**chauˊ·vin·isˊti·cal·ly** *adv.*

etymology within brackets []

By permission of *The American Heritage Dictionary of the English Language,* Fourth Edition.

---

etymology in brackets [ ]

chau·vin·ism \ˈshō-və-ˌni-zəm\ *n.* [F *chauvinisme*, fr. Nicolas *Chauvin* character noted for his excessive patriotism and devotion to Napoleon in Théodore and Hippolyte Cogniard's play *La Cocarde tricolore* (1831]] (1870) **1 :** excessive or blind patriotism — compare JINGOISM **2 :** undue partiality or attachment to a group or place to which one belongs or has belonged **3 :** an attitude of superiority toward members of the opposite sex; *also:* behavior expressive of such an attitude — **chau·vin·ist** \-v ə -nist\ *n. or adj.* —**chau·vin·is·tic** \ˌshō-v ə-ˌnis-tik\ *adj.* —**chau·vin·is·ti·cal·ly** \-ti-k(ə-)lē\ *adv.*

By permission of *Merriam-Webster's Collegiate Dictionary,* Tenth Edition.

The etymology may include an abbreviation of the languages a word was derived from, such as Middle English (ME), French (F or Fr), Latin (L), and so on; or it may tell how the word came into use. The abbreviation "f.," or the symbol < means "from" or "derived from." If a year is in parentheses, it is the earliest date the dictionary editors found the word in print. See the front matter of your dictionary for how to decode its etymology symbols. Don't memorize them; just get an overview and remember where to find them when needed.

MEMO FROM THE WORDSMITH—ETYMOLOGY EXAMPLES

Curfew	From an old French word *covrefeu*, which meant "cover fire," the hour the peasants of the Middle Ages had to extinguish their fires.
Sadist	Derived from an 18th-century French nobleman's name, the Count de Sade. He liked describing to friends how cruel he was to women.

Recap

Look up *sandwich* in your college dictionary.

1. Which dictionary did you use? <u>Answers will vary</u>
2. What symbols enclose the etymology? <u>Answers will vary</u>
3. Is the etymology before or after the definitions? <u>Answers will vary</u>
4. After what British nobleman was the word *sandwich* named?
<u>Earl of Sandwich</u>

Check answers to items 1–4 in your dictionary.

DEFINITIONS

If the meaning of a word has changed over time, some dictionaries list the earliest definitions first, right after the etymology, and continue in more or less chronological order. Other dictionaries give the central meaning first and end with older meanings, followed by etymology. Find out the system in your own dictionary by consulting the front matter.

Recap

Look up *nice* in your dictionary.

1. Are the current definitions listed first, or are the definitions in historical order; that is, early definitions before current ones: early first? <u>Answers will vary with dictionary being used</u> current first? ——— Which dictionary did you use? _____

Check answers in your dictionary.

USAGE, STYLE, OR FIELD LABELS

> **Usage, style, or field labels note something special about how the word is used.**

Certain words, definitions, and spellings are "labeled" according to systems that vary with the dictionary. Most words in the dictionary, however, are Standard English and don't require identifying "labels." A word labeled "archaic" or "arch" is used only in special contexts—although it was appropriate for general use in the past. Picture a new store in your neighborhood named YE OLDE SWEETE SHOPPE: The archaic word *Ye* is followed by three archaic spellings. Archaic differs from obsolete in that "obs." means the word is no longer used except historically. A word labeled "rare" is, as you can guess, rarely used. Certain other examples are known as field labels like *chemistry, biology, medicine;* regional labels like *British, Scottish, Southwest;* and style labels like *slang, colloquial, obscene, poetic, informal, taboo, technical, American.* The "labels" are usually abbreviated in dictionaries.

> **Consult your dictionary's "front matter" for the usage labeling system and translations of the labeling code; that is, how certain words are used.**

Some examples are archaic, technical, informal, humorous, slang, taboo, used mainly by children, regional, nonstandard, American, British, and many others, depending on the dictionary you're using.

Recap

1. In your dictionary, what usage label is shown for *ain't?* <u>Answers vary</u>

2. What usage label does your dictionary show for *mesa?*
   <u>Answers vary</u>

3. Give two examples of other words in your dictionary with a usage, field, or style label. <u>Answers vary; for example, *lefty, dandy, stoned*</u>

See page 388 to check your answers.

Although labeling systems and usage explanations differ, do not use *ain't* in business unless the intent is clearly to emphasize something jokingly.

OTHER INFORMATION

> Your college dictionary is probably the single most valuable reference source you own—and all in one portable book. Although much of the same data are online, it's often more convenient and faster to look something up in the paper dictionary beside you.

Read the table of contents or index, and scan the front matter and appendixes. See which types of information are in *your* college dictionary and where to find them. For example, in good, current dictionaries, you might find such assistance as A Guide for Writers, Avoiding Insensitive and Offensive Language, American Holidays, Science and Math Symbols, Currency (money) of Various Countries, Proofreader's Marks, Roman Numerals, Metric System, Foreign Language Expressions, Geographical and Biographical Information, Synonyms, Calendars used in various parts of the world, and even Commonly Misspelled Words.

If you use one of the newest dictionaries, you'll find hundreds of new words not in your old dictionaries. However, perhaps you have said, "How can I find a word in the dictionary if I can't spell it?" Well, one of the new dictionaries—*Microsoft Encarta College Dictionary*—lists about 700 frequently misspelled words and refers you to the correct spelling. The misspelled words have a line through them so that you can't mistake them for a correct spelling.

Synonyms are words with similar meanings. You can find them among the definitions of a word. Synonyms help a writer get the exact shade of meaning desired. If a synonym isn't labeled as such, you can discover one by reading the definition.

Word Power

Workplace Jargon

New vocabulary is often popularized in the workplace. You're not likely to find these in your old dictionary:

Barista   Maker and server of coffee in a coffee bar.

Tweenager   Someone between about 8 and 12 years old—no longer a small child but not yet a teenager.

Destination wedding   The couple travel to an exotic destination for their wedding ceremony.

Replay 16

Refer to your dictionary to answer these questions. Even if you already know the answer, look it up to get the practice of intelligent use of a dictionary.

EXAMPLE   What is the etymology of *smog?* SM(oke) + (F)OG Your answers may differ slightly from those on the answer page in the back of this book—but will mean about the same if you use a current college dictionary.

1. Give two synonyms for *idle*. __inactive, lazy, useless, futile__

2. What usage label does *yak* have when it means to talk too much? __slang__

3. Which definition of *quiz* is out of date or not used in American English? __queer or eccentric person, practical joke, hoax, to make fun of__

4. Divide *appendage* into syllables and give a synonym for it. __ap.pen.dage—adjunct, addition, accessory__

5. IRA is an abbreviation for __Individual Retirement Account__, but it's also an abbreviation for __Irish Republican Army__.

6. When was President Ulysses S. Grant born? __1822__

7. When did Guy Fawkes die? __1606__

8. Show the pronunciation (with diacritical marks) for *honorable* as it appears in your dictionary. __hon'r able__

9. What does *vacillate* mean? __to be indecisive__

10. The month January is derived from the name of the ancient Roman god of gates and doorways. What was his name? __Janus__

11. What does *graphology* mean? __the study of handwriting__

12. What does *stet* mean? __proofreader's mark meaning "let it stay; don't change it"__

13. According to your dictionary, in what year did Hawaii become a state in the United States? __1959__

14. Look in your dictionary for the information about writing Roman numerals. Write your year of birth in Roman numerals. __Answers will vary__

15. *Myxedema* is a disease of the __thyroid gland__.

Check answers on pages 388–389 and in your college dictionary.

Word Power

More Jargon

Digital divide   The division between people with unequal access to modern information technology.

HOAS   Abbreviation for **h**old **o**n **a S**econd—used in e-mails.

OPM financing   Financing a business using **o**ther **p**eople's **m**oney.

CHECKPOINT

Upon completing this chapter, you'll be familiar with the multiple kinds of information to be found in a comprehensive, up-to-date college dictionary. Be sure to explore the front matter and appendixes. Because a good dictionary is such a versatile reference book, keep a current one on your desk at home as well as at work.

Many of the new words are workplace-related and useful for your career. Technology, culture, and language change so rapidly that it is important to use the latest editions of textbooks and dictionaries.

WORD TO THE WISE

When checking in the dictionary for the meaning of a word, be sure to distinguish between the etymology and the definition.

Open your college dictionary to any page. Randomly point to any entry word, and then open your eyes. Write a check in the blank if you understand the dictionary code for interpreting the following:

_____ Spelling and variations of the spelling, if any

_____ Pronunciation

_____ Syllables

_____ Parts of speech

_____ Definitions

Write a check in the blank if you:

_____ Have read the table of contents or index of a college dictionary.

_____ Have looked through the front matter and appendixes of a college dictionary.

_____ Know the order in which definitions are listed in your college dictionary; i.e., oldest to current or current to oldest.

_____ Understand the labeling system in your college dictionary; for example, "labels" such as slang, vulgar, obsolete.

_____ Can locate and interpret the etymology of a word.

_____ Know what kinds of special information are in your dictionary's appendixes and front matter.

_____ Can find synonyms for *house* in your dictionary (abode, living quarters, etc.).

SPECIAL ASSIGNMENT

WORD SEARCH Test your vocabulary skill by filling in the blanks with a word that matches each definition. The first and last letter of a suggested word is shown.

A. Words About People

1. Person who looks on the bright side — o<u>ptimis</u>t
2. Person who looks on the dark side — p<u>essimis</u>t
3. Person who looks at reality — r<u>ealis</u>t
4. Person with a new and original approach — in<u>itiato</u>r
5. An organizer of a business — m<u>anage</u>r
6. A wealthy, powerful businessperson — m<u>agnat</u>e
7. One who judges art, wine, food, music, etc. — c<u>onnoiss</u>eur
8. An extremely wealthy person — z<u>illionair</u>e

B. Words About Places

9. Where money is legally produced — m<u>in</u>t
10. Where fish are kept — a<u>quariu</u>m
11. Where bees are kept — a<u>viar</u>y
12. A country in southern Africa — Z<u>air</u>e
13. A place with many trees — f<u>ores</u>t
14. Another place with many trees — a<u>rboretu</u>m

C. Words About Words

15. Words with opposite meanings — a<u>ntony</u>ms
16. Dictionary mark to show pronunciation — d<u>iacritica</u>l
17. Words with similar meanings — s<u>ynonym</u>
18. Words with same spelling but different meanings — h<u>omonym</u>
19. Shortened words — a<u>bbreviatio</u>ns
20. History or origin of a word — e<u>tymolog</u>y

PROOFREADING FOR CAREERS

The following press release was sent to a Seattle newspaper by a certified financial planner (CFP). However, I planted errors in it for your proofreading practice. Please proofread one word at a time, and correct the spelling and other errors. If you keyboard the corrected article, use standard format for a short report: Begin about one inch from the top, double-space, indent the first line of each paragraph, and use margins of about one inch. Use your spell checker, but correct other kinds of mistakes, too.

SOCIALLY RESPONSIBLE INVESTING

Two October work shops on socially responsible investing will be offered in Seattle and Bellevue by Certified Financial Planner Eric A. Smith and stock broker Wade Smith. The Humane society will sponsor the Bellevue workshop on Oct. 8. refreshments will be served, and admission is free.

Socially screened investing emerged as the result of churches and pension funds that wanted to make investments consistant with there ethicle values. In the last decade, however, socially responsible investing has groan from $40 billion two over $900 billion annually. this increase has ocured because the idea has sprad too many individual investers as well as other types of organizations.

Smith and Smith point out that most people carefully decides which organizations too donate time and money to each year. "Why," they ask, "shouldn't people consciously chose among financially sound investments that meat there ethicle, envirnmental, or social beleifs?"

Call 1 800 234 5678 to make your reservation for this informative and interesting workshop.

Some students want to spell out the contraction in the proofreading exercise. Explain that contractions are not errors and create a conversational tone, desirable in most business communications.

PRACTICE QUIZ

www.prenhall.com/smith

Take this Practice Quiz as though it were a real test. You'll find all the information in your college or unabridged dictionary or within Chapter 3.

1. In the blank, write the correct spelling of the two misspelled words: rediculous, personnel, occurred, ocassion, fulfilled. _____ ridiculous _____ occasion _____

2. What is the capital of the state of Georgia? _____ Atlanta _____

3. What does OPEC stand for? _____ Organization of Petroleum Exporting Countries _____

4. Illiterate expressions, vulgarities, and slang are not found in better dictionaries. True or False? _____ false _____

5. What does *colloquial* mean? _____ ordinary conversation or writing; informal language _____

6. In what year was Geronimo (the leader of the Apaches) born? _____ 1829 _____

7. What was the birth name of famous composer George Gershwin? _____ Jacob Gershvin _____

8. Give three synonyms for *small*. _____ miniature, tiny, minute (pronounced MY.NOOT) _____

9. In your dictionary, where is the etymology in relation to the definitions? _____ before or after, depending on what dictionary you're using _____

10. If you look up a word in a college dictionary and don't find a usage label, what does this mean? _____ The word is Standard English, and there are no special restrictions for its use _____

11. In college dictionaries, you can find a pronunciation key at the _____ bottom of each page or every other page _____.

12. The word *also* between two spellings of the same word means _____ the second spelling shown is less acceptable for general use _____.

13. Show with a dot how to divide *twinkling* between syllables. _____ twin.kling _____

14. The noun *pair* has two correct plural spellings. What are they? _____ pair or pairs _____

15. Correctly spell the seven misspelled words: accomodate, seperate, weird, persue, villain, bachlor, congradulate, superintendant, dilemma, pronunciation, priviledge
    accommodate, separate, pursue, bachelor, congratulate, superintendent, privilege

16. The *i* in the word *juvenile* is pronounced like the alphabet sound of *i*. True or False? _____ False _____

17. College dictionaries have a pronunciation key _at the bottom of each page or every other page_ and _in the front matter_ .

18. Where in a college or unabridged dictionary do you find detailed instructions and explanations for intelligent use? _in the "front matter" of the dictionary_

19. What is the plural of *addendum* and from what language does the word originate? _addenda, Latin_

20. What is a *tittle*? _the dot over the letter i (or other small marks used in writing and punctuation)_

21. A word with the same pronunciation as another but with a different meaning is a/an homonym (b) synonym (c) antonym. _____ a _____

22. Some college dictionaries have "back matter" that may be called an appendix or a supplement. Other dictionaries have supplementary information in what is usually called "front matter." Name three different types of information included in the appendix of your college dictionary.
    Answers will vary; sample answers are weights and measures, monetary units, chemical elements, Roman numerals,
    geographical information, populations of world cities (Webster's New World College Dictionary, Fourth ed.)

23. If your dictionary does not have "back matter," name three types of information in the "front matter."
    Answers will vary; sample answers are How to Use the Dictionary, Subject Labels for Specialist Areas, Abbreviations
    and Symbols, Usage in Crisis, The Internet as a Research Tool (Microsoft Encarta College Dictionary, First ed.

24. What is the capital of Spain? _____ Madrid _____

THE GLOBAL MARKETPLACE—ISRAEL

Hebrew	Spell	Say	Currency
GREETING	shalom	sha.lom	New Israel Shekel (NIS)
DEPARTING	shalom	sha.lom	
PLEASE, THANK YOU, YOU'RE WELCOME	bavakasha	ba VAK a sha	
I DON'T UNDERSTAND	Male: Ah-**nee** lo may-**veen**		
	Female: Ah-**nee lo** m'vee-**nah**		

Courtesy of Jenny Glass and Debi Rowe

Most dictionaries have charts showing currencies of the world. Such a chart might be near the listing for currency, money, or monetary. For example, *Microsoft Encarta College Dictionary* has an alphabetic list of currencies throughout the world, alphabetically arranged by the currency's name.

Grammar for Grownups

4

Identify Tools of the Trade

After Completing Chapter 4, You Will

> Apply your knowledge of the parts of speech to understand the English principles that will help your career.

"Even a brief survey of grammatical issues leaves one somewhat in awe at the extraordinary variety of patterns that exist in the languages of the world. Repeatedly the lesson is brought home that there is nothing sacrosanct or superior about the grammar of any one language."

—David Crystal,
Cambridge University Press

Chapter 4 Pretest is in the *Instructor's Resource Kit*.

Simplified, expedient definitions of parts of speech help students understand the overall structure of English; the essential details follow in Chapters 5–8. Only grammar terms and principles that enable students to improve language skill are included. For additional grammar information, see "Grammar for the Expert" in your *Instructor's Resource Kit*.

In some classes, it is useful to discuss facts of African American, Hispanic American, and other dialects vis-à-vis English/American grammar. A dialect is neither "bad" nor "incorrect" English and is all right to use in appropriate settings, such as with family and friends. However, it's important to know and use Standard English when your objective is success in a career. Not using Standard English handicaps career progress.

Read **17**

Grammar for grownups means reviewing what you learned in the past about the basic tools of the English "trade"—the parts of speech. This review will help you communicate correctly and effectively for your career. Unless you plan to be an English teacher, it isn't essential to study grammar in depth. What you *do* need is to be able to identify, so that you'll use correctly, the basic tools of the English "trade," the parts of speech.

The parts of speech are the system for organizing words into the following categories: nouns, pronouns, verbs, adjectives, adverbs, prepositions, conjunctions, interjections. Every word fits into one of those categories—depending on how the word is used in the sentence. If you've learned the parts of speech in the past, this chapter should be a pleasant review for you.

Uh oh; I just heard somebody say, "Oh no, not that; I never *could* understand those things." Well, the *English for Careers* method is different. You'll discover an enjoyable and efficient way to acquire language style that helps your career and enables you to communicate correctly and effectively for your career. Try it with an open mind.

WHO, WHOM, WHAT, OR WHERE

NOUNS

> Nouns name persons, animals, things, ideas, places, times, activities; they may tell who/whom or what.

Proper nouns name specific persons, animals, things, ideas, places, times, or activities—and begin with capital letters: Business Department, Professor Alicia Holmes, Toronto, Fido, Republican Party, Halloween.

Common nouns begin with lowercase letters: business, woman, city, dog, organization, holiday. Here are some nouns shown as individual words and then in sentences:

PERSONS	professor accountant woman Twileen members brother
	My **brother,** who is an **accountant,** is also a **professor.**
	Twileen and the **woman** with her are both **flirts.**
ANIMALS	goldfish Fido eagle dog flea dinosaur
	Fido has **fleas.**
	Dinosaurs were prehistoric **animals.**
THINGS	piano sandwich *People Magazine* subway seat pen
	A **piano** and a **pen** were advertised in ***People Magazine.***
IDEAS	integrity Buddhism modesty beauty success business
	My **beauty** is exceeded only by my **modesty.**
	Integrity in **business** contributes to **success.**
PLACES	home company factory city New Zealand porch
	Her **home** in **New Zealand** has a purple porch.
	The **company** will build a **factory** in this **city.**
TIMES	birthday holidays week today Thanksgiving Day Monday
	Today is my **birthday** and **Thanksgiving Day.**
ACTIVITIES	drinking running sleeping driving eating crying
	The baby spent the entire day **sleeping, eating,** and **crying.**
	Drinking and **driving** or **running** are dangerous.

Recap

Two of the preceding example sentences have nouns not printed in bold type.
Find these nouns and write them in the blanks.

1. _____porch_____ 2. _____baby_____ 3. _____day_____

4. Circle the nouns in the following sentence:

   When (interviewers) speak with (applicants), they look for (energy), (competence), (loyalty), (skill), (ambition), and (flexibility). (8 nouns)

5. Look around the room and write three common nouns that name what
   you see:
   Answers will vary: ____window____   ____sofa____   ____carpet____

6. An example of a proper noun is *Katherine.* Some common nouns that
   could refer to Katherine are mother-in-law, friend, sister, wife, mathematician. Write one proper noun and three common nouns that may
   be used for someone you love:

   Possible Answers: Proper Noun: ____Seymour____

   Common Nouns: ____principal, pilot, pal____

See page 389 for the answers.

RECOGNIZING NOUNS

Most nouns have two forms: singular for one and plural for more than one:

Singular	Plural
country	countries
child	children
loaf	loaves
boy	boys

Certain nouns—such as honesty, ethics, helpfulness—do not have plurals.

PRONOUNS

Pronouns substitute for nouns.

We often substitute pronouns for nouns to avoid repeating the nouns. Like nouns, pronouns tell *who, whom,* or *what*. For example, the pronouns *she, her,* and *herself* could substitute for any of the following nouns.

Sue   woman   stockbroker   lady   Ms. Gilchrist   daughter   student

NOUNS **Jerry Sider** has finished **Jerry Sider's work.**

PRONOUNS **He** has finished **his** work. **He** has finished **it.**

NOUNS **Dick Tracy** won the **prize.**

PRONOUNS **He** won **it. Who** won **it? Somebody** won **that.** Did **anyone** win **this?**

Chapter 6 provides answers to questions about specific pronouns like **I/me** and **who/whom.** For now, just know a pronoun when you see one.

"Personal" pronouns come in three categories—first person, second person, and third person. Each person has a singular and a plural form.

FIRST-PERSON PRONOUNS · The person or people speaking or writing:

Singular	I	me	my	mine	myself
Plural	we	us	our	ours	ourselves

SECOND-PERSON PRONOUNS · The person or people spoken or written to:

Singular or Plural	you	your	yours
Singular only	yourself		
Plural only	yourselves		

THIRD-PERSON PRONOUNS · The person(s), people, or thing(s) spoken or written about:

Singular Masculine	he	him	his	himself
Singular Feminine	she	her	hers	herself
Singular Neutral	it	its	itself	
Plural Neutral	they	them	their	theirs	themselves

Indefinite **pronouns don't specify whom or what they substitute for:**

who	this	everyone	everybody	everything
whoever	that	someone	somebody	something
whom	these	anyone	anybody	anything
whomever	those	no one	nobody	nothing
whose	each			

Pronouns substitute not only for nouns but also for other pronouns.

Everyone is responsible for keeping **his** own locker secure. [Personal pronoun *his* substitutes for indefinite pronoun *Everyone.*]

RECOGNIZING PRONOUNS

A pronoun may replace a noun or pronoun previously mentioned or understood. For example, "**It** is hot today" really means "The **air** is hot today." We understand that *it* substitutes for *air.*

Replay
17

A.   Circle the 24 nouns in the following five sentences. If a noun is made up of two or more words, use one circle.

EXAMPLE   Don't tell ethnic jokes while on the job at Abernathy Jones, Inc. (3)

1. A smile is likely to help even the most challenging situation. (2)

2. Well-written correspondence in the United States is less formal than it is in Asia, South America, Europe, and Africa. (6)

3. It is important to your career to accept people of various cultures and to judge them as individuals. (4)

4. Avoid using slang when you speak with co-workers for whom English is a second language. (4)

5. The multicultural neighborhood, classroom, and workplace are typical in the United States, Canada, Great Britain, and many other parts of the world. (8)

B. Insert pronouns in the blanks to replace the words in brackets.

EXAMPLE [My brother and my father] __They__ found [a strange item] __it__ in [the person I'm writing to] __your__ house.

1. [The temperature] __It__ is too hot today for running in the park.

2. [This man] __He__ will be the next president of the club.

3. [Which person] __Who__ will be the next governor?

4. [Ashley] __She__ talks to [Ashley's] __her__ friends on the phone for hours.

5. [These friends] __They__ urged [Twileen] __her__ to marry [George] __him__.

C. Underline the 19 pronouns in the following sentences.

EXAMPLE   My luck improves—everyone's does—whenever I work at it.

1. His report is better than hers.

2. He spent more time organizing it than she did.

3. She herself baked these and sold them at the fair.

4. Somebody said, "Everyone who was anyone was invited to it."

5. We were paid nothing for our work, but we learned almost everything about the job.

D. Fill in the blanks with the pronouns that represent each of the following:

1. The person speaking or writing (first-person singular):
   __I__   __me__   __my__   __mine__   __myself__

2. You and I together (first-person plural):
   __we__   __us__   __our__   __ours__   __ourselves__

3. The person or people you are speaking or writing to (second-person singular):
   __you__   __your__   __yours__   __yourself__   __yourselves__

4. Nina Beth (third-person singular, feminine):
   __she__   __her__   __herself__   __hers__

5. Steve (third-person singular, masculine):
   __he__   __him__   __his__   __himself__

6. The Hummer:
   __it__   __its__   __itself__

7. Joyce, Ron, and the cat Piglet (third-person plural):
   __they__   __their__   __them__   __theirs__   __themselves__

After verifying your answers on page 389, check one of these blanks:

How did you do? Wonderful_____ So so_____ I'd rather not say_____

VERBS—DOING, HAVING, BEING, AND HELPING

Verbs are action, being or sense, or "helping" words. Every sentence has at least one verb. Action or being verbs are sometimes preceded by one or more "helping" verbs, thus creating a "verb phrase."

ACTION VERBS

Most verbs express some kind of action; for example:

work   worry   invite   write   dance   receive   have   own   love   pay
read   call   run   think   relax   proofread   do   hop   work   play

Circle the eight action verbs and underline the pronouns in this paragraph.

He (wrote) the application letter in Word and (proofread) it. When the manager (received) it, she (looked) it over. After she (read) all the applications, she (invited) a few applicants for interviews. She (has) an assistant who (helped) her.

Check your answers on page 390.

BEING VERBS

"Being" verbs (also called *existence* or *linking* verbs) include forms of the verb *to be*, verbs of the senses, and a few other verbs. *Being* verbs "link" the subject of the sentence to a word or words that tell something about the subject.

be*   being   been   am   are   is   was   were   become   seem
remain   appear   feel   sound   taste   smell   look

The "being" verbs are highlighted in the following sentences:

You **seem** calm and **sound** happy.
She **is** glad that you **feel** good.
The soup **looks, smells,** and **tastes** good.
The pie **was** good.

Circle the ten "being" verbs in the following sentences:

She (is) capable. He (appears) efficient. I (am) glad the program (sounds) interesting. All employees (were) on vacation last week. It (seems) as though Gene (was) at the beach all summer. Fresh apple pie (smells) good. The raincoat doesn't (feel) wet. Eric (remains) on the Dean's List.

Check your answers on page 390.

---

*Avoid "dialect style" *be*—as in "We be going." If you think you speak in a dialect that makes you sound poorly educated, ask a friend or an instructor to remind you when you do.

Some "being verbs" are also used as "action verbs."

Action verb:	Did you **taste** the pasta?
Action verb:	Did you **smell** the stew?
Action verb:	She **looks** at everything in the store.
Being verb:	The stew **tastes** good to me!
Being verb:	They **smell** good too.
Being verb:	That cat **looks** healthy.

HELPING VERBS

To show time, possibility (maybe), or emphasis, a helping verb may precede the main verb—either action or being. Some *being* verbs also function as *helping* verbs.

Notice the *helping* verbs in the following sentences; each one precedes a main verb. The helping verbs may show such ideas as emphasis, possibility, or time (now, before, or later).

Underline each helping verb and main verb in the following sentences:

Judge Hoover <u>is reading</u> all my reports and <u>should finish</u> soon.

He <u>had danced</u> with the princess, but the Queen <u>did</u> not <u>like</u> it.

They <u>were enjoying</u> their vacation, but they <u>are returning</u> home today.

She <u>did love</u> him, and he <u>does visit</u> her often.

She <u>has seen</u> him every Sunday afternoon but <u>does</u> not <u>greet</u> him.

He <u>had selected</u> the ring. They <u>have dined</u> at the Ritz often.

Terms like *infinitive* and *participle* discourage some students at this point. They probably do better with terms such as "to plus a verb" and "*ing* words."

RECOGNIZING VERBS AND WORDS THAT LOOK LIKE VERBS BUT AREN'T VERBS

Ing Words

- Most verbs can add *ing*: *see seeing, have having, eat eating, be being,* and so on. When *ing* words express action or existence, they are verbs and are always preceded by a helping verb.

  Recent studies show that honesty **is paying** off in business.

- When an *ing* word *names* an activity, however, it's a noun called a *gerund* and is **not** preceded by a helping verb.

  Noun (Gerund): Paying bills has never been my favorite activity. (gerund—*paying*; verb—*has been*)

To + Verb

- When *to* precedes a verb—such as *to eat, to work, to swim*—the result is a noun naming an activity, not a verb. *To* plus a verb is called an *infinitive*. Infinitives are used as **nouns.** That sounds hard to believe; so just look at the sentence below to see how it works:

  To know me is **to love** me. *To know* and *to love* are *infinitives*; the verb is **is.** Notice that *to know* and *to love* name activities and are therefore infinitives, which are used as nouns, as stated in the paragraph above.

Replay 18

A. Circle the nine being verbs in the following sentences, and underline the one action verb.

She (is) capable. He (appears) efficient. I (am) glad the programs (are) interesting. All employees (were) on vacation last week. We <u>think</u> Gene (was) at the beach all summer. Fresh apple pie (sounds) good. Nina (feels) good because she will (be) on the Dean's List.

B. Circle the complete verbs—helping and main.

I (do dance) every Wednesday, but I (haven't met) the president. Judge Hoover (may read) my reports. The jury members (are reading) his reports. He (has danced) with the princess often. Professor Dowd (will be going) to Cairo. In June you (will have been working) here for five years. He (might have danced) with Diana in London. Ms. Moultry (does sign) all the letters herself.

C. Fill in the blanks with verbs that make sense in the sentence.

EXAMPLE   A word to the wise __is__ enough.—Laurence Sterne

1. From time to time most office professionals __seek__ more computer training, but they may not __have__ time for regular college classes.

2. Four out of five companies __spend__ a great deal of money training their employees.

3. We __are asking__ for your opinion. (helping verb and main verb needed)

4. Workplace etiquette __is__ important to employers.

5. Many firms __spend__ thousands of dollars for etiquette seminars for their employees.

D. First, underline the verbs in the following sentences. When a main verb is combined with a helping verb, underline both. (See *could help* in the example.) Second, circle the nouns and draw a triangle around the pronouns.

EXAMPLE   (Paul) <u>could help</u> (Ms. Adams) with △her△ work.

	NO. OF MARKS
1. (Applicants) <u>are judged</u> on △their△ (behavior) as well as △their△ (knowledge).	6
2. (Etiquette) <u>is</u> (part) of all (activities) on the (job).	5
3. △No one△ <u>has</u> much (data) on that (topic).	4
4. <u>Do</u> not <u>call</u> (clients) or (colleagues) by △their△ first (names) until △you△ <u>are</u> sure △it△ <u>is</u> customary in △your△ (organization).	12
5. (Sheela Danielle) <u>was</u> the (winner) of the (scholarship).	4
6. The (secretaries) <u>were keyboarding</u> the (answers) in the (blanks).	4
7. △Someone△ <u>should have completed</u> the (job) by (Tuesday).	4
8. △Who△ <u>should go</u> to the (conference) in (Las Vegas)?	4

9. The husband of the manager is known for his courtesy.   5

10. He will ask for the mail.   3

11. These have been sent to Southeastern Community College.   3

12. They know a lawyer in the building who will represent me.   7

13. Barbra Streisand rescheduled three performances in Anaheim.   4

14. Larry should deliver the tickets to you.   4

15. It is in the drawer next to the computer.   4

16. Their receptionist will mail these on Monday.   5

17. The Aldrich Company received everything.   3

18. Our auditors checked everyone's books for accuracy.   6

19. Frankenstein built a factory in a city on the moon.   5

20. The children dressed themselves.   3

Check your answers on pages 390–391. Great_____ Good_____ Fair_____ Ugh_____

Read
19

THE MODIFIERS

Modifiers—adjectives or adverbs—are words that explain or describe.

ADJECTIVES

> Adjectives modify (or describe) nouns and pronouns. They add information about which, what kind, or how many.

Articles—A Kind of Adjective Often Called "Noun Markers"

Articles tell "which one." You can easily memorize them because there are only three: *a, an, the.* Articles are a "warning" that a noun is coming up. In Chapter 8 you'll review when to use *a* and when to use *an.*

> **An** apple **a** day keeps **the** doctor away.

Pointing Adjectives

Also easy to memorize are the four adjectives that point at nouns—just as you might do with an outstretched finger. If *this, that, these,* and *those* "point" at nouns following them, they are adjectives that tell "which one."

> **this** book   **these** books   **that** jacket   **those** jackets

We just heard an alert student protest, "Hold on a minute, Leila: In Read (17), you told us *this, that, these,* and *those* are pronouns. Now you're telling us they're adjectives!" *This, that, these,* and *those* **are** adjectives—when they precede a noun, but are pronouns when used *instead* of a noun:

adj   n
> **This** book is the one I want to read. [*This* points at *book*.]

However, *this, that, these,* and *those* are pronouns *if:*

1. The noun is left out.

    pron
    These are interesting. [The noun *books* is understood.]

        pron v   n pron v
2. **That** is the book I want to read.

        pron v   pron pron v
3. **This** is the one I want to read.

You may be questioned about the pronoun *this* in item 3. Refer students to "pointing adjectives" in Read 19.

Limiting Adjectives

Words that "limit" nouns in the sense of quantity are also adjectives. They tell "how many." Limiting adjectives are words such as these:

more	enough	most	several	few	all	each
any	many	some	50 or fifty	no	every	numerous

               adj   n   adj   n
I bought **several** reams of paper at Cheapo Depot a **few** months ago.

            adj   n   adj   n
We have **enough** employees to pack **50** boxes properly.

Descriptive Adjectives

Descriptive adjectives tell "what kind." To tell what kind of *house,* you might choose *yellow, brick, contemporary, shabby, two-story, luxurious,* and so on. By carefully selecting descriptive adjectives, your words create a picture for the reader or listener. In the following sentences the descriptive adjectives are in bold print. The noun or pronoun they describe is underlined.

Respected <u>employees</u> work well with people from **diverse** <u>backgrounds</u>.

Upscale gourmet <u>dinners</u> are served in that **elegant** <u>restaurant</u>. (*that* is a pointing adjective)

The **new work** <u>environment</u> emphasizes **individual** <u>responsibility</u>.

Recap

A word that describes, explains, or restricts a noun or a pronoun is an **adjective.** The adjectives below are in bold print. Underline nouns or pronouns they describe.

1. Adjectives usually precede nouns:

    He has **an** <u>idea</u> that Twileen likes **rock** <u>music</u>, **red** <u>dresses</u>, and **hot** <u>potatoes</u>.

2. Adjectives often follow "being" verbs:

    You seem **smart.** <u>Ethics</u> is **important.** <u>Ms. Parks</u> appears **happy.**

3. An *ing* word is an adjective if it describes a noun or pronoun:

    He folded **multipurpose** <u>paper</u> to make **a flowering** <u>blossom</u>, **a flying** <u>bird</u>, and **a kneeling** <u>nun</u>.

Check your answers on page 391.

ADVERBS

Both adjectives and adverbs describe or explain. Adjectives describe or explain nouns or pronouns. Adverbs, however, describe, limit, or explain verbs, adjectives, or other adverbs. Adverbs add information about when, where, how, or how much. Many adverbs—but not all—are formed by adding *ly* to an adjective.

Recap

Fill in the blanks with adverbs; be sure to spell them correctly.

	ADJECTIVE	ADVERB
EXAMPLE	peaceful	peacefully
	quiet	quietly
	exceptional	exceptionally
	intelligent	intelligently
	attractive	attractively
	final	finally
	real	really

See answers on page 391.

Adverbs Describe Verbs

adv   v   adv

Always prepare invoices **carefully.** [The adverb *always* tells "when" about the verb *prepare*. The adverb *carefully* tells "how" about the verb *prepare*.]

Adverbs Describe Adjectives

adv   adj   adv   adj

This **extremely expensive** book is required for an **especially important** course.

    [The adverb *extremely* tells "how" about the adjective *expensive*. The adverb *especially* tells "how" about the adjective *important*.]

Adverbs Describe Other Adverbs

adv   adv   adv   adv

Ms. Wanielista works **so efficiently** that she **almost never** makes a mistake. [The adverb *so* describes the adverb *efficiently*; *efficiently* describes the verb *works*; the adverb *almost* describes the adverb *never*; *never* describes the verb *makes*.]

WORD TO THE WISE

A word that "modifies" (tells about) a verb, an adjective, or an adverb is an adverb. While many adverbs end in *ly*, some words often used as adverbs don't end in *ly*:

almost   more   never   so   very   even   much   not   too   well

Replay 19

A. Circle the nouns: underline four limiting and five pointing adjectives and one article:

EXAMPLE   Please pay this (bill) within five (days).

1. These <u>rooms</u> don't have any (air-conditioning).
2. They have <u>several</u> (windows), however.
3. That (office) is mine; <u>this</u> is his.
4. <u>Some</u> (homes) on <u>this</u> (street) are (tri-level).
5. Ten <u>attorneys</u> left the (firm) <u>this</u> (year).

B. The following paragraph is from a sales letter meant to bring business to a resort hotel. In the blanks, write the adjectives (descriptive, pointing, or articles) that modify the ten nouns in bold type. Numbers in parentheses tell how many adjectives to look for. For item 1, write just one adjective *(an)* in the blank to modify the noun *individual*. For item 2 write the three adjectives describing the noun *hotel*. Continue until you complete all ten items.

Surveys show you are an **individual** who would be interested in visiting an exclusive resort **hotel** on the beautiful **Pacific,** where the crystal blue **waters** meet the white sandy **beaches.** You will enjoy exquisite guest **villas** in Mediterranean **decor** with luxurious Jacuzzis. The **climate** is smogless, sunny, and mild. You will want to stay in this **paradise** forever.

1. _____an_____ individual (1)
2. ___an exclusive resort___ hotel (3)
3. ___the beautiful___ Pacific (2)
4. ___the crystal blue___ waters (3)
5. ___the white sandy___ beaches (3)
6. ___exquisite guest___ villas (2)
7. ___Mediterranean___ decor (1)
8. ___luxurious___ Jacuzzis (1)
9. ___The smogless, sunny, mild___ climate (4)
10. ___this___ paradise (1)

C. Write three different sentences. In each sentence, use an article, a limiting adjective, a describing adjective, and a pointing adjective. Circle each adjective.

EXAMPLE   (The) woman bought (those) (two) (red) dresses.

Answers will vary.

1. (Some) people in (that) room have (a) (good) attitude.
2. (This) morning I found (two) dimes near (a) (red) phone.
3. (These) companies bought (a) (new) computer for (every) manager.

This is a good activity for students working in small groups.

Read 20: FAN BOYS (**f**or, **a**nd, **n**or, **b**ut, **o**r, **y**et, **s**o) is an acronym for memorizing coordinate conjunctions. Another learning technique is for students to practice saying aloud quickly— *and, but—or, nor, for— yet, so. Yet* may be either a coordinate conjunction or a conjunctive adverb. It may be preceded by either a comma or a semicolon. However, I suggest not mentioning punctuation unless someone asks. The coordinate conjunction *so* is too informal for written business communication—except for e-mails or memos. However, *so that* is a good subordinate conjunction.

D. In each blank, insert a one-word adverb to describe or explain the word in bold type. The adverb you select will tell when, why, where, how, or how much.

EXAMPLE   Dolores Denova **speaks** _clearly_ . Possible Answers

1. He **was hired** _yesterday, today, immediately_ .
2. The children in that class **read** _well, poorly, quickly_ .
3. Do you think that I **drive** _carefully, recklessly, fast_ ?
4. You **added** the figures _correctly, accurately, slowly_ .
5. They **should** _always, never_ **use** good quality paper for resumes.
6. When you **go** _home, away, there_ you will meet our dog Fred.
7. A new computer **would cost** _more, less_ than upgrading the old one.
8. Although Dan Druff is on the team, he _rarely, seldom, never_ **plays** in important games.
9. We didn't believe Delilah when she said she **had** _really, carefully, never_ **cut** Sampson's hair.
10. This _elegantly, tastefully_ **designed** home was featured in *Architectural Digest.*
11. Those are _extremely, very, such_ **expensive** restaurants.
12. Nutrition is one of the _most, least_ **important** aspects of health.
13. Good word processing software is _not, very, sometimes_ **hard** to find.
14. Employees who ignore etiquette while at work **are** _usually, often, always_ **disliked.**
15. I think Billie Clubb is _much, even_ **sexier** than her sister.

Check answers on page 391.

Turn to page 338 and take a Pop Quiz on Reads 17–19.

THE CONNECTORS

The connecting words—conjunctions and prepositions—help us understand how words within a sentence relate to one another to create meaning.

CONJUNCTIONS

> Conjunctions connect two words, phrases, or clauses. A relationship between the words or groups of words is shown by the choice of conjunction.

Coordinate Conjunctions

> Coordinate conjunctions—and, but, or, nor, for, yet, so—join equals, such as two or more nouns, adjectives, phrases, or sentences. Study the following examples so that you can identify conjunctions. An easy way to memorize them is to think of the words "FAN BOYS"—**f**or, **a**nd, **n**or, **b**ut, **o**r, **y**et, **s**o.

FOR   Twileen decided not to become a movie star, **for** she feared the money and fame would go to her head. (Coordinate conjunction *for* joins two sentences into one.)

AND   Studies show that honesty, ethics, **and** social responsibility pay off in business. (A series of nouns are joined with *and*.)

NOR   He will neither attend the convention **nor** visit the showroom. (Two verbs—attend and visit—are joined with *nor*.)

BUT   George said it's hard to get a job as a movie star, **but** the pay is good. (*But* joins two sentences.)

OR   Workplace ethics include not violating civil law **or** company policy. (Two nouns—law and policy—are joined with *or*.)

YET   English is the international language of business, **yet** we should all know how to say "please," "thank you," and "hello" in several languages. (*Yet* joins two complete sentences into one.)

SO   Adventures of Twileen and George are spread throughout this book, **so** be sure to read every word. (*So* joins two complete sentences into one.)

Recap

Say the seven coordinate conjunctions quickly now to help you remember them. Then write them here: and, but, or, nor, for, yet, so

Dependent Conjunctions

> **Dependent—also called subordinate—conjunctions precede a word group that cannot stand by itself as a sentence. Further details regarding conjunctions are in the punctuation chapters.**

The words below are often used as dependent conjunctions. The sentences following the lists show the dependent conjunctions in bold type.

after	although	as	because	before
even though	if	since	so that	than
unless	until	when	which	while

Business letters in the United States and Canada are less formal **than** in most other parts of the world. [*Than* is used as a dependent conjunction.)

The new branch office will be successful **if** Petrina manages it. [*If* is a dependent conjunction preceding the noun *Petrina* and the verb *manages*.]

If the order of the sentence is reversed, the dependent conjunction moves to the beginning of the sentence.

Because Petrina manages it, the branch office will be successful.

Unless it is repaired this week, we can't use the copier for the report.

Conjunctive Adverbs

> **Conjunctive adverbs, also called transitional expressions, join two complete sentences into one. Examples are:**

however	therefore	consequently	moreover	furthermore
also	for example	nevertheless	yet	in addition

My friend in Spain works from 8 A.M. to 6 P.M.; **however,** she has an hour and a half for lunch.

Recap

Observe how the conjunctions are used in the following sentences. Circle the dependent conjunctions, underline the coordinating conjunctions, and draw a heart around the conjunctive adverbs:

1. What kind of car would you drive if you could have any one you want?

2. The European Summit in Nice was supposed to last two to three days, but it ended up lasting a full week.

3. Some people don't accomplish their goals because they either don't set specific goals or don't keep them in mind.

4. Dinner in Turkey often begins with soup and ends with dessert and fruit.

5. In Turkey it is customary to have three sit-down meals a day; therefore, household members are together at mealtime and share events of the day.

Answers are on page 391.

PREPOSITIONS

> A preposition begins a word group called a prepositional phrase. The preposition shows how the noun or pronoun ending the phrase relates to a word elsewhere in the sentence. A prepositional phrase NEVER HAS A SUBJECT OR A VERB but does have an object.

Look through this list of words often used as prepositions:

about	among	beside	from	of	to	with
above	around	between	in	off	toward	within
across	at	by	inside	on	under	
after	behind	during	into	over	until	
against	below	except	like	since	up	
along	beneath	for*	near	through	upon	

A preposition cannot usually operate by itself; it is likely to be meaningless without the noun or pronoun—called the **object of the preposition**—that ends the word group. The preposition and its object plus any words between them are called a **prepositional phrase.** Only the preposition and object are required to make the phrase. IMPORTANT: If the word group has a **verb,** it is not a prepositional phrase. In these examples the prepositions are in bold type, and the objects are underlined.

under the antique table   **below** the newly painted roof

with Victoria's little sister   **through** the air

on the decrepit plane   **across** the street

from Twileen's window   **in** the leather briefcase

to you and me   **after** lunch

However, you can sometimes use a preposition without stating its object, which may be understood. The old saying, "Never end a sentence with a preposition" has been updated so that it sometimes is all right, especially in conversation. When the

*For is a conjunction when it means "because." Otherwise, it's a preposition.

Read 20 | The Connectors   75

Here are examples of typical PREPOSITION-RELATED ERRORS:

1. Where are you going **to**?

2. Where have they been **at**?

3. Don't take any money **off of** him.

4. Divide the work **between** the three of you.

5. After waiting **on** him for three hours, she left the church and sobbed.

6. To **who** should she return the gifts?

7. Send them to Donna and **I.**
Answers:

1. going? (delete **to**)

2. delete **at**

3. from

4. among

5. for

6. whom

7. me

object of the preposition can be easily understood, it is all right to omit it. Here are some examples in which the preposition is "understood":

Please come **in.**   Don't jump **out.**   She went **under.**   I'll go **along.**

Some prepositions are also used as dependent conjunctions—but only when a noun or pronoun and a verb are part of the word group that follows. For now, just be able to recognize prepositions in a sentence.

> conj   pron   v
> He hasn't been here **since** he spoke to the manager. [*Since* is a dependent conjunction.]

> prep
> He hasn't been here **since** last month. [*Since last month* is a prepositional phrase.]

Study the prepositions in the following sentences. The prepositions are in bold type, and the prepositional phrases are underlined.

(a) Ozzie hid the diamonds **under** the bed **near** the door.

(b) Victoria will go **with** you **in** the single-engine plane **to** a small village **beside** a mountain.

(c) They went **into** the city **by** bus and then dined **at** a coffee shop.

(d) George saw the newly painted house **from** Twileen's window.

Recap

1. What are the objects of the prepositions in sentence (b) above?

   ____you____   ____plane____   ____village____   ____mountain____

2. How do you know "hid the diamonds" in sentence (a) and "newly painted house" in sentence (d) in the preceding sentences are not prepositional phrases?

   ____Hid____ is a verb and ____newly____ is an adverb. A prepositional phrase begins with a preposition, ends with a noun or pronoun, and never includes a verb.

Answers are on page 392.

Although prepositions are usually small words, they are important for showing the exact meaning of a sentence. Notice how the preposition shows the relationship between *pen* and *desk* in the sentences that follow:

The pen is **on** the desk.   The pen is **in** the desk.

The pen is **under** the desk.   The pen is **near** the desk.

The pen is **beside** the desk.   The pen is **behind** the desk.

RECOGNIZING CONNECTING WORDS

Conjunctions connect (or join) words, groups of words, or complete sentences.

Prepositions are connecting words that show the relationship between an object and another word in the sentence. Say the preposition aloud and then ask "whom" or "what"; if you see an answer, that word is the object. A verb cannot be part of a prepositional phrase.

Replay 20

A. Insert a suitable coordinating conjunction, dependent conjunction, or conjunctive adverb in each blank. Possible Answers

1. Neither Leianne _____nor_____ Michael wants to move to New York.

2. Ning has led many tour groups through China, _____but_____ he enjoys teaching Chinese brush painting more than anything else.

3. I was there to meet Sue _____and_____ Steve _____when_____ they arrived.

4. Victoria has worked for Pacific Telephone Company for many years, _____although_____ she would prefer to work at George's company.

5. _____Since_____ Victoria was an administrative assistant _____while_____ she took management classes, she understands the problems of office support personnel.

6. Management consultant Peter Drucker said the better employees are, the more mistakes they make _____and_____ the more new things they try.

7. We won't know whether this department will be profitable _____until_____ we check the records.

8. _____"If_____ you want the rainbow, you must put up with the rain," said Dolly Parton.

9. She did not make personal phone calls at the office _____because_____ she wanted to be known for her good work ethic.

10. "Do you prefer Kona _____or_____ Kauai for our honeymoon?" asked Twileen.

B. Write a preposition in each blank to show a relationship between the verb *walked* and the noun *mountain*. The choice of preposition controls the meaning of the sentence. Possible Answers

EXAMPLE   Twileen and George walked _____around_____ the mountain.

1. _____into_____
2. _____by_____
3. _____to_____
4. _____through_____
5. _____over_____
6. _____across_____
7. _____below_____
8. _____under_____

C. Underline the prepositional phrases and circle the prepositions.

EXAMPLE   Victoria looked (at) the return address and threw the envelope (into) the trash.

1. Courtesy (in) the workplace means the practice (of) kindness and consideration (toward) other employees and customers. (3 phrases)

2. Advertising has a profound influence ⟨on⟩ the behavior ⟨of⟩ people and ⟨on⟩ their lifestyles. (3 phrases)

3. Do you know the difference ⟨between⟩ a dream and a goal? (1 phrase)

4. Fax the letter ⟨from⟩ your office ⟨to⟩ my home. (2 phrases)

5. Radio advertisers are experts ⟨at⟩ producing spot announcements ⟨for⟩ their customers. (2 phrases)

Check your answers on page 392.

VERSATILE TOOLS

Tools are versatile when they do more than one kind of job. So it is with words. That is, most words may be more than one part of speech. You can't be sure of the part of speech of a word until you see it in a particular sentence. The idea is simple. Here's what you need to remember:

> **The part of speech is determined by how a word is used in a sentence.**

Consider the word *dancing*. It looks like a verb, doesn't it? After all, it represents *action*. It sounds like a verb too—because it ends in *ing*. As a matter of fact, it *is* a verb . . . sometimes.

VERB   He **is dancing** with Fergie. [The complete verb consists of the helping verb *is* and main verb *dancing*.]

> **The *ing* form is a verb only when a helping verb precedes it.**
>
> **Otherwise, the *ing* word is a noun or an adjective.**

NOUN   **Dancing** is fun. [*Dancing* is a noun because it names an activity. *Is* is the verb.]

ADJECTIVE   The child attends a **dancing** class. [*Dancing* is an adjective because it describes the class; *attends* is the verb. Some words called verbals can function as verbs, adjectives, or nouns. An *ing* word, such as *dancing*, can't be a verb unless a helping verb precedes it.]

For more about verbals, see "Grammar for **VERBALS** the Expert" in your *Instructor's Resource Kit.*

Some words called *verbals* can function as verbs, adjectives, or nouns. For example, what part of speech is *brown*? It can be a:

VERB   Did you *brown* the onions?

ADJECTIVE   She has beautiful *brown* hair.

NOUN   Is *brown* your favorite color?

A. The word *fish* may be a noun, an adjective, or a verb. What part of speech is *fish* in each of these sentences?

EXAMPLE   Let's go to the fish market.   _____adjective_____

1. We'll fish every day while we're on vacation.   _____verb_____

2. We'll have a fish dinner every evening.   _____adjective_____

3. Dietitians say that fish is a healthful food.   _____noun_____

B. The word *trade* may also be a noun, an adjective, or a verb. What part of speech is *trade* in each of the following sentences?

EXAMPLE   Would you like a career in international trade?   ___noun___

1. That was not a fair trade.   ___noun___

2. I will not trade that stock.   ___verb___

3. He reads several trade papers.   ___adjective___

C. *Reading* is another word that readily changes part of speech. Write a sentence using *reading* as each part listed. Possible answers

1. verb Ozzie is reading a report from the president .

2. noun Reading is my favorite activity .

3. adjective She needs new reading glasses .

D. Write two sentences using *plant*—first as a verb and next as a noun. Possible answers

1. (verb) Let's plant roses and gardenias this year .

2. (noun) In our Seattle plant, we hire many environmentalists .

E. Please respond to the following:

1. What part of speech are *city* and *country* in the sentence below? Do you prefer **country** life to **city** life? ___adjectives___

2. Write a sentence with *city* and *country* as nouns.
    Young children are probably better off in the country than in the city.
    (Answers will vary.)

Answers to this replay are on page 392.

DICTIONARY DILIGENCE

In Replay 22, with the assistance of your dictionary, you're asked to identify parts of speech of some words in the following business letter. Of course, many words are more than one part of speech; for example, *fish*—verb, adjective, or noun, depending on how it's used in the sentence.

The following letter and others in this book, especially in Chapter 14, are examples of the business writing style, tone, and format most likely to achieve your objectives. Although I changed the names, addresses, and details, this is basically a letter I wrote about a vehicle my husband and I purchased. The first reply was from the customer services manager, who wrote she was investigating the situation. Later we received a check and an apology from the dealer.

2301 Garnet Place
Sioux Falls, SD 57107
July 1, 0000

Customer Services Manager
Rollex Motors Ltd.
33 Edgemont Drive
Presque City, ME 04769

Dear Customer Services Manager:

My husband and I are pleased with the new Humvee we purchased on June 25. The dealer, Goniff Imports, Inc., provided courteous and prompt service, which we appreciated.

We believe this dealer and your organization conduct business fairly and honestly. However, yesterday we paid $115 (copy of bill enclosed) for balancing the wheels, even though we had driven the vehicle only 32 miles. Although we were unaware of it, the need for this service apparently existed when we drove the vehicle out of your showroom.

When we returned the vehicle, we were billed for this adjustment because of the exclusion provisions on page 3 of the warranty. The third paragraph reads, "Normal maintenance services . . . such as . . . wheel balancing . . . " are not covered. A condition existing at the time of purchase cannot be considered "normal maintenance" and surely should be adjusted without charge.

Although we assume a misunderstanding occurred, the dealer would not release our Humvee without payment. Will you please instruct Goniff Imports, Inc., to send us a check for $115.

Sincerely,

Leila R. Smith

Leila R. Smith

enclosure
c: Goniff Imports, Inc.

Word Power

One of the earliest dictionaries was written in the 1600s by Robert Cawdrey, an Englishman. He called it *A Table Alphabeticall* [sic], in which he defined "hard words for benefit and helpe [sic] of ladies, gentlewomen, or any other unskillful persons."

Replay 22

After reading the letter on page 80, answer the following parts-of-speech questions. Use your dictionary along with what you've learned in this chapter. Remember that although a word may be used as several different parts of speech, only one of them is correct for a particular meaning.

A. First Paragraph: For each of the following words, write the part of speech in the blanks:

1. My ___pronoun___
2. are pleased ___verb___
3. on ___preposition___
4. The ___adjective (article)___
5. and ___conjunction___

6. prompt ___adjective___
7. service ___noun___
8. which ___conjunction___
9. appreciated ___verb___

B. Second Paragraph: Write the verbs in blanks 1–7 and the pronouns in blanks 8–15.

1. ___believe___
2. ___conduct___
3. ___paid___
4. ___had driven___
5. ___were___
6. ___existed___
7. ___drove___
8. ___We___

9. ___your___
10. ___we___
11. ___we___
12. ___we___
13. ___it___
14. ___we___
15. ___your___

C. Third Paragraph: Write the prepositional phrases in the blanks.

1. ___for this adjustment___
2. ___of the exclusion provisions___
3. ___on page 3___
4. ___of the warranty___

5. ___at the time___
6. ___of purchase___
7. ___without charge___

Check your answers on pages 392–393.

Take the Pop Quiz on Reads 20–22, page 339.

CHECKPOINT

You've reviewed the "tools of the trade": seven of the eight parts of speech. The eighth is the **interjection,** an exclamatory word or phrase such as *No! Cool! That's great!* Perhaps you have special ones for certain occasions, such as when you've made a mistake or are angry.

You should now recognize parts of speech well enough so that you can learn and apply the information provided in the chapters that follow.

Check each blank after you know the definition and understand the examples.

_____ **NOUN** Names something, somebody, or someplace.

Beauty is in the **eyes** of the **beholder.**

Keyboarding is a basic **skill** needed by most **employees.**

_____ **PRONOUN** Substitutes for a noun. To decide whether a word is a pronoun, see if it can substitute for a noun.

Give **that** to **me.** substitutes for Give the **tape** to **Debra.**

They were selected to fill **them** out. substitutes for **Students** were selected to fill out **forms.**

_____ **VERB** Expresses action or being.

Long paragraphs **look** difficult and **discourage** concentration. Well-written paragraphs **are** about one subject.

_____ **ADJECTIVE** Modifies—tells something about—a noun or pronoun. The four kinds of adjectives are pointing, descriptive, articles, limiting.

This clever accountant has **a** job with **many** responsibilities. [*This* is a pointing adjective; *clever* is a descriptive adjective; *a* is an article; *many* is a limiting adjective.]

_____ **ADVERB** Modifies—tells something about—a verb, an adjective, or an adverb. Many, but not all, adverbs end with *ly.*

Yesterday I walked **really fast.** [Adverbs *yesterday* and *fast* tell when and how regarding the verb *walked.* Adverb *really* modifies adverb *fast.*]

_____ **CONJUNCTION** Joins—or connects—two words or groups of words. Coordinate conjunctions are *and, but, or, nor, for, so,* and *yet.* Dependent conjunctions precede a word group made up of a noun or pronoun plus a verb and may either begin a sentence or join parts of a sentence. A few examples are *when, as, if, since, because,* and *although.*

Although he's in love with Twileen **and** Victoria, he has to choose one **or** the other.

_____ **PREPOSITION** Introduces a prepositional phrase. A prepositional phrase begins with a preposition and ends with a noun or pronoun that is the object of the preposition. Describing words may come between the preposition and its object. A verb cannot be part of a prepositional phrase. Prepositional phrases are underlined in the following sentence, and prepositions are in bold type.

During the past month he took each woman **to** dinner **at** his parents' home **near** Dallas.

_____ **PART OF SPEECH** depends on how a word is used in a sentence. You may look up the part of speech in a dictionary; however, many words have more than one part of speech—depending on the definition. The pronunciation may even differ.

He wound the bandage around the wound.

They will service the vehicle, but the service manager doesn't give very

good service.

SPECIAL ASSIGNMENTS

A. Compose sentences using each of these words as an adjective, a verb, and a noun: *running, program, radio.*

B. Compose a sentence using *after* as a preposition and another sentence using *after* as a conjunction.

Complete this Special Assignment by _____ (date).

PROOFREADING FOR CAREERS

Everyone makes mistakes: I once thought even **I** had made one; but, of course, I was mistaken. If you want good grades or career advancement, why advertise your mistakes! Instead, proofread, *proofread,* **proofread.** Special symbols called "proofreaders' marks" are shown in college dictionaries, keyboarding textbooks, and reference manuals—and the inside back cover of this book.

Proofreading tests alertness and knowledge, is slower than other types of reading, and requires more than one reading.

Proofreading is much more than checking for spelling and typographical errors—often called "typos." Don't let anything get past your desk that doesn't make sense to you or is incorrect. Correct the errors in the following:

CORPORATE ORGANIZATION: A VIEW FORM THE TOP

The owners of a corporation are called stock holders or shareholders. each stockholder has one vote for each share of stock he or she owns. At the annual meeting, the stockholders elect a bored of directors, Which usually has between 7 and seventeen members. Because this board represents the stock holders, it ellects the high-ranking coorporate oofficers, or at least the president. The officers are responsable for the day-by-day operation the of corporation.

The board of directors, the officers, and the stockholders, Are required by law to meat at least once eacv year. The corporate secretary notify the stockholders of the meeting by male. It is common practis to enclosre a proxy form. With the notice of the meeting. Use of such a form allows stock holders to submit there votes in liese of attending the meeting.

After this error-filled article is corrected, circle one of the following to show how you did with the corrections: Excellent, Good, Fair, Ugh.

PARTS OF SPEECH SUMMARY

NOUNS naming words—such as *Joe, doctor, table, books.*

PRONOUNS substitution words—such as *he, she, it, them, who, everybody*—enable us to avoid needless repetition of nouns.

VERBS action words, such as *eat, love, count;* **being** and **helping** words, such as *be, am, seem;* **"possession"** words, such as *have, has, had,* which double as helping verbs.

ADJECTIVES modifying words—tell something about nouns, such as *comfortable* shoes, *yellow* banana, *sexy* guy, *expensive* homes, *the* mouse, *an* alligator—or about pronouns, such as *it* is expensive, *she* was polite, *they* are yellow, *we* are happy.

ADVERBS modifying words—tell something about:

* **verbs** run *fast* (*fast* tells about the verb *run*)
* **adjectives** *extraordinarily* graceful child (*extraordinarily* tells about adjective *graceful*)
* **other adverbs** such as ran *very* fast (adverb *very* tells about adverb *fast*)

CONJUNCTIONS joining words such as *or, and, but, however, therefore;* conjunctions join words or parts of sentences.

* Would you prefer to attend Harvard *or* Yale?
* Start with a rough draft; *then* do a great deal of revising. (*then* joins the sentence parts)

PREPOSITIONS linking words such as *under, through, with, beside, between, to;* a preposition is the first word of a "prepositional phrase." A prepositional phrase begins with a preposition and ends with a noun or pronoun, which is called the *object* of the preposition.

* The name "Jeep," according *to the captain* came *from the abbreviation* used *in the army for the* **G**eneral **P**urpose Vehicle.

EXCLAMATIONS express emotion and are usually followed by an exclamation mark.

* Never!
* Wonderful!
* You don't say!

PRACTICE QUIZ

www.prenhall.com/smith

A. Fill in the blanks with the name of the part of speech.

1. To name persons, animals, things, ideas, places, times, or activities, use a/an ____noun____.
2. To show or suggest action or existence, use a/an ____verb____.
3. To describe a verb, adjective, or adverb, use a/an ____adjective____.
4. To begin a prepositional phrase, use a/an ____preposition____.
5. To describe a noun or pronoun, use a/an ____adjective____.
6. A prepositional phrase may not include a/an ____verb____.
7. The coordinate conjunctions are ____and, but, or, nor, for____.
8. *Because* is an example of a ____dependent____ conjunction.
9. A word that substitutes for a noun is called a/an ____pronoun____.
10. *I, me, my, mine,* and *myself* are examples of ____first-person pronouns____.

B. Insert the part of speech for each word.

While Sarala slept soundly, she dreamed about snowboarding with Jonathan at the new Utah resort. Wonderful!

11. While _____conjunction_____
12. Sarala _____noun_____
13. slept _____verb_____
14. soundly _____adverb_____
15. she _____pronoun_____
16. dreamed _____verb_____
17. about _____preposition_____
18. snowboarding _____noun_____

19. with _____preposition_____
20. Jonathan _____noun_____
21. at _____preposition_____
22. the _____adjective_____ (also called article)
23. new _____adjective_____
24. Utah _____noun_____
25. resort _____noun_____
26. Wonderful! _____exclamation_____

THE GLOBAL MARKETPLACE—JAPAN

Japanese	Spell	Say	Currency
Greeting	konnichi wa	ko.NICH.ee.wa (name)-san	yen (y)
Departure	sayonara	SY.oon.ara	
Please	kudasai	KOO.di.si	
Thank you	arigatoo	ar.ee.GAT.oo	
I don't understand	Wakarimasen		

Courtesy of Alex Paul

Apples, Tigers, and Swahili

5

Tell Who, Whom, or What—Nouns

After Completing Chapter 5, You Will

> Apply Standard English principles to form plurals of both regular and irregular nouns, write compound nouns correctly, and capitalize proper nouns.

> Correctly spell, pronounce, and use the words presented in this chapter.

> Use bias-free language.

"For there be women fair as she whose verbs and nouns do more agree."

—*Bret Harte*

A pretest is in the *Instructor's Resource Kit.*

Nouns name people, animals, places, things, ideas, and actions. When you think about exotic faraway places or your hometown . . . movie stars, kings, and the person next door . . . life, health, peace, success . . . monopoly, jogging, and homework . . . apples, Swahili, or a tiger—you are thinking **NOUNS**. Look around and you see nouns. Reading, listening, and life experience enable you to use many nouns correctly. With a few simple rules and intelligent use of the dictionary, you'll avoid the common errors.

Sometimes a noun error is simply choosing the wrong word because of inadequate vocabulary, carelessness, or poor spelling—like the actual noun bloopers that follow:

Recap

Please correct the noun errors below, since your spell checker is unlikely to do so:

Starlings ate the farmer's grain and spoiled his ~~corpse~~. [crops]

The ship was bound for ~~Nausea~~ in the Bahamas. [Nassau]

My sister writes in her ~~dairy~~ every night. [diary]

When we went to Canada, we had to pass through ~~costumes~~. [customs]

The focus is on usage rather than on grammar terminology and rules. If your students would profit from information about abstract and concrete nouns, direct and indirect objects, complements, etc., see "Grammar for the Expert" in your *Instructor's Resource Kit.*

Check your answers on page 393.

Other noun errors, such as in the poem below, occur because of the many irregular forms, or exceptions, which are especially confusing to adults learning English as a second language and to young children with English as a first language:

The plural of box we all know is boxes,
Yet the plural of ox is oxen, not oxes.*
A goose is a goose, but two are called geese,
But why isn't more than one mouse quoted meese?
A mouse and his family are mentioned as mice,
But the plural of house is houses, not hice.
You can readily double a foot and have feet,
But try as you might, you can't make root reet.
If the singular's this and the plural, these,
Should the plural of kiss ever be keese?

Some irregular nouns are less obvious than those in the poem, even to those who have used English all their lives. In Chapter 5 you'll spot these troublesome irregulars, interpret the dictionary symbols for them, improve spelling and vocabulary, and learn the latest capitalization style.

Read 23

SAFETY IN NUMBERS

A singular noun means just one of whatever person, place, or thing the noun names; a plural noun means more than one. Spelling most plural nouns is easy. We usually just add *s* to the singular noun to make it plural: *one check* but *three checks*. When the singular noun ends with *s, z, x, ch,* or *sh,* add *es* to spell the

---

*Sometimes acceptable—but not preferred.

plural: *one box* but *three boxes*. Most English rules, however, have exceptions. Three types of exceptions are the plurals of nouns ending in *y*, *o*, and *f*.

NOUNS ENDING IN *Y*: IS THE PLURAL *Y-S* OR *I-E-S*?

MORE EXAMPLES
journeys   monkeys

If a vowel (*a, e, i, o, u*) precedes *y*, simply add *s*.

valle*y*, valle*ys*   attorne*y*, attorne*ys*   turke*y*, turke*ys*

MORE EXAMPLES
boundary, boundaries . . . festivity, festivities . . . vacancy, vacancies . . . apology, apologies . . .

If a consonant (all letters other than vowels) precedes *y*, change the *y* to *i* and add *es*.

industr*y*, industr*ies*   compan*y*, compan*ies*   hobb*y*, hobb*ies*

NOUNS ENDING IN *O*—IS THE PLURAL *O-S* OR *O-E-S*?

Parent: What did you learn in school today? **Child**: The teacher taught us how to make babies. **Parent**: Ohh?? What did the teacher say? **Child**: You change the Y to I and add ES.

If a noun ending in *o* relates to music, add just an *s*.

banj*o*, banj*os*   sol*o*, sol*os*   alt*o*, alt*os*

MORE EXAMPLES
piccolos, contraltos, bassos, duos

If a vowel precedes the *o*, add just an *s*.

studi*o*, studi*os*   rode*o*, rode*os*   radi*o*, radi*os*

pintos, autos, provisos or provisoes, vetoes, tobaccos, tornadoes or tournedos, volcanos

If a nonmusic noun ends in *o* preceded by a consonant, look it up to find out whether to add *s* or *es*.

mem*o*, mem*os*   domin*o*, domin*os*, domin*oes* (both are correct)

Spelling Tip: tomatoes and potatoes: both have toes!

Silly mnemonics are effective: Former Vice President Quayle had an embarrassing moment when the teacher corrected him during a visit to an elementary school classroom. When a child spelled *potato* correctly, VP Quayle said that was wrong; it's p o t a t o **e**.

WORD TO THE WISE

For plurals of nouns, you sometimes find two spellings in the dictionary. This means both are correct, but it's better to use the one appearing first.

In most dictionaries the plural is not given if all that's needed is *s* or *es*. This means if you don't find the plural in the dictionary, spell it in the "regular" way; it is not an exception.

NOUNS ENDING IN *F*—IS THE PLURAL *F-S* OR *V-E-S*?

EXAMPLES
half/halves, calf/calves, loaf/loaves

If a word ends with *ff*, just add *s* to make it plural. Otherwise, no useful rules determine how to spell plurals of nouns ending in *f*. When in doubt, check the dictionary because some simply add *s* and others change the *f* to *v* and add *es*.

rebu*ff*, rebu*ffs*   chie*f*, chie*fs*   belie*f*, belie*fs*   wi*fe*, wi*ves*   kni*fe*, kni*ves*

APOSTROPHES? NO!

Chapter 9 is about apostrophes. Chapters 5 and 9 are separated to help students avoid the common error of adding apostrophes to nonpossessive plurals.

In general, apostrophes (') do not make a noun plural.

The occasional exceptions are in Read 49 of Chapter 9, which is all about apostrophes. Most nouns become plural by adding *s*, not by adding apostrophe and *s*.

Recap

Correct the plurals by crossing out incorrect apostrophes.

Founded in 1926, the Globetrotter's were known worldwide as the "Magi-cian's of Basketball." Former Secretary of State Henry Kissinger said, "I have particularly admired both the athletic skills of the Trotter's and the sheer joy and good will they have created over the year's."—Metromedia, Inc.

See page 393 for the solution.

Replay 23

Have your dictionary handy. This is not a spelling test. It's a test of alertness to spelling errors and efficiency in dictionary use.

A. Spell the plurals of these nouns. One of these nouns has two correct spellings.

1. ally ___allies___
2. accessory ___accessories___
3. itinerary ___itineraries___
4. proxy ___proxies___
5. facility ___facilities___

6. injury ___injuries___
7. money ___moneys or monies___
8. authority ___authorities___
9. ferry ___ferries___
10. survey ___surveys___

B. Spell the plurals of these nouns. For words with two correct spellings, write both in the blank.

1. tattoo ___tattoos___
2. domino ___dominos, dominoes___
3. cargo ___cargoes, cargos___
4. piano ___pianos___
5. hero ___heroes___

6. alto ___altos___
7. potato ___potatoes___
8. zero ___zeros, zeroes___
9. portfolio ___portfolios___
10. memento ___mementos, mementoes___

C. Spell the plurals of these nouns.

Say *chifs*, not *cheefs* or *cheevs*.

STUDENT ALERT
tariff: 1 *r*, 2 *f*s

1. thief ___thieves___
2. handkerchief ___handkerchiefs___
3. knife ___knives___
4. tariff ___tariffs___
5. wife ___wives___

6. half ___halves___
7. safe ___safes___
8. wolf ___wolves___
9. plaintiff ___plaintiffs___
10. chief ___chiefs___

D. Correct the plural errors in this paragraph.

During tour's of the United State's and Canada by the Globetrotter's, the team played before more than 1,400,000 fan's in 263 game's. These basketball player's amused and delighted audience's with a combination of incredible ballhandling, seemingly impossible shooting, and classic comedy routines'.

Answers are on page 393.

The etymology of *school* is the Greek word *skhole*, which means "leisure" or "spare time." In ancient Greece, only the leisure classes could afford an education.

Read 24

THE ECCENTRIC "S"

An eccentric person or thing is odd or strange and doesn't conform to an expected pattern. The following nouns are "eccentric" because of the *s* or absence of it at the end of a word.

For plurals of words like the nouns below, change the spelling, and do not add *s*.

SINGULAR	man	woman	mouse	tooth	child	foot
PLURAL	men	women	mice	teeth	children	feet

Below are examples of another group of "eccentrics." Spell them the same whether they are singular or plural.

SINGULAR OR PLURAL   aircraft, British, corps,* deer, Dutch, fish, Japanese, salmon, series, sheep, statistics

Check some of the preceding words in your dictionary; notice that each word is respelled following the abbreviation *pl.*, which means the plural spelling is the same as the singular. For some nouns, however, dictionaries show two correct plurals—either without an *s* or with it—like "deer n. pl. deer, also deers." Usually choose the first spelling for workplace communication as that is the more commonly seen and heard.

Writing for your Career

Provide clues for the reader. For example, the words *a, one, few, several,* or *this* are possible clues before a noun spelled the same for both singular and plural forms.

A deer stopped at the pond. (*Deer* is singular.)

Several deer munched on our yellow roses. (*Deer* is plural.)

Some nouns ending in *s* are singular or plural depending on the meaning. Your dictionary shows which meanings are singular and which are plural. If one of these nouns is the subject of a sentence, be sure to choose the correct verb form.**

---

*See the dictionary for pronunciation of singular and plural *corps*.
**In Chapter 7 you'll review using a singular or plural verb form to agree with a singular or plural subject.

politics   mathematics   economics   measles

The **politics** of affirmative action **have** not always been clear. (plural)
Politics is the focus of his life. (singular)

> Some nouns ending in *s* are always singular, while others are always plural.

ALWAYS SINGULAR   news   aeronautics

ALWAYS PLURAL   scissors   proceeds*

The latest **news is** that the **scissors were** (not *was*) lost.
Aeronautics is the principal industry in that town, and the **proceeds** from it **have** (not *has*) doubled.

> Spell plurals of proper nouns (those beginning with capital letters) in the "regular" way; simply add *s* or *es*.

Do not change the spelling of a name (proper noun) to form a plural even though it may end in *y*, *o*, or *f*. If you apply the rule for plural nouns ending in *y* to MARY, you end up—incorrectly—with MARIES—instead of MARYS. To form the plural of *wolf*, if you change the *f* to *v* and add *es*, you get *wolves*. If you mean Mr. and Ms. Wolf, you wouldn't say "The Wolves are coming for dinner," but rather "The Wolfs. . . ."

> To make a proper noun plural, just add *s* or *es*; do not add an apostrophe.

My family name on greeting cards reads *The Smiths*—not *The Smith's* nor *The Smiths'*. Notice that the plural proper nouns in the following sentences do not have apostrophes:

Four hundred *Taylors* are listed in the telephone directory.
The *Lopezes* have new Web sites.

A.   In the blanks write plurals of the following nouns. See your dictionary if in doubt about common nouns (those beginning with lowercase letters). Some of these plurals are spelled the same as the singular form.

EXAMPLE   Perkins   Perkinses

Awkward plurals like Perkinses or Floreses may be expressed as the *Flores family, the Flores sisters, Mr. and Mrs. Flores*, etc.

1. corps _____corps_____   7. Jones _____Joneses_____
2. economics ___economics___   8. aircraft _____aircraft_____
3. deer __deer (occasionally deers)__   9. fish ___fish (or fishes)___
4. Larry _____Larrys_____   10. stepchild ___stepchildren___
5. series _____series_____   11. foot _____feet_____
6. Chinese _____Chinese_____   12. Flores _____Floreses_____

---

*See your dictionary for the **noun** *proceeds* and distinguish it from the pronunciation and meaning of the **verb** *proceeds*.

B. Write either **S** or **P** in the blank to show whether the noun is singular or plural. If a noun may be used either way, write **S/P.** Consult your dictionary as needed.

EXAMPLE   premises ___P___

1. trousers _____P_____   4. statistics _____S/P_____

2. corps _____S/P_____   5. news _____S_____

3. mumps _____S/P_____   6. politics _____S/P_____

C. Use the dictionary or word clues to decide whether the noun subjects (shown in bold type) require a singular or a plural verb. Then circle the correct verb for that subject.

SINGULAR VERBS   is,   was,   has

PLURAL VERBS   are,   were,   have

EXAMPLE   The new **pants** (was/were) shortened.

1. The **scissors** (have/has) sharp edges.

2. **Mathematics** (is/are) my favorite course.

3. These **statistics** (is/are) accurate.

4. The **documents** (was/were) faxed yesterday.

5. All **earnings** from this show (are/is) being given to charity.

6. **Clothes** (is/are) all over the floor.

7. **Genetics** (is/are) an important field in modern science.

8. Each day's **news** (was/were) carefully edited.

9. Do you think **politics** (is/are) a subject to avoid discussing at a party?

10. Several lecture **series** (was/were) offered in anthropology.

Please check your answers on pages 393–394.

PLURALS OUT OF UNIFORM

Plurals that don't require adding *s* are a bit like soldiers out of uniform: You can't recognize their occupation by their appearance, although their behavior might reveal military training. When a noun doesn't need *s* to make it plural, it seems to be out of uniform; but its "behavior" in the sentence reveals its plural status.

Certain plurals that don't require adding *s* have been incorporated into English from other languages. These words follow the spelling rules of the original language. Sometimes an English plural spelling has been adopted as well; then we have two plurals to choose from. It's helpful to know both forms as they may have different uses. Following are examples of these unusual plurals.

SINGULAR	ORIGINAL LANGUAGE PLURAL	SINGULAR ENDING	PLURAL ENDING	ENGLISH PLURAL
formula	formulae	a	ae	formulas
vertebra	vertebrae	a	ae	**or** vertebras
alumnus	alumni	us	i	alumni
stimulus	stimuli			stimuli
analysis	analyses	is	es	analyses
diagnosis	diagnoses			diagnoses
criterion	criteria	on	a	criterions
phenomenon	phenomena			phenomenons
medium	media	um	a	mediums
curriculum	curricula			curriculums
datum	data			data
memorandum	memoranda			memorandums
addendum	addenda			addendums

Pronounce the last syllable of words like *analyses, diagnoses, parentheses, bases* with a long *e* sound.

See usage notes in the newest dictionaries regarding *criteria, data,* and *media*.

USAGE ALERTS

Check your dictionary for pronunciation of foreign plurals. *Formulae* is used in scientific and technical communication. *Criteria*, not *criterions*, is preferred in American usage. A common, but serious, grammar error is using *criteria* as though it were singular. Remember **one criterion is** but **two criteria are.** *Media* is the preferred plural for newspapers, radio, etc. *Mediums* is preferred in expressions like "mediums of exchange" to refer to various kinds of money, and "mediums" who claim to communicate with "other entities." *Data* and *media* are now widely used as singular or plural forms and are accepted as singular in some newer dictionaries. Many business and professional people, however, still consider the singular form wrong if the plural is required for correctness. Don't risk having a colleague, supervisor, or client silently think you're grammar-ignorant. Instead either use the singular and plural according to the chart, or choose another word.

NEVER *memorandas, phenomenas, curriculas!!*

AVOID This data *is* too technical for quick investigation.

USE These data *are* too technical for quick investigation. OR

These statistics (figures, etc.) are too . . . . OR

This information is . . . .

MEMO FROM THE WORDSMITH

Curriculum originally meant a racecourse in Latin; some people still believe a curriculum is a runaround.

Careful communicators recognize nouns that require a change other than adding *s* to become plural. Yet some words look as though the plural might be irregular when it isn't. For example, just add *es* to make *sinus* plural; it's not *sini.* Develop an "instinct" for noticing words like the examples in the preceding chart, and refer to the dictionary when in doubt. If you're sure a word has two plurals that mean the same, use the one appearing first in the dictionary entry.

A. With the help of your dictionary, spell the plurals of these words. If two plurals are correct, write them both in the blank.

EXAMPLE   bureau bureaus, bureaux

1. formula ___formulas, formulae___
2. alumnus ___alumni___
3. basis ___bases___
4. census ___censuses___
5. criterion ___criteria, criterions___
6. axis ___axes___
7. parenthesis ___parentheses___

8. crocus ___crocuses, croci___
9. appendix ___appendixes, appendices___
10. concerto ___concertos, concerti___
11. index ___indexes, indices___
12. analysis ___analyses___
13. medium ___media, mediums___
14. diagnosis ___diagnoses___

B. Write **S** or **P** to show whether these nouns are singular or plural. Use your dictionary.

EXAMPLE   nucleus ___S___

1. alumna ___S___
2. criteria ___P___
3. alumnus ___S___

4. data ___P or S___
5. hypotheses ___P___
6. kibbutzim ___P___

C. Circle the correct form.

EXAMPLE   The (media/(medium)) is sometimes as important as the message.

1. Television is the (media/(medium)) that would be preferred.
2. She broke several (vertebra/(vertebrae)/vertebraes) in the accident.
3. How many (criterion/(criteria)/criterias) did they consider?
4. Many people forget to type the closing (parentheses/(parenthesis)).
5. Several (alumnus/(alumni)/alumna) attended the opening game.

You'll find the answers on page 394. Then review Reads and Replays 23–25 before taking the Pop Quiz on pages 340–341.

COMPOUNDS AND PROPERS

TOGETHER, SPLIT, OR SEPARATED?

MORE EXAMPLES
blowout, halfback, sugarcane, won ton or wonton

A noun made up of more than one word is a **compound noun,** such as *high school.* Compound nouns are written in three ways: as one word, split with hyphens, or as separate words:

ONE WORD	dollhouse, checkbook
SPLIT WITH HYPHEN	tie-in, follow-up
SEPARATED	doggy bag, time clock

Follow-up as a noun or an adjective is hyphenated. As a verb+ an adverb, it's two words, no hyphen: *Please follow up on that account.* A follow-up letter was sent. The follow-up was too late.

Some compound expressions may be written in more than one way; therefore, check the dictionary to be sure of whether your compound noun requires one solid word, a hyphenated word, or two separate words.

> If a dot or an accent mark (but no space) appears between the parts, write the expression as one word. The dot or accent mark indicates syllables, not separations or spaces.

DICTIONARY ENTRY	brick.lay.er
IT'S ONE WORD	bricklayer

> If the word parts have a space between them or the expression is not shown in a good college dictionary, write each part as a separate word. Use a hyphen only if you see a hyphen between the parts of the dictionary entry.

DICTIONARY ENTRY	swiss cheese (two words)
DICTIONARY ENTRY	bric-a-brac (one hyphenated word)

PLURALS OF COMPOUND NOUNS

> To form the plural of a compound noun spelled as one word, usually add *s* or *es* to the end of the word—unless the noun is irregular.

bookcases	spoonfuls	headlines	businesswomen (irregular)

> For compound nouns with a hyphen or space between the parts, the dictionary shows which part to make plural.

sister**s**-in-law	letter**s** of credit	write-off**s**	trade-in**s**

> If the dictionary shows two ways to form the plural, choose the first for business writing.

DICTIONARY ENTRY	notaries public or notary publics
USE THE FIRST	notaries public

 Recap

Use proofreading symbols from the inside back cover of this text, from your dictionary, or from your computer to correct the compound noun errors in the following paragraph.

My brotherinlaw asked me to design a letterhead for his stockcar company, which accepts tradeins. Because his cars have high pricetags, he wants the letter head to be classy.

Check your answers on page 394.

Compound Sports

When basketball, baseball, and football were invented, they were separate words: *basket ball*, *foot ball*, and *base ball*. As each game became more popular, it became hyphenated: *basket-ball*, *foot-ball*, *base-ball*. Eventually all became the one-word compounds they are today. Many new compound words eventually progress from separate words to hyphenated compounds to one-word forms. What about *volley ball*? or is it *volleyball*?

A GUIDED TOUR OF THE CAPITAL

A noun beginning with a capital letter is a proper noun. Nouns that do not begin with capitals are common nouns. A complete guide to capitalization is in the "Mini Reference Manual" Appendix D. Following, however, is a review of noun capitalization principles most often needed in the workplace.

If more on capitalization rules and practice would help your students, assign Appendix D, Read and Replay "Capitalization."

Capitalize official titles used directly before the person's name OR used in "direct address" instead of the name. Generally do not capitalize an official title that follows the name.

Did you know that **P**resident Jimmy Carter owned a peanut farm?

A letter of recommendation was received from **S**ister Mary Margaret.

We hope, **M**adam **P**resident, that you will like this gift. [Direct address means you call a person by name or title in speech or writing.]

Jimmy Carter was **p**resident of the United States from 1977–1981. [The title follows the name.]

The **c**olonel, Rachel Rothstein, is a West Point graduate. [No capital is used for *colonel* because a comma separates it from the name.]

EXCEPTION

Be alert to corporate and government customs regarding capitalizing titles of high-ranking company officials.

Capitalize titles that follow the name—in addresses and in letter closings.

In address—Mr. Steam Gene, Head Honcho

Typed name at the end of a letter—Rosalyn Amaro, Vice President*

Capitalize a family title when used as part of the name or instead of the name. Do not capitalize, however, if a noun or pronoun precedes the title.

I attended **U**ncle Jack's 90th birthday party at the Town Club.

I asked **U**ncle about the old days.

Do not capitalize it if a noun or pronoun comes before the title.

I asked my **u**ncle about the old days.

My **c**ousin Billy's **w**ife, Janice, was the flowergirl at my wedding.

---

*Avoid "honorifics"—Mr., Ms., Dr., Mrs., Professor, and the like—in signatures. Official titles, however, that **follow** signatures, such as Web site Manager, Vice President, are appropriate and helpful to readers.

Words such as *company, college,* and *association* are usually capitalized only when used with the name of the organization.

Microchip **C**ompany is on 5th Street. **BUT** The **c**ompany is on 5th Street, near the college.

Capitalize the official name of a department or a committee if you are a member of the company or organization.

Prepare a requisition and send it to our **P**urchasing **D**epartment.
Our company's **S**ocial **C**ommittee is planning the company picnic.

BUT

He works in the **p**urchasing **d**epartment of our competitor and is a member of its **c**onservation **c**ommittee.

Capitalize words like *town, city, state,* and *county*—when they follow the name, or if a governing body uses the geographic term officially.

New York **S**tate, Kansas **C**ity
The **s**tate of Iowa is a great place to live, and he moved to Sioux **C**ity.
Mayor Wimbledon said the **C**ity is requesting bids from building contractors.

This principle may require knowledge of geography and culture; on Real Quizzes, obvious examples are used.

Capitalize definite geographic regions with compass point names. Do not capitalize directions or general locations.

He lives in the **E**ast, but he talks like a **W**esterner. [The East, North, South, West, and Midwest are considered definite geographic regions in the United States.]
Disneyland is **e**ast of Los Angeles and **n**orth of the **S**outh Pole. Drive three blocks **e**ast on Disney Lane.

Usually use lowercase letters for the names of seasons.*

The office will close for a week this **s**ummer.

Always capitalize names of languages; do not capitalize names of courses unless they are languages or official course names.

He is studying **B**usiness **E**nglish, **b**ookkeeping, **A**ccounting 101, **S**panish, and **G**reek at Gonzaga University. Next year he will take a **s**ociology course, **B**usiness **L**aw 230, and **F**rench. He already speaks **H**indi and **F**arsi.

Races named by color begin with lowercase letters, but sociological names of races are capitalized. Religions are also capitalized.

white/**C**aucasian   **b**lack/**A**frican **A**merican
Islam   **C**hristianity   **J**udaism   **B**uddhism

---

*Exceptions are explained in Appendix D.

Replay 26

A. Use your dictionary for help with these compound nouns. Show whether the expression is one solid word, a hyphenated word, or two or three separate words. Some dictionaries give more than one spelling for certain plurals; if yours does, include both.

	SINGULAR	PLURAL
EXAMPLE notarypublic	notary public	notaries public / notary publics
1. followup	follow-up	follow-ups
2. textbook	textbook	textbooks
3. tradein	trade-in	trade-ins
4. editorinchief	editor in chief	editors in chief
5. runnerup	runner-up	runners-up
6. spaceflight	spaceflight	spaceflights
7. headhunter	headhunter	headhunters
8. bushleague	bush league	bush leagues
9. chiefofstaff	chief of staff	chiefs of staff
10. volleyball	volleyball	volleyballs

B. Complete Read and Replay Capitalization on pages 373–376.

C. Write **C** in the blanks below if the capitalization is correct, or make the corrections.

EXAMPLE   If their Sales Manager calls, let me talk to her.
                            s   m

_____ 1. The General Manager of Horizon telephone likes Tostadas and Tacos.
                     g   m

_____ 2. We hire former Business Teachers to work in our office.
                                b   t

___C___ 3. The secretary of the ski club is taking Art 41.

_____ 4. Lacy's expects sales associates to speak spanish when needed.

_____ 5. The president of the United States has just entered the White House.

_____ 6. Did the Cardinal discuss the issue with other catholic leaders?

___C___ 7. The typed signature at the end of the letter should be capitalized like this: Patty Killman, Professor of Office Technology.

_____ 8. We drove East last Summer until we reached Kansas city.
                     e   s

_____ 9. The City of Azusa, which is in the State of California, is named after everything from A to Z in the USA.
                     c   s

___C___ 10. The clerk in our Credit Department needs to take the Business English course.

_____ 11. My sister has taught anthropology 101 at several Colleges.
                                                                        c

_____ 12. Until the 1960s African American, Mexican American, Asian American, and Jewish office workers were not offered employment in this Company.
                                                                                                                                                                                    c

<u>C</u> **13.** Some African Americans and Chinese Americans did factory work but had to eat in a separate lunchroom.

<u>C</u> **14.** I believe Governor Shawn A. Taylor joined his famous sisters, Krista and Ashley, at the Inauguration Ball.

_____ **15.** Krista won the <u>academy</u> <u>award</u> for her performance in "The Iron Magnolia," and Ashley was awarded the <u>nobel</u> <u>prize</u> for her efforts on behalf of world peace.

Answers are on pages 394–395.

Read 27

BANISHING BIAS FROM BUSINESS ENGLISH

Learn to sense words and expressions that may offend, result in lawsuits, arouse ridicule, or detract from the grace of the English language. We have no hard-and-fast rules; students should be reminded to use good taste and good judgment. Examples: Instead of "deaf and dumb," use hearing and/or speech impaired. Instead of "ladylike," use well-mannered, polite, cultured. Replace "repairman" with technician, mechanic, plumber, etc.

Businesspeople know—or should know—that offending customers or clients is bad business. So is offending co-workers; people don't work well when they're angry. Equally or more serious are legal problems resulting from biased language. What follows is simple advice about using language that banishes bias and encourages businesslike behavior.

> **Avoid language in the workplace that inappropriately draws attention to someone's age, color, body characteristics, disability, nationality, race, religion, sex, or sexual orientation, or political beliefs, unless clearly required. Do not use these descriptions to identify individuals unless clearly required.**

In your place of employment or when with co-workers or clients elsewhere, do not use terms that could suggest bias. Avoid allowing political and religious beliefs to influence your choice of appropriate workplace language. Whether the reference is complimentary, disparaging,*or even neutral, avoid stressing differences or making one group sound inferior or superior to another. Using nonbiased-sounding language helps a person appear well-educated, intelligent, and current.

AVOID An African American networking specialist is the keynote speaker at the Computer Hardware Convention.

AVOID A networking specialist—a black woman—is the keynote speaker at the Computer Hardware Convention.

USE A networking specialist is the keynote speaker at the Computer Hardware Convention.

AVOID Two gay men came into the store today, and each bought a complete wardrobe for a Hawaii vacation.

USE Two men came into the store today, and each bought a complete wardrobe for a Hawaii vacation.

AVOID Honey, that blouse shows off your best features.

USE Jenny, that's a good-looking suit.

AVOID We went to Alaska to buy Eskimo art to sell in our gift shop at the museum.

USE We went to Alaska to buy Inuit art to sell in our gift shop at the museum.

*If you're unsure about this adjective, why not look it up now?

The term *Eskimo* is considered offensive by some, as it has been incorrectly defined as "eater of raw meat." Canadian officials use the term *Inuit* or *Innu* (literally *the real people*) for Native Americans of Canada and Alaska. Oddly, however, the term *Eskimo* is still used in some academic contexts.

> **Most people like to be thought of as individuals, rather than have a minority status stressed. Therefore, focus on the person as an individual, not as a member of a certain group. In addition, replace terms used in derogatory and hurtful ways with tactful, but honest, words.**

Although the following two AVOID sentences are more concise, the longer one is preferable.

AVOID   Jean Chung, our blind transcriber, needs a new lighting fixture.

USE   Jean Chung, the transcriber, needs an improved lighting fixture.

AVOID   His daughter, who's a retard, goes to this school.

USE   His daughter, who has Down syndrome, attends this school. OR
His daughter, a special-needs student, attends this school.
Preferable if there's no good reason to include the extra information: His daughter attends this school.

AVOID   Please give your application to the elderly man at the desk in the corner.
Please give your application to the senior citizen at the desk in the corner.
Please give your application to the old gentleman at the desk in the corner.

USE   Please give your application to the man at the desk in the corner.

> **Use woman/women (not gal/girl) and man/men (not boy) or other appropriate identifying terms to refer to adults in the workplace. *Gal* is an out-of-date word, and *girl* shouldn't be used for a female over 18. Avoid lady/ladies or gentleman/gentlemen except to stress a caring attitude or to refer to British nobility.**

AVOID   That's the (gal, girl, lady) who recommended the new client to our firm.

USE   That's the woman who recommended the new client to our firm.

USE   She is truly a lady in every sense of the word, and her husband is a gentleman.

USE   Queen Elizabeth had three new ladies-in-waiting that year.

AVOID   man and wife

USE   husband and wife or man and woman

USE   The psychologist discussed differences between men and women.

AVOID   This reception is for the executives and their wives.

USE   This reception is for the executives and their spouses.

> **Do not use terms like *honey, sweetie,* or *dear* to address women at work or in public institutions or women you don't know.**

Use the woman's name if you know it; otherwise, *ma'am* or *Ms.*, just as you would use *Sir* for a man whose name you don't know. Addressing women is simple; if you are not on a first-name basis with the woman, call her *Ms.* followed by her last name—just as you would address a man as *Mr.* plus his last name.

You don't have to know a person's marital status to choose Mr. or Ms. (pronounced *Miz*). For workplace writing, use *Ms. Myrna Fox* to address the communication; for the greeting, write *Dear Myrna* or *Dear Ms. Fox*, depending on your relationship with her.

Avoid attaching *ess* or *ette* to a noun to create a feminine word for historically masculine roles:

ESS/ETTE WORDS	REPLACE WITH
actress [still often appropriate]	actor [preferred by some females]
poetess, authoress	poet, author
proprietress	proprietor
stewardess	flight attendant or steward (on a ship)
waitress/waiter	server

Choose a form without sex identification unless it is relevant. Many terms that include the word *man* can be expressed in neutral gender without sounding clumsy. If an expression sounds awkward, rephrase it.

AVOID	USE
mankind	human beings, society, human race, humanity
manmade	synthetic, artificial, constructed, factory-made, plastic
manpower	workers, employees, crew, laborers, staff, workforce, etc.
workmen's compensation	workers' compensation
businessmen	If an inclusive term is preferable, try executives, managers, businesspeople, etc.
cameraman	photographer, cinematographer
chairman	chairperson, chair,* leader, moderator, coordinator, facilitator, etc.
clergyman	member of the clergy, minister, rabbi, priest, etc.
congressmen	congressmen and congresswomen, congressperson/s, representatives, legislators
fireman	firefighter
insurance man	insurance agent
mailman	mail carrier, letter carrier, postal worker
male nurse	nurse
male secretary	secretary
policeman	police officer

AVOID He's studying to be a male nurse.

USE He's studying to be a nurse.

---

*Yes, it's in the dictionary. Chair is often used instead of the alternatives and has been used in the sense of *leader* since the 1600s, particularly in British English.

USE The client requests a male nurse as a live-in for two weeks. [*Male* has a workplace purpose in this sentence.]

AVOID A lady policeman said the area is off limits.

USE A police officer said the area is off limits.

> **When communicating in the workplace, avoid language that stereotypes or belittles either sex.**

Avoid expressions like *the fair sex, the little woman,* or *ball and chain* (referring to a wife), *bachelorette, old maid, spinster, gal, girls* (for adult women), and *coed* for a female college student. The terms *gay, lesbian,* and *partner* or *same-sex partner* are currently preferred for men and women who are homosexual. On the job, do not call grown men *boys,* and avoid sexist jokes that make either all men or all women seem stupid or clumsy.

> **Be alert to changing terms for nationalities, races, and religions.**

Hyphenated nationality designations are no longer appropriate. Avoid *Hispanic* except in government documents. Use Chicano or Chicana only if you're sure the individuals involved choose those terms; political ramifications make them offensive to some.

AVOID	USE
Hispanic	Latina, Latino/s, Mexican American, Cuban American, Hispanic American
Indian (use for someone from India)	Native American, American Indian, Inuit, Alaska native, etc.
Jew (as a verb)	Jew (noun only) or Jewish (adjective) for a male or female ("Jewess" is an offensive term.)
Oriental	Asian, Asian American [preferably Chinese, Chinese American, Vietnamese, Vietnamese American, etc.]
negro, colored, Afro-American	African American, black [keeping in mind that not all black Americans are direct descendants of Africans]
Christian name	first name, given name

Notice that I don't suggest silly sounding "politically correct" terms, such as "visually challenged" for blind, "personhole" for manhole, "pregnant persons" for pregnant women, and "vertically challenged" for short. Although stereotyping has decreased considerably in the United States, businesspeople should not allow personal, religious, or political beliefs to result in human relations or legal problems. Keeping workplace communication nonbiased contributes to your career growth.

MEMO FROM THE WORDSMITH

Detail man is an old term for a pharmaceutical company representative who introduces medications and medical supplies to doctors. Now both men and women have this job and are simply referred to by the name of the pharmaceutical company + *representative* or *rep*. For example, "Anna is an Eli Lilly Corporation rep."

Replay 27

Adapt the language in these sentences to reflect current attitudes, not yesterday's stereotypes. Draw a line through four sentences to show they should be eliminated altogether. Write C beside the two correct items.

average person
EXAMPLE The ~~common man~~ wants peace.

Students with weak vocabulary skills often improve in collaborative learning situations. In teams of about four, students read the sentences aloud to one another and look up meanings or pronunciations as needed during the team session. Each team selects one member to pronounce and define some of the words to the rest of the class.

_____ 1. *Society or Humanity*
  ~~Mankind~~ can rejoice when a way is found to settle conflicts without war.

_____ 2. *flight attendant*   *Delete this sentence*
  The ~~stewardess~~ is bringing drinks now. ~~She looks old enough to be my grandmother.~~

_____ 3. *women*
  Some of the men as well as several ~~girls~~ in this company play golf at the Braemar.

_____ 4. *chairperson*
  The ~~chairman~~ needs a good understanding of parliamentary procedure.

_____ 5. *spouses*
  The president invited the managers and their ~~wives~~ to a formal dinner.

___C___ 6. Because Wisconsin Pizza Kitchen hired only male servers, it was investigated by the Fair Employment Practices Commission.

_____ 7. *police officers*
  Several ~~lady policemen~~ are guarding against intruders.

_____ 8. *End this sentence after "intelligent."* *Delete the rest of this sentence*
  My new assistant is so reliable and intelligent. ~~that it's hard to believe she's only 19.~~

_____ 9. *Photographers*
  ~~Cameramen~~ loved taking pictures of Jacqueline Kennedy Onassis.

_____ 10. *women*
  The ~~girls~~ in my office go to lunch at 12.

___C___ 11. "Full inclusion" means children with disabilities are in regular classes, and extra help is provided by trained teaching aides.

_____ 12. *Delete first part of sentence.*
  ~~Despite being confined to a wheelchair,~~ he is a caring and capable counselor.

_____ 13. *an excellent*
  She is doing a ~~man-sized~~ job.

_____ 14. *delete male*
  My husband is a ~~male~~ nurse.

_____ 15. *Delete this sentence.*
  ~~Asian Americans are poor drivers.~~

_____ 16. *Delete this sentence.*
  ~~African Americans are good dancers.~~

_____ 17. *synthetic*
  We prefer garments made of natural rather than ~~manmade~~ fabrics.

_____ 18. *agent*
  Can you recommend a good insurance ~~man~~?

See page 395 for answers.

Read 28

EXPANDING YOUR VOCABULARY

Extensive, varied, and current vocabulary is needed for better jobs in business, technology, or the professions. Vocabulary expansion involves not only learning new words and adding meanings to words previously acquired, but also spelling, pronouncing, and using more words correctly. Below are the usual vocabulary building stages, although we are often unaware of them as they occur:

1. We recognize a written word; that is, it looks familiar to us.
2. Next we understand the written word in context.
3. Later we understand the meaning when we hear the word.
4. Eventually we're ready to use the word correctly in our own writing.
5. Finally, we use the word correctly in conversation, which means the word has been mastered.

The process continues as long as we read books, magazines, and newspapers; listen to radio, films, live plays, and TV; and communicate with educated people. The result is a continuous increase in vocabulary.

We speed up vocabulary growth by conscious effort. As you read the 15 sentences that follow, give special attention to the underlined nouns for spelling, definition, capitalization, and/or pronunciation. If you look carefully at the words and say them, you'll probably spell them correctly when you do the Replay. Even any you're not sure of will look "funny" if misspelled, prompting you to refer to a dictionary. Use the dictionary to check on both singular and plural spellings, definitions, and pronunciation.

1. My allies found the cargo mentioned in both memoranda.
2. Please put the bills of lading in the portfolio, as they will be a good tie-in with the data.
3. The alumni were given copies of the itinerary as a memento.
4. The notaries public gave the banjos to the editors in chief.
5. An importer must know the criteria for determining the amount of the tariff as well as how to get the letters of credit.
6. The embargo on dynamos and dominos affects our earnings and causes several weird phenomena.
7. An addendum to the contract states that the attorneys distribute the proceeds from the sale according to these criteria.
8. The plaintiffs are technicians in our Aircraft Division but do not understand the phenomenon.
9. A survey of the co-owners shows that two speak Swahili and three, Hindi.
10. According to the media, proxies from 1,000 stockholders were received at these facilities.
11. According to the alumnae, Native Americans used the shells and clay pots stored on the premises.
12. When my proposal was rejected in all three memoranda, I accused the authorities of bias.
13. The altos sang solos for the Civilian Conservation Corps.
14. The author of the textbook about college curriculums received the Nobel Prize.
15. Her credentials show she worked for the Vietnamese in Saigon and studied aeronautics, mathematics, and monkeys' vertebrae.

MEMO FROM THE WORDSMITH

Ancient Greek athletes trained naked. The word *gymnasium* entered the English language through the Greek word *gymnos* which means *naked*.

Replay
28

Test your spelling and vocabulary knowledge of Chapter 5 nouns. The first letter of each word is given.

EXAMPLE   My sister's husband: **b**rother-in-law

1. Nongender-specific words for *businessmen*:
   businesspeople   **b**usinesspersons

2. One who starts a legal action: **p**laintiff

3. A compound noun meaning programs for use in a computer:
   software

4. The place where you work or live: **p**remises

5. Plural of newspaper executive: **e**ditors in chief

6. Freight carried by a ship: **c**argo

7. The singular and plural are spelled the same but pronounced differently: **c**orps

8. Carrying case for holding papers or a list of investments:
   portfolio

9. Written authorization to act for another: **p**roxy

10. A plural noun that means profits from a commercial or other venture; when the same word is a verb, the accent is on the second syllable:
    proceeds

11. A travel plan: transportation, times, dates, hotels:
    itinerary

12. An object that is a reminder of the past, such as a souvenir:
    memento

13. Persons authorized to guarantee signatures on legal documents:
    notaries public

14. The plural of chassis: **c**hassis

15. Something we want more of at income tax time (compound noun):
    write-offs

Check your answers on page 395. Then review Reads and Replays 26–28 before taking the Pop Quiz on page 342.

CHECKPOINT

A regular noun becomes plural by adding **s**—or by adding **es** to nouns ending in *s, x, z, sh,* and *ch.* However, irregular nouns (the exceptions) don't follow this principle. In this chapter you've reviewed irregular plural spellings and pronunciations. In addition you studied compound nouns, capitalization, and bias-free language. You also may have increased your vocabulary.

SPECIAL ASSIGNMENTS

Word Game from the Wordsmith

A. The words represented by the following definitions rhyme with BIRD. Fill in the blanks. The first letter of each word is shown.

   EXAMPLE   Slang for a person regarded as clumsy and dull   **n**erd

1. Listened ___**h**eard___
2. Next after second ___**t**hird___
3. You should put one in each blank ___**w**ord___
4. A group of large animals ___**h**erd___
5. Unreasonable and ridiculous ___**a**bsurd___
6. The thick part of sour milk ___**c**urd___
7. Went astray or made a mistake ___**e**rred___
8. Postponed ___**d**eferred___
9. Came together to discuss ___**c**onferred___
10. Agreed; we're of the same opinion ___**c**oncurred___

Most students get items 1–4, but many do not know 5–10. Divide class into several teams and provide a few minutes to see which team gets the most words. Follow with applause or grade points for the winning team.

B. Use the proofreader's mark (≡) to show which words should be capitalized. See "Mini Reference Manual" Appendix D if you want help with capitalization.

   EXAMPLE: susan is an occupational therapist in new york state.

1. joseph flew via united airlines to uganda on veteran's day.
2. world war II preceded the era known as the fabulous forties.
3. our english professor met reverend perez in this city last fall.
4. Did senator charles west from pennsylvania give a speech in northern maine?
5. to get to monroe college, go north on jerome avenue until you reach fordham road.
6. we ate a bag of yum yums and drank koka kola while we watched the film "father's day."
7. my uncle was the first to do accounts payable auditing for department stores.
8. if you speak german and take business 101, you might qualify for that job.
9. although she has a bs and an ma degree, she does not have a doctorate.
10. The salutation is "dear credit manager"; the complimentary close is "sincerely."

Special Assignment B may be an open-book quiz. In addition, see "Capitalization Quiz" and "Open-Book Reference Manual Quiz" in the *Instructor's Resource Kit*. Capitalization questions on Real Quiz, Midterm, and Final are based on Chapter 5 only.

PROOFREADING FOR CAREERS

Proofreading is challenging because it requires concentration on so much at the same time: errors in spelling, word choice, meaning, noun usage, numbers, capitalization, and just plain carelessness.

If you proofread on your screen, start by running your spell checker, and make needed corrections. Next, whether the copy is on the screen or in your book, use a pointer or closed pen to "underline" slowly each line of type for sense, correctness, and clearness. The "underlining" changes to pointing when you catch a possible error. This underlining–pointing process focuses your attention on proofreading.

The letter to Ms. Gonzalez has noun errors—capitalization, spelling, and typographical. Use your dictionary and spell checker as needed. In addition, be sure you end each sentence with a period, and begin the next sentence with a capital letter. Your instructor will tell you whether to rekey or to use proofreaders' marks (see inside back cover).

Dear Ms. Zonzalez:

We can help your Sales Staff by providing you firm with a new dimention of client service—an inferior design tiein with the sale of office space and studioes.

clients looking for new office facilitys often ask about desks, filling cabinets, chairs, and carpeting. They are interested in prices and availability of these items this is were we come in to help close the sale for you. At no charge to you, we can furnish a complete plan to fit any offices you offer. this will be an important sales aid to you that will pay off in faster and increased sales.

Alert Reality Firms like yours are always looking for new and creative consepts in selling. The next time you have a client with an office furnishing need. Give us a call or reffer your client to us we will prove that this service will work for you.

May we get together with you or one of your key sales representatives soon to discuss precise arrangemts.

Yours Very Truely,

Manny Errata

What's your score? Excellent? Good? Fair? or Ugh??

Have you completed and reviewed "Read and Replay Capitalization" in Appendix D?

PRACTICE QUIZ

www.prenhall.com/smith

Write the letter of the best answer in the blank. It's all right to use your dictionary.

_____a_____ **1.** Marie and Jennifer are the two new (a) ushers (b) usherettes (c) female ushers (d) ush-eresses (e) usher's. (Read 27)

_____d_____ **2.** My sister is a (a) woman sculptor (b) sculptress (c) sculpterette (d) sculptor. (Read 27)

_____a_____ **3.** The birds fly (a) south (b) South in (Read 26)

_____a_____ **4.** the (a) winter (b) Winter. (Read 26)

_____d_____ **5.** Stockholders mail in their (a) proxys (b) proxi (c) proxyes (d) proxies. (Read 23)

_____b_____ **6.** Aeronautics (a) were (b) was my favorite subject. (Read 24)

_____c_____ **7.** They are our (a) alleys (b) allys (c) allies (d) alloys in the controversy. (Read 23)

_____c_____ **8.** Her (a) sister-in-laws (b) sister in laws (c) sisters-in-law (d) sisters-in-laws manage the office. (Read 23)

_____c_____ **9.** Our treasurer wants the fund-raisers to deposit the (a) dynamos (b) rebuffs (c) proceeds (d) mementos (e) itineraries in the bank. (Read 28)

_____b_____ **10.** To import from Thailand, we need two (a) letters-of-credit (b) letters of credit (c) letter of credits (d) letters of credits (e) letter's of credit. (Read 26)

_____b_____ **11.** Our office in the (a) midwest (b) Midwest (c) Mid west will be (Read 26)

_____c_____ **12.** closed on (a) labor day (b) Labor day (c) Labor Day. (Read 25)

_____a_____ **13.** The Wyoming (a) senator (b) Senator voted for a million dollar grant to Casper College. (Read 25)

_____b_____ **14.** Both (a) secretarys (b) secretaries (c) secretarys' (d) secretary's seemed confident as they interviewed for the data control position. (Read 23)

_____c_____ **15.** The inventory indicates we have three (a) celloes (b) cello (c) cellos (d) cello's (e) cellos' in our Music Department. (Read 23)

_____d_____ **16.** Although James McCarthy is the president, three other (a) McCarthy's (b) McCarthies (c) McCarthys' (d) McCarthys are on the Board. (Read 23)

_____c_____ **17.** The (a) Jones's (b) Jones (c) Joneses (d) Jones' invited us to dinner. (Read 23)

_____d_____ **18.** Our (a) Advertising department (b) advertising Department (c) advertising department (d) Advertising Department is (Read 26)

_____d_____ **19.** headed by the (a) vice President (b) Vice president (c) Vice President (d) vice president. (Read 26)

_____c_____ **20.** The nouns (a) cargo and embargo (b) vertebra and chassis (c) addenda and appendix (d) tariff and bill of lading have almost the same meaning. (Read 25)

_____c_____ **21.** The (a) diagnosis (b) diagnosises (c) diagnoses (d) diagnosis's are accurate. (Read 25)

_____a_____ **22.** These (a) criteria (b) criterion (c) criterias (d) criteriae were developed for us. (Read 25)

_____a_____ **23.** Which word is incorrect in the following? An alumnae and an alumnus broke their ver-tebrae when they placed the bric-a-brac on top of the chassis. (a) alumnae (b) alumnus (c) vertebrae (d) bric-a-brac (e) chassis (Read 25)

_____a_____ **24.** Which word is incorrect in the following? We sent two memorandas about the new for-mulas for dealing with the two nuclear crises publicized by the media. (a) memorandas (b) formulas (c) nuclear (d) crises (e) media (Read 25)

_____b_____ **25.** The new curricula (a) seems (b) seem more challenging than the old. (Read 25)

108 Chapter 5 | Apples, Tigers, and Swahili

The Global Marketplace—Italy

Italian	Spell	Say	Currency
GREETING	Ciao	Chow	Italian Lira (Lit)
DEPARTING	Ciao	Chow	
PLEASE	Per favore	Per fa.VOR.ay	
THANK YOU	Grazie	GRA.tzee	
I DON'T UNDERSTAND	I o non capisco		

Courtesy of Sebastian

Be Kind to Substitutes

6

Pronouns Substitute for Nouns

> "Who would succeed in the world should be wise in the use of pronouns. Utter the **You** twenty times where you once utter the **I.**"
>
> —*John Milton Hay*

After Completing Chapter 6, You Will

> Use pronouns according to Standard English principles.

See the *Instructor's Resource Kit* for Chapter 6 Pretest.

Chapter 6 grammar terms are explained with as few grammar terms as possible. Students with weak English skills are more likely to improve usage with fewer intimidating terms. If your students profit from more traditional rules and terms, see "Grammar for the Expert" in the *Instructor's Resource Kit.*

Some students like to reminisce about junior high school experiences with substitutes. The appearance of a substitute signaled the class clowns to go into action. Even usually well-behaved students sometimes joined in the fun. When the substitute lost control of the class, students knew they had achieved the ultimate in success. As an adult, you may feel remorseful about how mean your classmates (not *you,* of course) were. Chapter 6 is a chance to be kind to substitutes—noun substitutes, that is—by using them correctly.

> **A word that substitutes for a noun is a pronoun. A pronoun refers to someone or something previously named by a noun.**

For example, after using the noun *pencil,* you could later use the pronoun *it;* the reader or listener understands *it* means *pencil.* To substitute for the plural noun *pencils,* the pronoun *them* or *they* might be used. We substitute pronouns for nouns to avoid repetition and awkward constructions.

When a noun or pronoun tells who or what a sentence is about, it is identified as a *subject* or as being in the "subjective case."* However, when a noun or pronoun is used in a sentence as an *object,* it is in the "objective case" as the object of a verb or a preposition. Nouns or pronouns in the "possessive case" show possession (or ownership). Here's a chart to help you select correct pronouns:

Handy Personal Pronoun Reference Chart

Person	Singular			Plural		
	Subject	Object	Possessive	Subject	Object	Possessive
First	I	me	my, mine	we	us	our, ours
Second	you**	you	your, yours	you	you	your, yours
Third	he, she, it**	him, her, it	his, hers, its	they	them	their, theirs
	who	whom	whose	who, whoever	whom, whomever	whose

Self/Selves	First Person: myself, ourselves
	Second Person: yourself, yourselves
	Third Person: himself, herself, itself, themselves,*** oneself
Indefinite Pronouns	• everyone, anyone, no one, someone
	• anybody, everybody, nobody, somebody
	• anything, everything, nothing, something
	• some, several, both, few, any, one, all (when not followed by a noun)

*also known as the *nominative case*
You can't go wrong with **you or **it:** They are used as either subjects or objects; **you** is singular *or* plural.
***Never use the "nonword" *themself*!

JUST BETWEEN YOU AND ME

More Examples: They sent a letter to Richard and (I/**me**). The report is for (she and I/**her and me**). They want (you and I/**you and me**) at the party. Dave and (**I**/me) believe the president and (he/**him**).

Or should it be "Just Between You and I"? **Definitely not!** Well, how about "Just Between Us"? You know it can't be "Just Between We"! With **two** pronouns—*you* and *me* or *you* and *I*—it's harder to tell which is right; this is a good time to review the subjects and objects on the chart on the preceding page. The correct title of Read 29 could be either "Just Between You and Me" or "Just Between Us." Objects of prepositions—such as *between, among, under, below, over, with, to, in, into*—must be nouns or object pronouns, not subject pronouns. Objects of action verbs must also be nouns or object pronouns—Please telephone either Janet or me (not subject pronouns like *I* or a *self* pronoun like "myself").

IMAGINE IT OMITTED

> Most pronoun errors occur when two or more are together, such as "he and I," or when a noun is used with a pronoun, such as Jonathan and I/me/myself. You will usually make the right choice if you imagine one of the nouns omitted; then decide whether the sentence "sounds right."

NO Jonathan and me went to the concert. Jonathan and myself went to the concert. [Leave Jonathan home, and you know that "Me or myself went to the concert" sounds **terrible.**] You would change the sentence to:

YES Jonathan and **I** went to the concert.

Try Some More

NO Give the report to Pat Garner and I tomorrow. [Omit Pat, and then you know "Give the report to I before noon" can't be right.] So . . .

YES Give the report to Pat Garner and **me** before noon.

NO Professor Boone will discuss the problem with he and I. [Wrong, of course; change *he* to *him* and *I* to *me*—change subject pronouns to object pronouns—objects of the preposition *with*.]

NO Him and me attended the concert. [Both wrong; you wouldn't say "Him attended the concert." or "Me attended the concert." Therefore, change object pronouns—*him/me*—to subject pronouns He and I.]

RESULT He and I attended the concert.

Marg, Larry, and (me, I, myself) lost all our quarters at the casino. [Just imagine the nouns (Marg and Larry) omitted, and you know the correct pronoun would be *I*—Marg, Larry, and I.]

Please give the instructions to the president and (he, him, himself). [If you imagine the noun *president* omitted, you easily choose the pronoun *him*.]

NO Bob and me went to the concert. [Imagine Bob out of the sentence, and you'll use the subject pronoun *I*, not the object pronoun *me*.]

No Conjunction Between the Noun and the Pronoun?
Imagine Omitting the Noun!

Sometimes the noun and pronoun have no conjunction between them. This happens when we combine a noun and a pronoun with the same meaning for emphasis or clearness. Which is right?

we clerks **or** us clerks   we boys **or** us boys

The four combinations above are Standard English. You can tell which to use when you see the rest of the sentence. You'll make the right choice if you imagine omitting the **noun;** then decide whether the sentence sounds right:

NO **Us students** need a longer lunch break. [Omit the noun *students* and you know "**Us** need a longer lunch break" is wrong.]

YES **We students** need a longer lunch break. [Therefore, **We** need a longer lunch break.]

Try Another

NO Please give **we students** longer lunch breaks. [Omit the noun *students*, and you immediately know that "Please give **we** longer lunch breaks" can't be right.]

YES Please give **us students** longer lunch breaks.

> **The subject of a verb tells who or what is doing or being. Subject pronouns are listed in the "Handy Personal Pronoun Reference Chart's" two "Subject" columns (Singular and Plural). Use a noun and/or a subject pronoun as the subject or object of an action verb or a preposition.**

Joan and **he** are majoring in business administration. [*Joan* and *he* are the subjects—NOT *him.*]
(subj / v)

He and **she** both graduated with honors. [not *him* and *her*]
(subj / v)

> **Use subject pronouns after "being verbs" (defined in Read 18, Chapter 4).**

The best students **are he** and **she.** [*are* is a being verb; therefore, use subject pronouns *he* and *she*]

It was Josh and (me/I/myself) who made the highest scores. [being verb—*was*/subject pronoun—*I*]

> **The object of an action verb or of a preposition tells what or whom; if it is a pronoun, it must be an *object* pronoun, not a *subject.***

She gave the **pizza** to **Joan** and **me.** [The object of the verb *gave* is *pizza*; the objects of the preposition *to* are *Joan* and *me.*]

My sister bought the theater **tickets** for **Jordan** and **us.** [*tickets* is the object of the verb *bought*; *Jordan* and *us* are objects of the preposition *for*]

Between **you** and **me,** we need a lunch break. [*You* and *me* are objects of the preposition *between.*]

> **The subject of an action verb is the "doer" of the verb's action.**

You and **I** plan to attend graduate school and study for our MBAs. [subject pronouns are subjects of the action verb *plan*]

The object of a preposition or a verb must be an object pronoun.

Let's divide the pie **between you** and **me.** [objects of the preposition *between*]

Give it to my **sister** and **him.** [objects of the preposition *to* are *sister* and *him*; object of the verb *give* is *it*]

We kissed **Marty** and **her** goodbye. [objects of the verb *kissed* are *Marty* and *her*]

NO   Bob and me went to the concert. [Leave Bob home, and you immediately know that "Me went to the concert" sounds **terrible.**] That's why you change the sentence to:

YES   Bob and **I** went to the concert.

Try Another

NO   Send a check to Ms. Dahlberg and I immediately. [Omit Ms. Dahlberg, and you immediately know that "Send a check to I immediately" can't be right.] So . . .

YES   Send a check to Ms. Dahlberg and **me** immediately.

Imagine Omitting the Pronouns One at a Time

Just consider the pronouns one at a time. Here's an example of incorrectly used pronouns: Professor Boone will talk to **he and I.** Note that *he* and *I* (subject pronouns) are both used incorrectly as objects of the preposition *to*.

NO   Professor Boone will talk to **he.** [*he*, a subject pronoun, can't be the object of *to*]

YES   Professor Boone will talk to **him.** [*him* is the object of the preposition *to*]

NO   Professor Boone will talk to **I.** [*I*, a subject pronoun, can't be an object of *to*]

YES   Professor Boone will talk to **me.** [*me* is the object of the preposition *to*]

ALSO CORRECT   Professor Boone will talk to **him and me.**

NO   Please give **we students** a longer lunch break. [Omit the noun *students*, and you know that "Please give we a longer lunch break" can't be right.]

YES   Please give **us students** a longer lunch break.

Recap   Write a different subject pronoun in the subject blank of each sentence and an object pronoun in the object blank. If in doubt, see the chart at the beginning of this chapter. Notice the last item in each column ends with a question mark.

	SUBJECT		OBJECT OF VERB		SUBJECT		OBJECT OF PREPOSITION
1.	I	love	you	.	I	am in love with	him	.
2.	He	loves	her	.	We	are in love with	it	.
3.	She	loves	him	.	He	is in love with	her	.
4.	Who	loves	them	?	They	are in love with	whom	?

Possible answers are on page 395.

> **If a subject is a pronoun, use a subject pronoun, not an object pronoun.**

YES **He** and **I** love Twileen. [*He* and *I* are the subjects.]

NO **Him** and **me** love Twileen. [*Him* and *me* are object pronouns.]

> **If the object of a verb or a preposition is a pronoun, use an object pronoun.**

OBJECTS OF VERBS

To find out if a verb has an object, say the subject and verb and then ask *whom?* or *what?* If you get an answer, the answer is the object. The following sentences show objects of verbs in bold type:

I like **peaches.** [I like what? peaches]

George loves **Twileen** and **her.** [George loves whom? Twileen and her]

If you don't get an answer to *whom* or *what,* the clause has no object.

I dance well. [*well* answers *how,* not whom or what]

OBJECTS OF PREPOSITIONS

Prepositions, usually direction or position words, begin prepositional phrases. A prepositional phrase ends with an object, which is a noun or an object pronoun. If you say the preposition followed by *whom* or *what,* the answer is the object. In the following sentences, the prepositions are underlined, and the objects of the prepositions are in bold type.

Twileen is <u>in</u> **love** <u>with</u> **him.**

Victoria wrote <u>to</u> **me** <u>about</u> the **secret.**

Just <u>between</u> **you** and **me,** the meeting was dull.

You can now answer the question on the first line of Read 29. Since *between* is a preposition, an object pronoun is needed. "Just Between You and **Me**" (not **I**) is right—*me* is an object of the preposition *between.*

Use the methods given in Read 29 to select and circle the correct pronoun.

1. If (she and me/her and I/(she and I)/her and me) study, we'll do better work.

2. Please tell Twileen and (I/(me)) Victoria's secret.

3. ((They)/Them) as well as Carla are involved in the Texas City project.

4. ((We)/Us) attorneys need more training in Excel.

5. Ms. Garcia and ((I)/me) should work together frequently.

6. Joining the union would be good for (we/(us)) pasta chefs.

7. Deborah of Denmark, South Carolina, phoned Suzy and (I/(me)).

8. Mr. Tawa authorized George and (she/(her)) to go to Fort Lauderdale.

9. Do ((we)/us) students have a vote on whether to have a final exam?

10. Ms. Denova and ((he)/him) could travel to the sales meeting together.

11. Should Luann show Mr. Byrnes and (I/me) the new outlines?

12. Christopher asked both you and (we/us) to visit her.

13. This committee needs Terry as well as (they/them).

14. Everyone except Ms. Rosenblatt and (he/him) works in Omaha.*

15. (Him and me/He and I) have an advantage in this situation.

16. The money should be divided between Mr. Park and (I/me).

17. Jamal and (him/he) will share the Pulitzer prize.

18. (We/Us) Americans transfer the fork to the right hand after cutting food.

19. We sent Ms. Papachristos and (her/she) to London.

20. All the responsibility was given to the auditor and (they/them).

21. The director told Ms. Ray and (I/me) about downtown Lancaster.

22. Were you and (he/him) preparing a PowerPoint presentation that day?

23. Ms. Lightle invited you but not (he/him).

24. Professor Newsom wants Yee and (I/me) to work on the project.

25. You and (them/they) should devise a new production schedule.

26. Let's keep Victoria's secret just between you and (I/me).

(Item 23) both *you* and *him* are objects of the verb *invited; not* is an adjective modifying the pronoun *him.*

Check your answers on pages 395–396.

ME, MYSELF, AND I

As a child, you probably asked your parents questions like "Can Johnny and me go to the movies?" Your mom or dad may have replied "Johnny and **I**" and wouldn't give you the money until you changed that evil word *me* to *I.* Although they were right in that instance, a number of incidents like that through your growing-up years may have made you a firm disbeliever in the word *me.* It really is quite a respectable word and does not deserve an X rating. Oddly enough, it's people who try to be correct who use *I* where *me* belongs.

UNDERSTOOD WORDS

If you're not sure whether to use a subject or an object pronoun when "understood words" are missing, decide whether "understood" words could complete the thought. "Understood words" are verbs or subject–verb combinations that complete comparisons but are often left out to avoid wordiness.

Imagine it Completed

> If you're not sure whether to use subject or object pronouns when understood words are missing, just complete the expression or imagine it completed.

*Subject is **everyone; except** is a preposition.

The words *than* or *as* are clues that a sentence may have "understood" words omitted from it.

> Did Mr. Ngueyen sell more tickets than (she/her)? [Than **she** sold or than **she** did is understood—not *her* sold or *her* did.]

NO No one wants to please you more than the manager and me (want to please you).

YES No one wants to please you more than the manager and I (want to please you).

Try Another

> Do you like him better than (I or me)? [Again, if you add the understood words *like him* or *do* after the questionable pronoun, you easily choose *I*, not *me*.]

NO Do you like him better than me (do)?

YES Do you like him better than I (do)?

But

In some sentences either a subject pronoun or an object pronoun is correct. The choice depends on what you want the sentence to mean. Let's take another look at the preceding example sentence:

> Do you like him better than (I *or* me)? [If you add the understood words *you like* **before** the pronoun in question, the meaning changes—and you need a different pronoun.]

NO Do you like him better than (you like) I?

YES Do you like him better than (you like) me?

To be certain you choose the right pronoun, complete in your imagination any sentence that omits understood words in a comparison. If more than one meaning is possible, use the words that deliver the meaning you intend.

Circle the correct answers for items 1 and 2.

    1. Mr. Swanson sold more tickets than (her/she).

    2. Marilyn is just as tall as (him/he).

Items 3 and 4 are both correct but have different meanings. Circle the number of the statement that sounds like a marriage in trouble. In the blanks below, write the understood word(s) in parentheses.

    3. My husband likes golf better than I. _____(do)_____

    ④. My husband likes golf better than __(he likes)__ me.

Check your answers on page 396.

COMPOUND PRONOUNS—SELF, SELVES

The *self* and *selves* pronouns that follow are compound personal pronouns but are also called *reflexive pronouns*—because they may be used to reflect back on a noun or another pronoun with the same meaning. They're also used to add emphasis or clarity. Other spellings of these words, which are in parentheses below, are unacceptable:

myself	itself
yourself, yourselves	oneself
himself (never hisself)	ourselves (not ourself)
herself	themselves (never theirself, theirselves, or themself)

Using Reflective Pronouns Correctly

I did that job **myself.** (emphasis)

You yourself know better than that. (emphasis)

They took care of **themselves.** (reflects *They*)

He frequently talks to **himself.** (reflects *He*)

We ourselves are to blame. (emphasis)

He corrected **himself** immediately. (reflects *He*)

She does the easy work by **herself.** (object of preposition *by*)

She **herself** does the easy work. (emphasis)

She placed **herself** at great risk. (reflects *She*)

The most common *self* error is using *myself* when *I* or *me* makes sense.

NO Ms. Trentham and myself will visit the client.

YES Ms. Trentham and I will visit the client.

Replay 30

A. Fill in the parentheses with the understood completion word or words.

1. He loves his wife more than she. (_____does_____)

2. He loves his wife more than (_____he loves_____) her.

3. I know the vice president better than he. (_____does_____)

4. I know the vice president better than (_____you know_____) him.

5. Ms. Guthrie can operate the device as well as he. (_____can_____)

B. Draw a line through the incorrect words, and make the necessary pronoun corrections.

Write *C* in the blank if the sentence is already correct.

____C____ 1. The president himself will attend the meeting.

__themselves__ 2. The twins learned how to dress ~~theirselves~~ at an early age.

__he can__ 3. She can do the work faster than ~~him~~.

__he__ 4. Neither his sister nor ~~himself~~ is willing to take care of the situation.

_____me_____ 5. Always give copies of memorandums to Joyce Moore and ~~myself~~.

_____we (do)_____ 6. Victoria usually leaves earlier than Twileen and ~~us~~.

_____I leave (or I do)_____ 7. The others usually leave as soon as Ms. Hixon and ~~me~~ leave.

_____C_____ 8. Ms. Walsh runs just as fast as he.

_____themselves_____ 9. Rhoda Bike and Isabelle Ringing often ask ~~themself~~ that question.

_____you_____ 10. Only Ms. Englehart and ~~yourself~~ know how to create bulleted lists.

_____I_____ 11. Isabelle Ringing and ~~myself~~ listened to them with great interest.

_____C_____ 12. I may find myself looking for a new manager.

_____C_____ 13. We ourselves are excluded from the contract.

_____C_____ 14. They felt like themselves again after the crisis had passed.

_____C_____ 15. She injured herself while they were in Oshkosh.

_____themselves_____ 16. The Indiana team members voted ~~theirself~~ a pay increase.

_____C_____ 17. Good table manners require that you spoon soup away from yourself.

_____himself_____ 18. He gave ~~hisself~~ a raise.

_____myself_____ 19. I gave ~~me~~ a perm and a manicure.

_____C_____ 20. Gomez himself should have known better than to buy 100 shares of GYPCO in a bear market; don't blame yourself.

Check your answers on page 396.

A TALE OF A LIZARD'S TAIL

If, as a child, you never tried to pull a lizard, tail first, out of a crevice in a rock, you've missed the shock of a lifetime—that of discovering that although its tail is in your hand, the rest of it didn't come along. This is not as horrible as it seems, for the lizard manages to grow a new one.

Had you wanted to tell your buddy, you might have said in amazement, "The lizard lost its tail!" Although you didn't know it as a child, that sentence skillfully uses a possessive pronoun to substitute for a possessive noun. Without

English experts label possessive pronouns as pronouns, adjectives, or pronominal adjectives (depending on the source book). Although they have adjective functions, most students find it easiest to identify them as pronouns—and the label doesn't affect correct usage. For the same reason, the term "possessive noun" is used instead of "possessive adjective."

that substitution, the sentence would be, "The lizard lost the lizard's tail," using a possessive noun* and sounding rather silly.

> **Do not use apostrophes with possessive pronouns—pronouns showing ownership—as listed below. Do use apostrophes with possessive nouns and contractions.**

That is *my* idea.   That idea is *mine.***

This is *his* idea.   This idea is *his.*

That is *her* businesss.   That book is *hers.*

These are *our* jeeps.   Those jeeps are *ours.*

This is *your* dollar.   That money is *yours.*

These are *their* jeans.   Those jeans are *theirs.*

The lizard lost *its* tail.   *Whose* tail is that?

> **Some *possessive pronouns* and *contractions* sound alike but are spelled differently.**

POSSESSIVE PRONOUNS	CONTRACTIONS REQUIRE APOSTROPHES
your—**Your** job pays well.	**you're** (you are)—**You're** a rich man.
its—**Its** wing was injured.	**it's** (it is)—**It's** a cockatoo.
whose—**Whose** bird is singing?	**who's** (who is)—I know **who's** here.
their—**Their** problem is serious.	**they're** (they are)—**They're** in debt.

> ***One*** and ***body*** words are often called *indefinite* pronouns. To make them possessive, add *'s.*

That could be **anybody's** calliope.   This isn't **anyone's** trumpet.

It must be **somebody's** saxophone.   **Someone's** guitar is on the piano.

That is **everybody's** banjo.   **Everyone's** song sheets are here.

One should mind **one's** own business.

> ***Body*, *one*, or *thing* pronouns may also be part of a contraction.**

Everyone is   **Everyone's** going to the movies.

Everything is   **Everything's** fine.

Somebody is   **Somebody's** in the kitchen with Dinah.

---

*Possessive nouns are explained in Chapter 9.

**Never add *s* to *mine* unless you mean the underground sources of valuable materials:

☹ **Mines** is better than hers.

☺ **Mine** is better than hers. My diamond **mines** have made me too rich.

Replay 31

Correct the pronoun and contraction errors. Write C beside the only correct sentence.

Everybody's   **1.** Everybody's going to the Dodger game today.

Nobody's   **2.** Nobody's here to mind the store.

its it's   **3.** Reserve the apostrophe for it's proper use, and omit it when its not needed.

C   **4.** Its color has faded.

ours yours   **5.** The new printer is our's, not yours'.

Yours   **6.** Your's is on the 18th floor of the New Otani Hotel.

hers   **7.** The one on the right is her's.

mine theirs   **8.** The CDs are mines; the floppies are their's.

You're   **9.** Your to use your own books today.

you're your   **10.** If your running behind schedule, let you're sister help.

Everybody's   **11.** Everybody's notebook is closed.

No one's   **12.** Noones work was checked.

anyone's   **13.** Whether or not we'll get that account is anyones guess.

Who's   **14.** Whose going to do the graphics?

Whose   **15.** Who's work do you prefer?

Answers to this Replay are on page 396. Your Pop Quiz is on page 343.

Word Power

Student is a word derived from the Latin *studium* meaning "zeal" or "eagerness."

Sophomore literally means "wise fool." It is an invented word combining two Greek words: *sophos* (wise) and *moros* (foolish).

Pedagogue, which now means *teacher,* originally meant a slave in charge of children—from the Greek *paidagogos.* The Romans, who respected the learning of the Greeks, assigned Greek slaves to train their children.

A WHODUNIT

Most of us have seen TV and movie whodunits; and if you're a mystery fan, you've probably read them. The best known English-language whodunit is the Mystery of Who and Whom. (Although it qualifies as a mystery, it's not exactly a thriller.)

Often those who truly care about their English err by using *whom* where *who* belongs. Some people correct themselves like this: "Who ah whom ah who ah is the treasurer of this company?" Then we have who/whom cowards who mumble the mystery word and hope the listener won't notice. Many well-educated people, however, do use *who* and *whom* traditionally. Now here's the good news: Some language experts recommend eliminating *whom* from **speech** to avoid awkward pauses and mumbles.

> **BUT**

For career-type **writing,** we recommend taking some care in choosing *who* and *whom* "correctly." It takes just a few seconds if you know the "clue." In **speech,** just use *who* unless you're certain about your choice of *whom.*

CLUE: IMAGINE HE OR HIM

> Imagine replacing *who* or *whom* with he or him, regardless of whether the person is male or female. If *he* fits, use *who/whoever;* if *him* sounds right, use *whom/whomever.* Both *him* and *whom* end with *m,* making it a good memory device.

- Mr. Agresta is the man (who/whom) gave me the package. [*He* gave me the package—not *him* gave me the package; therefore, choose *who.*]
- Jim Young, (who/whom) we understand visited you yesterday, is a database expert. [Jim Young is a database expert; we understand *he* visited your office yesterday—not *him* visited; therefore, choose *who.*]
- Dr. Perry, (who/whom) Prudential hopes to hire, has the highest qualifications. [Dr. Perry has the highest qualifications; Prudential hopes to hire *him*—not hire *he;* therefore choose *whom.*]
- Assign the report to (whoever/whomever) you wish. [Assign the report to *him;* therefore, choose *whomever.*]

> If either *he* or *him* fits in the same place, choose *who* or *whoever,* not *whom* or *whomever:*

- You should go with (whoever/whomever) is ready first. [*Him* fits after *with,* but *he* fits before *is.* Therefore use *who.* (Always allow a subject to have priority over an object.)
- Follow the same who/whom steps for **who**ever and **whom**ever.

Language experts agree it's better to use *who* for *whom* rather than the reverse. Use *who* in speech, therefore, to avoid risking "overrefinement."

Bergen Evans wrote, "The semiliterate, intimidated, and bewildered struggle to decide between *who and whom.* People who speak natural, fluent, literary English often use the nominative to begin questions even when the objective is technically correct." *New York Times* columnist and eminent linguist William Safire wrote, "My rule is when *whom* is correct, use some other formulation."

Underline the correct pronoun.

1. She is the expert (who/**whom**) we told you about.
2. He is the one (**who**/whom) will receive the Texas award.
3. George can invite (whoever/**whomever**) he pleases.
4. Give* the clothes to (**whoever**/whomever) you think needs them.
5. Send the money to (**whoever**/whomever) will use it wisely.
6. Jacqueline is the woman (who/**whom**) I chose for the job.
7. I want to hire (**whoever**/whomever) will do the job efficiently.
8. (Who/**Whom**) should she choose for the job?
9. To (who/**whom**) did George tell his innermost secrets?
10. Victoria, (who/**whom**) Jesse planned to marry, was Twileen's best friend.

Check answers on page 396.

I believe we still need the distinction for written business communication, but we're probably fighting a losing battle. The *New Yorker* magazine used to mock examples of the misuse of *whom;* e.g., "Whom did you say is coming?" These examples were discontinued because few readers understood what was wrong.

Item 4: *whoever* is the subject of the verb *needs.*

An atrocious, but grammatically correct, sports section headline in a Chicago newspaper read: "Whom's He Kidding?"

CLUE: IMAGINE A STATEMENT

> **If the sentence is a question, imagine it as a statement before making the who/whom choice.**

(Who/**Whom**) are you going with? [Change to statement—You are going with who/whom. *Him*, not *he*, sounds right after *with*. OR *You* is the subject of the verb *are going*; therefore, the object pronoun *whom* is needed as the object of the preposition *with*.]

Circle the correct answer.

(Whom/**Who**) do you think won the contest?

See page 396 for the answer.

The question of (**who**/whom) should go will be discussed. Bring anyone (**who**/whom) would like to come. Mr. Valdez, (who/**whom**) we expected, can't attend.

You is the "understood" subject of *give; you* is the subject of *think; whoever* is the subject of *needs.*

Replay 32

Choose *who/whoever* or *whom/whomever;* apply Read 32 clues. **Practice,** not guessing, makes perfect.

EXAMPLE   We referred a programmer to you <u>whom</u> we believe you will like.

The demise of *whom* has been predicted from about 1870 through present times. Grammarians' rules have had little effect on actual usage, except to encourage hypercorrect usage of whom and to make people unsure of themselves; e.g., "It is not known *who or whom* invited her." or . . . how to handle who/whom with your students is up to you.

1. You should go with ___whoever___ is leaving first.

2. I will give the money to ___whoever___ will take it.

3. People ___who___ never make a mistake never make anything else.

4. Mohammed is the one ___who___ should do the work.

5. ___Whom___ does Professor Serrano prefer for the job?

6. ___Whom___ would you like to join me?

7. ___Who___ do you believe will win the election?

8. ___Whom___ should we ask to investigate?

9. ___Whoever___ is willing to work hard will be given the responsiblity.

10. Each candidate will support ___whomever___ the convention chooses.

11. We think he is the professor ___whom___ you will want at Carl Sandburg College.

12. Give the scholarship to the one ___who___ needs it most.

13. She is the woman ___whom___ I took to be your sister.

14. You are the court reporter ___whom___ I requested for the deposition.

15. Professor Costner is the instructor ___who___ I believe could help you.

16. Give it to ___whomever___ you wish.

17. Ms. Ferguson is the one ___who___ helped me most in Seattle.

18. We selected Ms. Serrano, ___who___ we know is a Renton professor.

19. Ms. Faries-Tondi, ___whom___ we met yesterday, is giving a speech.

20. The prize will be awarded to ___whoever___ writes the best essay.

21. The board will approve ___whomever___ we select for vice president.

22. Ms. Stranix, ___whom___ I told you about last week, will speak on business education at Yuba College.

23. The question of ___who___ should do the art work will be discussed.

24. Give the package to ___whoever___ can identify it.

25. He ___who___ has courage and faith will never perish in misery.— Anne Frank

---

See if your answers match those on page 396.

EVERYBODY NEEDS MILK

This title was once a radio and TV commercial for the dairy industry. Did it mean *every body* or *everybody*? That's just one type of question about indefinite pronouns—pronouns that don't refer to a definite person or thing:

INDEFINITE PRONOUNS

each	any	some	none	both	most
everyone	anybody	somebody	nobody	few	several
everybody	anyone	someone	no one	all	many
everything	anything	something	nothing	others	more

One Word or Two

Except for *no one,* the compound pronouns (those combining two words into one) above are one word. In some sentences, however, these expressions are not pronouns; the first part is an adjective and the second, a noun. In that case, two words are used. Fortunately, you don't need to analyze the grammar to know whether to use one word or two. Just apply these two word tricks:

Any one and *every one* are two words when *of* follows.

Any one of you might go to Topeka.

Every one of you might go to Omaha.

Anyone may go to Houston.

Everyone might go to Brunswick.

Some of the preceding compound expressions are two words—if your own good judgment indicates a special meaning:

Dr. Cutup told the medical students at the morgue to examine **any body.**

The man I met at Muscle Beach has **some body.**

BUT

Phyllis Diller said, "(Every body/*Everybody*) needs milk."

Does (any one/**anyone**) want ice cream? (**Any one**/Anyone) of those flavors is delicious.

Everybody (or everyone, anybody, etc.) as an antecedent for *their/they* is considered OK for spoken and written British English and spoken, but not written, American English.

American students should treat one/body pronouns as singular for business and professional communication—so that their grammar meets the expectations of discriminating employers or clients. For a more complete explanation, see *The American Heritage Dictionary of the English Language.* (The short usage note in the Dell-published paperback is inadequate.)

Either one is correct: *Every body* stresses the body's need for milk, and *everybody* stresses that all people need milk. The double meaning is probably why the dairy industry once chose the words as part of a commercial.

The Singles Scene

The following words are often indefinite pronouns. However, when the words in the first column precede a noun they become adjectives. Good news: It isn't necessary to identify the part of speech—just remember they are all **singular.**

each	everyone	everybody
every	someone	nobody
either	anyone	somebody
neither	no one	anybody
	one	

Although *everyone* might include five hundred people, it's singular because it refers to each one acting individually. Use singular pronouns like *his, her,* or *its*—not the plural *their*—to substitute for words in the preceding list. In informal speech this rule is frequently ignored, but it is still important in workplace writing.

USE **Every** student in men's physical education needs **his** own locker. [*Every* means *every one.*]

AVOID **Every** student in men's physical education needs **their** own locker.

USE **Each** building has **its** own heating unit.

AVOID **Each** building has **their** own heating unit.

USE **No one** is consulting a lawyer about it. [No one is consulting **his** or **her** lawyer about it.]

AVOID **No one** is consulting **their** lawyer about it.

Indefinite Plural Pronouns

The following words are often adjectives but are also used as indefinite plural pronouns: all, any, both, few, more, most, none, some. When using these words as indefinite plural pronouns, be sure to use the plural verb form.

Both are beautiful.

Few tourists **go** there.

Some were better than others.

More seem interested this year.

Most have attended.

Many are here today.

Writing for Your Career

Careful business writers omit unnecessary words. Eliminate words that contribute nothing to clearness, smooth flow, or desired emphasis. Adding unnecessary words is a "skill" developed by some students when doing a report and not knowing much about the subject.

Make the corrections needed for written English. Use # (to show a space).

1. No#one from this office responded.

2. Every#one of the books was sold yesterday.

3. Each man has ~~their~~ his own access code.

4. In American business situations, keeping some one waiting is considered rude.

Check answers on page 396.

GENDER AND NUMBER IN WRITTEN COMMUNICATION

Gender refers to male, female, and neutral words. When both sexes are represented, avoid the shortcut of using only male gender—*his, he,* or *him. Number* refers to singular or plural; nouns and pronouns should agree in number.

> For business and professional writing, use singular pronouns to represent singular nouns or pronouns, and avoid using masculine pronouns to represent both males and females.

AVOID   **Each** employee did **his** work quietly. [No women on the job?]

WRONG   **Each** person did **their** work quietly. [A plural pronoun *their* represents singular words, each person.]

GOOD   Each person worked quietly. OR Everyone worked quietly. OR The students worked quietly.

WRONG   Each employee has their job to do.

GOOD   Employees have their jobs to do. OR An employee has his or her job to do.

A.   Correct the following sentences:

1. Everyone in this department should be sure their nouns and pronouns agree in number in their written communications. [When correcting this sentence, note that "Everyone" is a singular pronoun, but *their* is plural.] Members of this department should be sure their nouns and pronouns agree in number in their written communications. OR All department members . . . .

2. Every mechanic finished their work quickly. All the mechanics finished their work quickly. OR Each mechanic finished the work quickly.

3. Every boy and girl in the class needs (his or her/their) own book. Every child in the class needs a book. OR All the children in the class need their own book.

4. Each applicant should write their name in the blank. All applicants should write their name in the blank. OR Each applicant should write his or her name in the blank.

B. Draw a line through the incorrect form in the following sentences.

EXAMPLE (No body/~~Nobody~~) was found where the fatal accident occurred.

1. (~~Any body~~/Anybody) at this meeting may speak on the subject.

2. (Every one/~~Everyone~~) of you can do well in this class.

3. (No one/~~Noone~~/~~No-one~~) but Patci knows the combination to the safe.

4. Please distribute the flyers to (~~any one~~/anyone) who wants them.

5. (~~Every body~~/Everybody) should learn touch keyboarding.

6. Ms. Kato asked that (~~some one~~/someone) from this office visit her store.

7. (Any one/~~Anyone~~) of you is qualified to prepare the report.

8. Although we believe the pilot was killed, (no body/~~nobody~~) was found.

9. Senators must use (~~his or her~~/their) own funds for this project.

10. (Every body/~~Everybody~~) in the morgue was in (its/~~their~~) drawer.

11. Each building has (its/~~their~~) own security guard.

12. A few mechanics completed (~~his or her~~/their) work by 4 p.m.

C. In the blank below "Best," write the letter of the form best for written workplace communication. In the next blank, write the letter of the only incorrect form.

BEST	INCORRECT	
c	b	1. (a) A person can usually improve if he really tries.
		(b) A person can usually improve if they really try.
		(c) People can usually improve if they really try.
		(d) A person can usually improve if he or she really tries.
a	d	2. (a) When customers express their dissatisfaction, listen courteously.
		(b) When a customer expresses his dissatisfaction, listen courteously.
		(c) When a customer expresses his or her dissatisfaction, listen courteously.
		(d) When a customer expresses their dissatisfaction, listen courteously.
b	d	3. (a) Every one of the contractors submitted his or her bid today.
		(b) Every one of the contractors submitted bids today.
		(c) Each contractor submitted his bid today.
		(d) Every contractor submitted their bid today.
b	c	4. (a) Did anyone here lose his notebook?
		(b) Did anyone here lose a notebook?
		(c) Did anyone here lose their notebook?
d	b	5. (a) Everyone should write his name on the form.
		(b) Everyone should write their name on the form.
		(c) Everyone should write his or her name on the form.
		(d) You should each write your name on the form.

Check answers on pages 396–397.

Read 34

A GAGGLE OF GEESE

Gaggle (geese), herd (cattle), flock (sheep), pride (lions), fleet (ships), and faculty (teachers) are all examples of COLLECTIVE NOUNS.

Collective nouns name collections of people, animals, or things.

| audience | committee | crowd | group | Navy | staff |
| class | congregation | family | jury | squad | team |

USE PRONOUNS THAT AGREE WITH THE COLLECTIVE NOUNS

Use a singular pronoun—*it, its, itself*—to substitute for a collective noun if the members of the group act as one; that is, as a single unit. If the pronoun can be eliminated, do so.

An example of a collective noun is jury; when a jury announces a verdict, it acts as a single unit.

YES The **jury** announced **its** verdict. [Singular pronoun *its* refers to jury acting as a unit.]

The **jury** announced the verdict. [no pronoun]

NO The **jury** announced **their** verdict. [Plural pronoun *their* is incorrect when members of the collection act as one.]

Use a plural pronoun—*they, them, their/s, themselves*—to substitute for a collective noun if the members of the group act separately or disagree.

NO The **jury** put on **its** coats as it prepared to leave. [Jury acts as single unit arriving at a verdict but as separate individuals when putting on coats.]

OK The **jury** put on **their** coats as they prepared to leave. [technically correct but awkward]

BETTER The **members** of the jury put on **their** coats as they prepared to leave.

YES The **committee** believed **it** had been right to take strong action.

NO The **committee** disagreed about whether **it** should take strong action.

YES The **committee** disagreed about whether **they** should take strong action.

Use a singular pronoun to substitute for names of organizations—for example, companies, unions, stores, schools, governments, and government agencies.

Though the college named below has many students and employees, use a singular pronoun because it is **one** organization.

NO Davenport College opens **their** offices at 8 a.m.

YES Davenport College opens **its** offices at 8 a.m.

BETTER Davenport College offices open at 8 a.m. [better because the same idea is expressed smoothly and with fewer words]

Rephrasing so that the pronoun is omitted when referring to collective nouns improves the wording.

Wrong	The Internal Revenue Service revised **their** 1099 forms. (IRS is a collective noun requiring a singular pronoun (its), if a pronoun is used.)
Correct	The Internal Revenue Service revised **its** 1099 forms.
Better	The Internal Revenue Service revised the 1099 forms.

> **Use a plural pronoun to substitute for a plural noun—juries, companies, colleges, classes, teams.**

YES The **juries** announced **their** verdicts simultaneously in New York and in Chicago.

By recognizing collective nouns, you can avoid choosing incorrect pronouns to substitute for them. Your understanding of collective nouns will help with "subject–verb agreement" in Chapter 7.

Circle the better answer.

1. Nordstrom's will open (their/its) doors early for the sale.

2. The Navy recalled (its/their) ships from dry dock.

3. Committees often make (its/their) decisions too slowly.

Answers are on page 397.

USE SUBJECT PRONOUNS AFTER BEING VERBS

> **A pronoun following a "being" verb must be a subject pronoun.**

The preceding rule is important in **written** English. In 21st century **spoken** American English, however, either an object or a subject pronoun is all right after a being verb. Adjectives are also often used after being verbs—but not adverbs. A subject pronoun or an adjective after a *being verb* is called a *complement*. (Notice *e*, not *i*, before "ment.")

SUBJECT PRONOUNS	I, we, you, he, she, they, it, who; these are correct as *complements*.
OBJECT PRONOUNS	me, us, you, them, it, whom; these are incorrect as *complements*.
BEING VERBS	am, is, was, were, been, be
YES	It **is he,** not Jerry, who sold it. OR **It's he,** not Jerry, who sold it.
NO	It **is him,** not Jerry, who sold it. OR **It is them,** not Jerry, who sold it.
YES	The only experts in the group **were** she and **I.** [not *her* and *me*]
	The winners might have **been** Raul and **he.** [not *him*]

> Point of view is the reader's or listener's relationship to the information. Avoid using *they* in a vague sense and *you* when you mean people in general or those in authority. Use specific nouns instead, or rephrase to omit the need for a noun or pronoun.

NO *They* give *you* grants or loans if *you* can prove *your* income is at or below the poverty level. [Use nouns instead of *they* and *you* so that you communicate in a general manner.]

YES The Financial Aid Office arranges grants or loans for students who can prove their income is at or below the poverty level.

NO *You're* prohibited from driving in this state if *you've* been drinking.

YES State law prohibits driving after drinking alcoholic beverages.

Replay 34

A. List collective nouns in the six blanks below. See how many you can think of without referring to the preceding pages. [sample answers]

1. _____group_____   **2.** _____class_____   **3.** _____jury_____

4. _____team_____   **5.** _____staff_____   **6.** _____committee_____

B. Circle the correct answer within the parentheses.

1. When the members of the group named by a collective noun—such as team—act separately or disagree, the pronoun is (singular/**plural**).

2. When the group named by a collective noun acts as one, a pronoun substituting for it is (**singular**/plural).

3. When writing, use (an object/a **subject**) pronoun after a *being* verb.

4. Aerojet Corp. has (**its**/their) offices in Escanaba, Michigan.

5. The Sales Office is on the third floor; (they are/**it is**) open until five.

6. Segal Institute always pays (**its**/their) bills promptly.

7. Metro College revised (**its**/their) application form.

8. It was not (me/**I**/myself) who made the suggestion.

9. Her family is planning (**its**/their) vacation for August this year.

10. The Board of Directors will hold (**its**/it's/their) next meeting at Mount Hood.

11. The city should regulate (**its**/it's/their) hiring policies more carefully.

12. The committee will review the new data at (**its**/it's/their) next meeting.

13. Why can't (they/**the janitorial service**) keep this place clean?

14. We all know (you shouldn't throw trash/**trash shouldn't be thrown**) from a car window.

15. Our best district managers in Dallas are Ron and (**she**/her).

16. Next to join the department are Jill Edelson and (he/him).

17. The guests of honor were Janice Bragia and (me/I/myself).

18. I think it was (they/them) who requested the report.

19. It is (me/I) who should go to Auburn Hills.

20. He wants (we clerks/us clerks) to bear responsibility for the project.

Check your answers on page 397. Then review the entire chapter before taking the Pop Quiz on page 344.

CHECKPOINT

Along with rapidly changing technologies and lifestyles, language styles also change. As with all change, some people use their energy defending the old, and others lead the parade. For your career, being in the middle is advisable when it comes to language. Your objective is for others to concentrate on your message and not be distracted by something unusual you say or write.

Whether preparing a report for the president of your company, a letter to an important customer, or an informal *e*-mail message, you should not have to ponder over whether to use *I* or *me, who* or *whom,* and so on. You should **know** how to decide. When speaking with colleagues or clients, you should not have to be self-conscious about your grammar. You should be **confident** your grammar is correct.

Pronouns are frequent trouble spots for people otherwise comfortable with Standard English as well as for those who grew up using a dialect or a language other than English. Chapter 6 principles provide Standard pronoun usage for business, professional, and technical careers.

Place a check in the blank when you understand the principle.

_____ Use a subject pronoun as subject of a verb and an object pronoun as object of a verb or of a preposition.

_____ *Who* is acceptable in speech for subjects and objects. However, when writing, use *who* as a subject and *whom* as an object.

_____ Use a "self" or "selves" pronoun only when another pronoun will not make sense in that position in the sentence. Never use *hisself, theirselves, themself,* or *theirself!!*

_____ The pronouns *hers, ours, yours,* and *theirs* never have apostrophes. However, indefinite possessive pronouns (such as someone or everybody) do have apostrophes before the *s* when used in possessive forms. Distinguish between possessive pronouns and contractions that sound alike but are written differently, such as *its/it's* and *your/you're.*

_____ Pronouns should agree in number (singular or plural) and in gender (masculine or feminine) with the noun or other pronoun for which they substitute. Reword sentences to avoid pronouns suggesting sexual bias.

_____ Except for *no one,* compound indefinite pronouns are written as one word (*someone*)—unless "of" follows or the meaning indicates two words are required.

Some language experts say the battle for nominative case pronouns after linking verbs is lost and that it's not a loss to mourn. The rule is often ignored by even educated speakers and is often avoided in written media by rephrasing. Nevertheless, I recommend students follow traditional usage—to avoid criticism that could hurt career progress.

_____ Singular pronouns substituting for singular collective nouns are singular if the members of the collection act as a single unit. If the members act separately or disagree, use a plural pronoun. Names of organizations are always considered singular.

_____ In workplace writing, use subject pronouns after "being" verbs.

_____ Avoid the pronouns *you* and *they* when you refer to the general population.

SPECIAL ASSIGNMENT

Some years ago a British steamship company operating vacation cruises wrote to the British Admiralty asking permission for its officers to wear swords, as British naval officers did. The Admiralty did not like the idea but didn't want to refuse permission. The ambiguous reply from the Admiral's office granted permission, adding, "provided the swords are worn on the right side." The cruise company dropped the idea, feeling foolish about writing again to clarify "the right side." This was probably intentional lack of clearness, since the Admiral's office didn't like the idea but didn't want to refuse.

When you were a child, a relative may have said, "Honey, you put your left shoe on your right foot." After you looked at your feet in confusion, the double meaning of "right foot" was explained and everyone laughed.

These stories are to remind you of the importance of clear writing—especially for your workplace. Concentrate on writing this brief report clearly as well as correctly.

DOCTOR, LAWYER, MERCHANT, CHIEF?

Choose a career that interests you: fashion merchandising, administrative assistant, business management, computer technology, law, teaching, auto mechanics, or ??. Interview someone who works in the chosen field. Prepare questions, such as: What does the work consist of? What is a typical daily routine? What about salary and advancement possibilities? What are the advantages and disadvantages of this kind of work? What kind of training is needed to qualify? Ask only questions that interest you; that is, what do *you* want to know about this kind of work? Summarize the information, and prepare a 175 to 200-word report.

Start with a topic sentence that tells the reader in an interesting way about the subject of the report. However, don't begin "I am going to write about. . . . "

SAMPLE TOPIC SENTENCES

1. "Accounting is an excellent career for women," states my friend, Twileen Disheru, who works for the CFO* at Berring Aircraft in Seattle.
2. I learned a great deal about the work of a systems analyst in the aircraft industry by interviewing George Gorjus at Berring Aircraft.
3. Like other careers, a teaching career has advantages as well as disadvantages.

Language expert William Safire wrote, "My imperative, authoritative ruling (on usage) is based on what other usage writers say, as amended by what I think." Linguist Theodore Bernstein said, "Be neither a language slob nor a language snob." Safire, Bernstein, and Roy Copperud (author of *American Usage, a Consensus*) are among my sources for de-emphasizing certain traditional pronoun and verb rules. However, students shouldn't confuse the changing language patterns with "anything-goes" dialects or other non-Standard English.

---

*CFO is an abbreviation for chief financial officer. See abbreviations in Appendix D.

Follow through with what you promise in the topic sentence. The rest of the report for Topic Sentence 1 could show why accounting is a good career, especially for women. For Topic Sentence 2, the writer will tell about the work of an aircraft industry systems analyst. The report beginning with Topic Sentence 3 explains advantages and disadvantages of a teaching career.

Stay on track right to the end. Here are sample closing sentences. Effective closings conclude or summarize what went before and do not introduce new ideas.

SAMPLE CLOSING SENTENCES

1. Business management sounds like an exciting career filled with challenges and rewards.
2. A human resources department manager requires a vast knowledge of many subjects as well as excellent people skills.
3. Now that I've learned about the training a neurosurgeon needs, I realize I do not have the tremendous drive, ambition, and ability required to prepare for this career.

Submit this report on _____ (date)

NOTES:

PROOFREADING FOR CAREERS

Please correct sentence construction, spelling, capitalization, noun and pronoun usage, and keyboarding errors. Do not change wording except to correct errors of the type brought to your attention in preceding chapters.

Dear Dr. Costner:

What in the world is ~~you're~~ your money doing?

Good business practices, protection of the environment, and social responsibility can all work together hand-in-hand; however, many investor's disregard enviromental and social records of the Company's ~~he or she~~ they invests in. Why be one of ~~them~~ those investors when you could earn competitive returns investing in activitys you beleive in!

In 1986 when few had even heard of socially responsible investing. We ~~was~~ were devloping SRI (socially responsible investing) programs. Individuals like you as well as small businesses is benefitting from our experience in this field.

Please let Eric Smith and I know if you would like a complementary initial consultation. Eric Smith and me will discuss you're needs and objective's to determine how too assist you with profitable, yet socially responsable investing.

Sincerely,

Wade Smith

Wade Smith

NOTES:

Circle the correct pronoun—or word group for written English.

1. (Every one/Everyone) must make (his or her/their) own decision.
2. They make you stand in the registration line for hours. The applicants must stand in line for hours.
3. (No one/Noone) regrets this incident more than (I/me/myself).
4. Is Mike Terry better qualified than (her/she)?
5. They completed more of the programming than (I/me).
6. Yoshi has a better background in Hebrew than (they/them).
7. The Rock Springs plant is owned by Martinez and (me/I/myself).
8. Neither Ms. Yamomoto nor (I/me/myself) will visit the Maryland plant.
9. (Some one/Someone) left (his or her/their) keys at the reception desk.
10. Mine is faster than (your's/yours'/yours) and (her's/hers).
11. (Everybody's/Everybodys/Every body's) getting a raise.
12. (Who's/Whose/Whos) laptop computer is missing?
13. (Who/Whom) did you say will handle the new account?
14. Victor seated (hisself/himself) between Twileen and (me/I/myself).
15. George became angry with the man (who/whom) was harassing Twileen.
16. (Him/He) and (me/I/myself) will report it to (whoever/whomever) is in charge.
17. If (anyone/any one) of the men would like a ticket, (he/they) may have one.
18. If (it's/its/its') too late, give the information to (whoever/whomever) needs it.
19. We sent letters to all (who/whom) we thought might visit our showroom.
20. Each officer was issued (his/their/a) new uniform yesterday.
21. The guests of honor were Nancy Burnett and (I/me/myself).
22. Her family members are taking (it/it's/their) vacations at different times this year.
23. The Internal Revenue Service revised (its/it's/their) forms again this year.
24. The Maintenance Department is on the third floor; (it's/they are) open every day.
25. (It's/Its) (I/me) who should take the risk.

THE GLOBAL MARKETPLACE—GREECE

Greek	Spell	Say	Currency
GREETING	Yiasou	Ya'soo'	hrimata
DEPARTING	Adio	Ah. dee. o	
PLEASE	se para. kalo	Say.para.kalo	
THANK YOU	efharisto	ef.har'ees.toe'	

Courtesy of John Papadakis

Looking for the Action?

7

**Then Find
the Verbs!**

"I am not yet so lost in
lexicography as to
forget that words are
the daughters of earth
and that things are the
sons of Heaven."
—*Samuel Johnson*

**After Completing Chapter
7, You Will**

> Use verbs with time (or tense)
expressed correctly.

> Use subjects and verbs that
agree in number and person.

With minimal grammar terms, students find and correct commonly made tense, agreement, and mode errors. Practice with "perfect" tenses is provided without naming them or giving rules for their use. Time spent on how and when to use perfect tenses is counterproductive for many of today's students. Those who might benefit—principally international students who studied English grammar as a second language in their homeland—will find useful information in a copy of "Grammar for the Expert" from your *Instructor's Resource Kit*.

Every sentence has at least one verb. The verb tells what the subject **does** or **has** (action verbs) or what the subject **is** (being verbs).

Some verbs consist of one word:

The bank manager **interviewed** two loan applicants. [action verb]

Our systems analyst **has** a new printer. [action verb]

The Internet sales director **was** in the office. [being verb]

Other verbs consist of two or more words: one or more helping verbs and a main verb.

The personnel director **had been interviewing** those applicants all day. [*Had* and *been* are helping verbs; *interviewing,* an action verb, is the main verb.]

The software engineer **should have been** here today. [*Should* and *have* are helping verbs; *been,* a being verb, is the main verb.]

Infinitives are the basic forms of verbs preceded by *to*. An infinitive is not used as a verb; the verb is elsewhere in the sentence.

Examples are *to cook, to dance, to have, to be, to love, to work, to send.*

George wants to cook for Twileen.   Twileen seems to like his cooking.

Choosing a verb form depends on time, number, and person.

TIME When does the action or being take place—before, now, or later?

NUMBER Does the verb have a singular or plural subject?

PERSON Is the subject of the verb in first person, second person, or third person?

Learning about correct verb use requires concentration and practice. The principles presented enable you to avoid verb errors often occurring in the language of intelligent adults.

BUT

If you grew up in:

- a community where regional or ethnic English is usually used
- a non-English-speaking country
- a community where English is a second language for many residents

THEN

Applying these rules could mean the difference between success and failure in a career. Most of the non-Standard verb forms brought to your attention are noticeable in business or professional environments. For that reason I urge you to acquire this information about verbs so that using Standard English forms will be a habit.

Read 35

TIMELY TIPS

Since *tense* is
sometimes how
students feel about
grammar study, you
may prefer to
interchange the word
with the less emotional
word *time* in your
discussions.

Action or being occurs in three principal time periods: past, present, and future. (The usual grammar term for *time* is *tense.*) Time is not so simple as **before, now,** or **later.** Some action or being includes more than one period of time; that is, something may start in the past and continue into the present. Other action might start in the present and continue into the future. English can express a complex range of time. Here are some ways we express time:

I have been working on this for days. [I started in the past and am still working now in the present.]

I worked on it earlier. [I am no longer working on it.]

I'll work on this for days. [I will continue to do the work in the future.]

REGULAR VERB FORMS

A verb is regular if it changes form by adding *s, ed,* or *ing* to the basic form in the usual way.

walk walk**s** walk**ed** walk**ing**   call call**s** call**ed** call**ing**

Look up the verb *walk* now in your dictionary. Be sure to look at the **verb** *walk,* not the noun. In some dictionaries, *walks, walked, walking* are next to the entry word. This means all forms of regular verbs *are* included in that dictionary. Other dictionaries, however, don't show the changed forms of regular verbs, because it is assumed you will know how to add *s, ed,* or *ing.*

RULES FOR REGULARS

A man told his
psychiatrist he had
dreamed he was a
wigwam and the next
night he had dreamed
he was a tepee. The
psychiatrist said, "Don't
worry about it; you're
just tense." (When they
groan, blame Leila.)

PAST TENSE: If the action or being was in the past or if *has, have,* or *had* precedes the verb, add *ed* or *d* to the basic verb form:

The baby **crawled** to the table.

The baby **has crawled** to the table every day this week.

IN PROGRESS: For "in progress" verbs, always use a helping verb before a main verb ending with *ing.*

If you **are inviting** some co-workers to your wedding, invite them all.

They **were talking** about Twileen and George.

They **will be cooking** stews and pot roasts every day from now on.

I (use/**used**) to ride the
bus every day. *Use to*
and *Used to* sound
alike, resulting in written
errors. Point out that *to
ride* is an *infinitive,* not
a verb. If you mean
formerly, write *used to.*

PRESENT TENSE + *s*: For action or being happening in the present, use the basic verb form ending with *s*—if the subject is a singular noun or a singular pronoun—except *you* or *I.*

In Spain a **visitor** to a home often receive**s** a gift from the host. [singular noun]

This crawl**s** like a worm. [singular pronoun]

PRESENT WITHOUT *s*: If the action or being is happening now, use the basic verb form; do not add *s* if the subject is *you, I,* or plural.

Twileen and **George walk** home holding hands. [plural subject—2 nouns]

They seem happy. [plural pronoun subject]

I walk home alone. [subject is *I*]

Please **mail** the checks. [subject is understood to be *you*]

FUTURE: If the action or being will be in the future, use *will, shall, would,* or *should* before the basic verb form.

George **should walk** home today.   I **would talk** to him today if I could.

Shall I walk with Twileen today?   He **will call** his mother tomorrow.

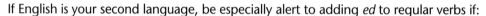

WORD TO THE WISE

If English is your second language, be especially alert to adding *ed* to regular verbs if:

- the past tense is required: call, called
- *have, has,* or *had* precedes the verb: have, has, or had called

*Using past when present tense is needed occurs mainly when two verbs are in a sentence. It also occurs in sentences with just one verb: He (**is**/was) a former student of this college. [*Was* would mean he had died.]*

IRREGULAR VERB FORMS

Irregular verbs do not follow a set pattern. Dictionaries, however, include the various forms.

The verb *begin* is an example of a highly irregular verb:

Basic Form . . . . begin—with plural subject

Add *s* . . . . begins—with singular subject

Instead of adding *ed*, change to . . . . began or begun—past tense with helping verb

Double final letter before adding *ing* . . . . beginning—happening now

*The children (clinged/clang/**clung**) to their story about the broken vase.*

Apply the same principles for *s* at the end of irregular verbs as for regular verbs.

Ms. Gina Hecht begins a new assignment today. [singular noun subject—verb ends with *s*]

Everything is fine at American Business Institute. [singular pronoun subject—verb ends with *s*]

The **boys begin** carefully. [plural noun subject—no *s* at end of verb]

These are useful. [plural pronoun subject—no *s* at end of verb]

You go to the stockholders' meeting every year. [you or I—no *s* at end of verb]

SENTENCES WITH TWO OR MORE VERBS

When a sentence has two or more verbs, we generally express them in the same "time"—or tense.

She **thinks** I **am** a millionaire. [both verbs are present tense]

Mr. Meek **wrote** me a note in which he **implied** I **passed** the accounting final. [3 verbs in past tense]

For a general truth or something still going on, use the present tense, even if a verb elsewhere in the sentence is past tense.

YES   Mr. Chung **told** us that Tokyo **is** larger than New York City.

NO   Mr. Chung **told** us that Tokyo **was** larger than New York City. [Even though he told us in the past, Tokyo **is** still larger.]

YES   Joan **demonstrated** in Binghamton that our printer **performs** better than any other.

NO   Joan **demonstrated** in Binghamton that our printer **performed** better than any other. [Since our printer still performs better **now**—*performed* (past tense) is wrong.]

YES   What **are** the titles of the books you **borrowed** from the library?

NO   What **were** the titles of the books you **borrowed** from the library? [Although you borrowed the books in the past, the titles are still the same.]

Replay 35

A.   Write in the blank a correct form of the verb in parentheses. Notice the **time** of the verb in parentheses at the end of the sentence. Add a helping verb if necessary.

EXAMPLE   They (talk) _talked_ every day on the telephone last week. (past)

1. He (work) ___works___ very accurately. (present)

2. The president (need) ___needs___ your decision now. (present)

3. Ms. DeVries (move) ___moved___ the books to Lakeport. (past)

4. They (sail) ___are sailing___ to Catalina this week. (*ing* form)

5. The floor is slippery because the custodian (wax) ___waxed___ it. (past)

6. Good manners in the workplace (be) ___are___ essential to success. (present)

7. The world (look) ___looks___ brighter from behind a smile. (present)

8. Dr. Perry (climb) ___climbed___ up the ladder to reach the carton. (past)

9. Left of the plate, you (find) ___will/should find___ an entrée fork and a salad fork. (future)

10. They (want) ___want___ something they cannot have. (present)

11. She (want) ___wants___ something she cannot have. (present)

12. He (want) ___wanted___ something he could not have. (past)

13. She (want) __will (or would) want__ something she cannot have. (future)

14. An employee's appearance (influence) __influences__ the way he or she is treated. (present)

15. He believes Ms. Oh (select) __selected__ a business career. (past)

16. Michael (consider) __will (or should) consider__ that problem tomorrow. (future)

17. Dennis (consider) __is considering__ that problem now. (*ing* form)

18. I (stay) __stay__ at that hotel every July. (present)

19. We (watch) __watched__ television all day. (past)

20. Marketing people (discuss) __discussed__ international sales. (past)

B. Each of the following sentences has a verb error. Cross out the incorrect verb and write the correct form in the blank.

EXAMPLE __has__ We knew that Orlando ~~had~~ a parade every Wednesday year round.

__is__ 1. The officer said that obeying traffic laws ~~was~~ necessary for accident prevention.

__is__ 2. Seth asked if New York ~~was~~ the biggest city in the USA.

__need__ 3. I brought the data with me that you said you ~~needed~~.

__are__ 4. Who ~~were~~ the authors of the books on yesterday's best-seller list? (Hint: Authors haven't changed since yesterday.)

__is__ 5. George said the old HP 150 ~~was~~ now in Twileen's office.

__are__ 6. The people meeting in the conference room ~~were~~ former employees of General Motors. [Hint: They haven't died.]

__flow__ 7. He taught us that rivers ~~flowed~~ into oceans. (They still do.)

__knows__ 8. We learned that no scientist ~~knew~~ with certainty how the universe originated.

__is__ 9. The Chicago Police have not discovered where the 5 million dollars ~~was~~.

__being__ 10. The CEO praised the office staff for ~~having been~~ so efficient. [Hint: The office personnel are still efficient.]

Check your answers on page 397.

These words have come into American English directly from Native American languages: raccoon, opossum, moccasin, skunk, moose.

Read 36

DELINQUENT VERBS

Delinquents—verbs that don't behave as they should—are called irregular verbs.

Assign two irregular verbs to each small group of students: *bid, spring, bend, kneel, tread, slay, wring, shrink, foresake, drag* (some students use *drug* for dragged). Teams look up principal parts of each verb and write a short, easy sentence for each form. Teams report to the rest of the class either orally or on the board.

> Recognize verbs that might be irregular; look them up in the dictionary to be sure of form or spelling.

Column 1		Column 2	Column 3
Present		**Simple Past (no helping verb)**	**Past Participle (use helping verb)**
A	B		
begin	begins	began	begun
break	breaks	broke	broken
choose	chooses	chose	chosen
do	does	did	done
drink	drinks	drank	drunk
freeze	freezes	froze	frozen
give	gives	gave	given
go	goes	went	gone
hang (suspend)	hangs	hung	hung*
ring	rings	rang	rung
rise	rises	rose	risen
run	runs	ran	run
see	sees	saw	seen
speak	speaks	spoke	spoken
stand	stands	stood**	stood**
swing	swings	swung	swung
take	takes	took	taken
wear	wears	wore	worn

Hang meaning a method of putting to death is regular: hang, hangs, hanged.
**Read the next Word to the Wise for information about *stood.*

WORD TO THE WISE

Notice **stand** in the preceding irregular verb chart. Using *stood* for *stayed* is a serious error; for example, "I should have stood in bed today." *Stayed* is the simple past and past participle of the regular verb *stay*.

For correct use of irregular verbs, refer to your dictionary to find the basic form—like those in column 1 in the preceding chart. If English is your second language, give special attention to the following:

ABOUT COLUMN 1

> The words in column 1A are the basic forms. Use them alone with *you, I,* or a plural subject. Also use them after helping verbs *will, shall, would, should, do, did, does, might, may, can,* and *could*—no matter what the subject is.

To Be or Not to Be in previous editions dealt with "black English," also called "Ebonics." Due to frequent moves from one part of the United States to another, Americans born in the United States sound more and more alike; and Ebonics and other dialects have decreased in significance. Non-Standard grammar is as prevalent as ever, of course, due to immigration and other sociological factors.

Without Helping Verbs—No *s*

You drink smoothies too fast. [Subject is *you.*]

I choose the beach for today's trip. [Subject is *I.*]

They break easily. [plural pronoun subject]

These **surfers wear** identical clothing. [plural noun subject]

With Helping Verbs

He **did choose** me.

I **should take** the best one.

Ms. Flood **does run** the business.

> For present tense (time), use the verb forms ending with *s*—column 1B in chart—with any singular subject except *I or you.*

Everybody begins at the same time. [singular pronoun subject]

Professor Seilo understands the problem. [singular noun subject]

ABOUT COLUMN 2

> Use Column 2 verbs—the simple past—for past time with any subject. Do not use a helping verb. (For some verbs, columns 2 and 3 forms are the same.)

She **ran** away.   The glass **broke.**   The bells **rang.**

ABOUT COLUMN 3

> Use Column 3 forms—past participles—with a helping verb, such as *have, has, had, been, was, were,* and with any subject.

SINGULAR	PLURAL
She has run away.	**They have** run away.
The **glass had broken.**	The **glasses had broken.**

WORD TO THE WISE

Although the irregular verbs in the preceding table are among the most commonly misused verbs, they aren't the only ones. Since other irregular verbs are used incorrectly also, careful writers refer to the dictionary when in doubt.

Replay 36

Write correct irregular forms in the blank. Change only the incorrect verbs.

EXAMPLE   <u>hung</u>   I hanged up all my clothes.

<u>broken worn</u>   **1.** I thought it was broke, but it had just wore out.

<u>began rang</u>   **2.** He begun to work when the bell rung. (or had just begun)

<u>chooses</u>   **3.** He choose the best one for himself. (present)

<u>chosen</u>   **4.** He had chose the best one for himself yesterday. (or delete had)

<u>does stands</u>   **5.** She do all the complaining, and he just stand around. (present)

<u>risen</u>   **6.** He had drunk all the tea before the sun had rose. (or delete had)

<u>have eaten</u>   **7.** The children have ate all the fresh strawberries. (or delete have)

<u>saw did</u>   **8.** We seen that Smith done a good job on the pricing. (or insert helping verbs)

<u>delete had</u>   **9.** They had flew to Taiwan last year. (or change *flew* to *flown*)

<u>quit</u>   **10.** They went to the convention before she quitted her job.

<u>run broken</u>   **11.** The child had ran away after he had broke the window. (or omit both hads)

<u>worn</u>   **12.** She had wore the new suit yesterday. (or omit *had*)

<u>stayed</u>   **13.** You shouldn't have stood in that job for two years.

<u>saw spoken</u>   **14.** I seen him before he had spoke. (or delete had)

<u>taken</u>   **15.** I had took it with me before you went away. (or delete had)

<u>wear</u>   **16.** They always wears suits to the office. (or *wore*)

<u>saw</u>   **17.** She seen that file before. (or has/had seen)

<u>doesn't</u>   **18.** "If a man don't know what port he's steering for, no wind is favorable," said the Roman philosopher Seneca.

<u>swung</u>   **19.** George swinged his tennis racket and then flung it down on the ground.

<u>gave</u>   **20.** Last week they give him three days to do the job. (or had given)

Please check your answers on pages 397–398.

Word Power

To use *lie/lay* correctly, see "Grammar for the Expert," and definitions in your college dictionary.

DICTIONARY DATA

The Read 36 chart lists just a few of over 100 irregular verbs. If English was your principal language during your growing up years, you've probably mastered most irregulars. However, a few of these verbs still cause even native English-speaking adults to hesitate and to use a few incorrectly. Since there are more irregular verb forms than are practical to memorize, it helps to know how to interpret a verb's "principal parts" in the dictionary.

PRINCIPAL PARTS OF A VERB

- **Basic form**—break, see, walk
- **Simple past form**—broke, saw, walked
- **Past participle form**—broken, seen, walked
- **Present participle form**—breaking, seeing, walking

Basic Form—Dictionary Entry

The basic form (infinitive without "to") is the dictionary entry for that verb. See Column 1A, Read 36. Column 1B shows the basic form ending in *s*.

Present Participle

The present participle is easy to recognize because it always ends with **ing.** Be aware, however, of other spelling variations.

- Usually the *ing* is simply added to the unchanged basic form: *see, seeing*
- Sometimes, however, a final *e* is omitted before adding *ing*—*love, loving*
- Or a final letter is doubled before adding *ing*—*win, winning*

Simple Past and Past Participle

> **Regular Verbs: Please place your dictionary beside you now.**

For regular verbs, the simple past and the past participle are the same: just add *ed*—as in earn*ed*. Use the *ed* form of any regular verb either with or without a helping verb. Both *I earned* and *I have earned* are correct, depending on the idea you wish to express.

> **Irregular Verbs**

- **Simple Past:** All column 2 verb forms in the "Delinquent Verbs" chart that begins Read 36 are simple past. This means use them **without** a helping verb. In the dictionary, the simple past either follows the entry word or follows the *ing* form. Look up *begin* now to see where *began* (the simple past) appears in your dictionary.

- **Past Participle:** The past participle follows the simple past and could be listed in Column 3 of Read 36. A helping verb **is required.** Notice how *begun* follows *began* in the dictionary.

- **Simple Past/Past Participle:** If only one *past* form appears, that form is *both* simple past *and* past participle. In that case use that form either **with** or **without** a helping verb. Now look up *bring*. You find only one past form: *brought*.

- **Or/Also:** If *or* or *also* is between two words in the dictionary, either word is correct—although often the meaning or customary usage differs. *Also* in a dictionary entry means the second word is less acceptable or less often used. Look up *broadcast.* You'll find that the simple past is either *broadcast* or *broadcasted.* Since no other form follows, the same two words are correct for the past participle.

- Now look up *show.* The first form after the entry (or after the *ing* word) is *showed,* meaning it's simple past—the one to use with no helping verb. Following *showed* is *shown* or *showed;* this means use either one **with** a helping verb. With that particular word, you can't go wrong with *showed;* however, *shown* without a helping verb is wrong. He shown it to me. (AWFUL!)

- Finally, look up *occur.* Notice the double *r* before adding *ed* or *ing.* When you're not certain whether to double the last letter before adding *ed* or *ing* or whether to drop a final *e* before adding *ing,* **look it up.**

Memo to the Conscientious but Restless Student

Did you take the necessary few minutes to look up each word suggested in the preceding dictionary-use hints? Did you also refer to the Delinquent Verbs chart in Read 36? If you did, you can now use a dictionary intelligently to get verb information. It's not necessary to memorize hundreds of forms; just know how to look them up.

Replay 37

A.   Fill in the blanks with the "principal parts" of these irregular verbs.

Consult your dictionary if you're unsure of the correct form.

BASIC VERB		PRESENT PARTICIPLE (*ing* ending)	SIMPLE PAST (no helping verb)	PAST PARTICIPLE (requires helping verb)
EXAMPLE	beat	beating	beat	beaten or beat
1.	be*	being	was	been
2.	bite	biting	bit	bitten
3.	blow	blowing	blew	blown
4.	come	coming	came	come
5.	cost	costing	cost	cost
6.	fall	falling	fell	fallen
7.	forget	forgetting	forgot	forgotten
8.	freeze	freezing	froze	frozen
9.	hide	hiding	hid	hidden
10.	lead	leading	led	led
11.	pay	paying	paid	paid
12.	shake	shaking	shook	shaken
13.	sink	sinking	sank	sunk or sunken
14.	sing	singing	sang	sung
15.	throw	throwing	threw	thrown

*Other "parts": *am, are, is.*

B.   Each sentence below has one or more incorrectly used irregular verbs. Make the corrections. When in doubt, consult your dictionary for spelling.

1.   Ms. Bogue has ~~beat~~ [beaten] all records and is ~~wining~~ [winning].

2.   Sean ~~payed~~ [paid] the bill because the job was done right.

3.   Ed has ~~broke~~ [broken] two rules and has ~~hid~~ [hidden] the evidence.

4.   Ms. Farrell ~~hanged~~ [hung] the picture but had ~~forgot~~ [forgotten] about it.

5.   Mr. James could have ~~stood~~ [stayed] another day if he had ~~wrote~~ [written] to me.

Check your answers on page 398.

IDENTIFYING SUBJECTS AND VERBS

Every sentence has at least one clause; that is, a subject–verb combination. The verb is the action or being word; the subject is the noun or pronoun that tells who or what is doing or being.

To find a subject, first find the verb. Then ask "who" or "what" before the verb. The answer is the subject.

The boys study geometry. (The verb is *study* and the subject is *boys*.)

An independent clause becomes a sentence when it begins with a capital letter and ends with closing punctuation such as a period: Boys study. Girls also study.

A related word group without both a subject and verb is a phrase.

Examples of phrases: in the morning; under the tree; an apple a day; men and women.

If a sentence has more than one clause, each clause has its own subject–verb combination.

Some students study and others daydream. [Who study? Students. Who daydream? Others. The subjects are *students* and *others*; the verbs are *study* and *daydream*.]

Questions are usually worded in either of two ways: (1) A helping verb may precede the subject, and a main verb may follow the subject, or (2) if a helping verb isn't needed, the verb simply precedes the subject.

1.   Do the girls study? [*Do* is the helping verb; *study* is the main verb; *girls* is the subject.]

2.   Are the girls smart? [*Are* is the verb; *girls* is the subject; *smart* is an adjective.]

Introductory or describing words may precede the subject.

In Schenectady, students study history. [The subject is *students*; the verb is *study*.]

Ambitious and diligent students study every day. [The subject is *students*; the verb is *study*.]

Sometimes a prepositional phrase separates the subject from the verb. The subject is never within a prepositional phrase.

Alice, as well as her husband Eric, lives in Ballard, Washington. [The verb is *lives*. Who or what *lives*? Alice *lives*. Therefore, *Alice* is the subject. The prepositional phrase is *as well as her husband Eric* and is not part of the subject.]

Word Power

A prepositional phrase begins with a preposition and ends with a noun or pronoun called an object. A prepositional phrase never includes a subject–verb combination. Some common prepositions are *in, to, with, by, for*.

Examples of two- or three-word prepositions are *alongside of, along with, in addition to, as well as*. You can review prepositions and prepositional phrases in Read and Replay 20.

Sometimes one clause "interrupts" another clause; that is, one clause may separate the subject and verb of another clause.

subj   subj   v   v
Alice, who has many skills and talents, is an RN. [The **independent clause** is *Alice is an RN*; the **dependent interrupting clause** is *who has many skills and talents*.]

A *compound subject* has two or more nouns or pronouns. A *compound verb* has two or more verbs.

┌————————compound subject————————┐   v
Laziness and irresponsibility impede success.

subj   ┌——compound verb——┐
Robin sings and dances.

Circle the subjects and underline the verbs in these correct sentences.

1. When (you) finish the soup, (you) understood leave the spoon in the plate. (2 subjects)

2. (Everyone) except your brothers was discharged.

3. (Lewis) and (Martin) told jokes and sang.

See page 398 for answers.

Although the subject is usually before the verb, sometimes the order is reversed. The order is almost always reversed in questions.

                   ┌——v——┐   subj
In the supply cabinet will be found six old floppy disks.

v   subj
Are six disks enough?

> **A subject is always a noun or a pronoun; however, it may look like a verb when it names an activity.**

subj   v
Studying is my favorite activity.

┌—subj—┐ ┌—v—┐
To run would be foolish.

> **The "understood" *you* may introduce a sentence that gives a command or makes a polite request.**

Put a new cartridge in the printer. [The subject is *you* understood; the verb is *put*.]

Please call before Friday. [The subject is *you* understood; the verb is *call*.]

Replay
38

Underline each verb and circle the subject. If a verb has a helping verb, underline the two or three words making up the "verb phrase." Remember that a sentence may have more than one subject–verb combination or may have a compound subject or verb.

EXAMPLE   A company may issue preferred stock only after common stock has been issued.

1. Do you enjoy the sunset?

2. The financial analysts are doing their jobs well.

3. Will she get a salary increase?

4. The clothing you select for the workplace should be appropriate.

5. (You) understood
   Ask yourself if your appearance sends a message that will benefit your career.

6. For a professional look, women should limit jewelry to a watch, necklace, earrings, and no more than two rings.

7. According to American-style table manners the diner puts the knife back on the plate, not the table, after he or she cuts a piece of food.

8. However, the knife remains in your right hand when you dine in most parts of Europe.

9. Career advancement is more likely for employees who get along well with co-workers, supervisors, and clients or customers.

10. Hard, steady work turns daydreaming into reality.

11. Playing Monopoly is fun.

12. During the past year our sales have risen dramatically.

13. In 1905 former president of the United States Grover Cleveland said, "Sensible and responsible women do not want to vote."

14. Everyone in the shop is working on the blueprints.

15. Xerox manufactured and sold copiers and other electronic equipment.

16. My assistant, who is very efficient, will gladly help you.

17. Would you like my assistant's help next week?

18. Considerable turnover of personnel is prevalent in the fast-food business.

19. Both poverty and riches are a state of mind as well as of pocketbook.

20. Good grammar and correct spelling are important in workplace e-mail.

Check your answers on page 398.

Read
39

MAKING THE SUBJECT AND VERB AGREE

NO: Here's the shoes and umbrella you forgot last week. **YES:** Here *are. . . .*

A subject and its verb should be compatible; that is, they must agree. Most agreement errors occur when a verb form is used that ends with **s** when it shouldn't or that does not end with **s** when it should. Although most **nouns** become plural by adding **s**, do **not** apply this principle to verbs.

SINGULAR SUBJECT

> **Singular subjects, except *you* or *I*, require an s verb form for the present tense and for "helping verbs" like *is, has,* and *was*.**

He runs 5 miles every day. **It was running** away. [singular pronoun subjects]

I believe that $100 (are/**is**) a fair price. (Point out that sums of money are singular.)

The **honor means** so much to me. Your **help has been** vital. [singular noun subjects]

> **You, I*, and plural subjects require a verb that does *not* end with s.**

You are the winner. **I work** across the street. [subjects: *you, I*]

We ship widgets everywhere on earth. [plural pronoun subject]

Faxes are ubiquitous.** [plural noun subject]

Neither his wife nor his friend (understand/ **understands**) why he refused the money.

> **If the words in a compound subject are joined by *or* or *nor*, make the verb agree with the noun or pronoun following the *or* or *nor*.**

Martha Jagel or **Petrina Noor has** the key. [singular noun—verb ends with *s*]

Christine or the **clerks** usually **sort** the mail. [plural noun—verb doesn't end with *s*]

Either Ms. Grover or **I am** holding the winning ticket. [I—verb doesn't end with *s*]

Neither Mr. Fairley nor **you are** to leave Blairsville. [you—verb doesn't end with *s*]

---

*Was is an exception. Although it ends with *s*, we use it with the subject *I*.
**No help here today; if you're not sure, look up *ubiquitous*. ☺

Pair *neither* with *nor*, not *or*!

Wrong Neither Charles **or** Dolores leaves at 5 p.m.

Right Neither Charles **nor** Dolores leaves at 5 p.m.

Neither Mary (or/**nor**) they do the work.
Either the land or the building (**is**/are) too expensive.
Every accountant in our offices (**has**/have) a different function.
Everyone (**is**/are) okay.
A **few** (is/**are**) better than none.

A compound subject joined with *and* is plural and, therefore, requires the verb form without *s*.

A **report and** a **letter provide** the information.
Victoria and he are guarding the secrets.

Except: If *each*, *every*, *many*, *a/an*, *one*, *either*, *neither*, *another*, or a pronoun ending with *one*, *body*, or *thing* precedes a subject, the subject is singular and requires the *s* verb.

Each man and woman needs an application form.
Many an applicant is denied an interview because of a spelling error.
Everyone dines after 9 p.m. in Spain.
Nobody likes this music.
Something has arrived from Bloomies.

For indefinite plural pronoun subjects—both, many, several, few—use the verb form without the *s*.

Several have been chosen.
Few take advantage of the opportunity.

When choosing a verb form, ignore words, phrases, or clauses separating a subject from its verb.

RIGHT The **box** of tools **is** on the table.

WRONG The **box** of tools **are** on the table.

WHY? The subject is the singular noun **box.** Therefore, use the singular verb form *is*. Ignore the prepositional phrase *of tools* when choosing the verb.

RIGHT The **supervisors,** as well as the CEO, **are** here today.

WRONG The **supervisors,** as well as the CEO, **is** here today.

WHY? Ignore the prepositional phrase *as well as the CEO* when deciding on the verb.

RIGHT The **reason** for his difficulties **seems** clear.

WRONG The **reason** for his difficulties **seem** clear.

RIGHT The **reasons** his job was difficult **seem** clear.

WRONG The **reasons** his job was difficult **seems** clear.

Recap

Correct the verb in the caption:

The Queen's English was fractured when the British Safety Council published a photo of Prince Charles with his arm in a sling after he had fallen off a horse. The caption read, "Ouch! One in three accidents are caused by falls." Change _____ are _____ to _____ is _____ .

Check your answer on page 399.

Although correct, there as an opener weakens a sentence. A box of tools was on the table is better than There was a box of tools on the table.

> **In a sentence introduced by *here*, *there*, or *where*, the subject usually *follows* the verb.**

RIGHT	There **were** several **boxes** of tools on the table.
WRONG	There **was** several **boxes** of tools on the table.
WHY?	Since the subject is the plural noun *boxes*, use the plural verb form *were*.
RIGHT	Here **come** my **sisters.**
WRONG	Here **comes** my **sisters.**
RIGHT	Where **are** your **sisters?**
WRONG	Where**'s** your **sisters?** [*'s* is a contraction of the **singular** verb *is*; **plural** subject *sisters* requires a plural verb.]

Use which or that for things, but never use which for people. That is now acceptable for people or things, but who or whom is preferable.

> **If the pronoun *who*, *which*, or *that* is a subject, the verb must agree in number (singular or plural) with the word for which the pronoun is substituting.**

RIGHT	This is the **man who talks** with you on the phone every day. [The pronoun *who* is substituting for singular noun *man*. Therefore, the verb *talks* ends with *s* to agree with *man*.]
WRONG	This is the **man who talk** with you on the phone every day.
RIGHT	These are the **men who talk** with you on the phone every day.
WRONG	These are the **men who talks** with you on the phone every day.

> **The word *number* is a singular subject if *the* precedes it. If *a* precedes *number*, it is a plural subject.**

RIGHT	**The number** of restaurants in this neighborhood **is** growing.
WRONG	**The number** of restaurants in this neighborhood **are** growing.
RIGHT	**A number** of books **have** been written on that subject.
WRONG	**A number** of books **has** been written on that subject.

of restaurants is a prepositional phrase.

SUBJECTS THAT GO EITHER WAY

> Subjects referring to "parts"—all, none, any, more, most, some, a fraction, or a percentage—may be singular OR plural depending on whether they refer to a plural or singular word.

SINGULAR   **All** of **it is** lost. [Use singular verb ***is*** because ***it*** is singular.]

PLURAL   **All** of **them are** happy. [Use plural verb ***are*** because ***them*** is plural.]

Here's a helpful trick to determine whether "part" subjects are singular or plural: If the item referred to can be counted, it's plural; otherwise, it's singular.

SINGULAR   At least **10 percent** of the **applesauce has** been eaten. [Applesauce can't be counted.]

PLURAL   **Half** the **apples are** in the refrigerator. [Apples can be counted.]

Writing for Your Career

For more concise writing, omit the preposition *of* when it's not needed. Notice **Half the apples** instead of **Half of the apples.** Write **All the software** instead of **All of the software.** If *of* can be omitted after the following—*all, half, none, any, more, most, some,* a fraction, or a percentage— do so. The preceding words are then adjectives describing the subject. It isn't important, however, to analyze the grammar. Just avoid prepositional phrases if you can express an idea more concisely without them.

CONTRACTIONS

> Another way to write more concisely is to consider use of contractions for *to be* verbs, *will* or *shall,* and *have.* Most business writing should be conversational in tone, and contractions often help meet that goal.

Many contractions are common and correct both in conversation and in career-type writing such as interoffice memos, e-mails, and most business letters.

YES   **I'm** being good today. **We're** always here to help you. **He's** an auditor. **You'll** always be welcome here. **We've** shipped the widgets. **It's** a good buy. **Here's** the document you requested.

Avoid contractions, however, in legal or other formal documents. Also, do not use *would* or *should* contractions (such as *I'd* for *I would* or *I should*) in workplace writing, although these contractions are fine in speech.

NO   **We'd** be pleased if **you'd** join us. [OK for conversation, informal e-mail, written telephone messages, etc.]

Replay
39

Decide whether the subjects and verbs agree in the following sentences. Draw a line through incorrect verbs and write the correct form in the blank. If the verb agrees with the subject, write C for *correct* in the blank.

was	1. Every one of the passengers ~~were~~ waiting in line quietly.
seems	2. Neither of your responses ~~seem~~ satisfactory.
were	3. The report and the letter ~~was~~ on my desk.
C	4. Either the report or the letter was on my desk.
was	5. Neither the report nor the letter ~~were~~ on my desk.
C	6. Both Twileen's story and Victoria's story sound true.
C	7. Neither Twileen's story nor Victoria's story sounds true.
is	8. Every battery, radio, and antenna ~~are~~ missing.
is	9. Everything that was in the garages ~~are~~ gone.
has	10. Many a floppy disk ~~have~~ been formatted and then lost.
are	11. About half the papers ~~is~~ gone from the file.
ride	12. All employees except the manager ~~rides~~ the elevator.
C	13. Any one of us is willing to help with the report.
C	14. A number of members are able to contribute to the fund.
C	15. When in Sweden, don't touch your drink until the host says "skoal."
has	16. Only one of the books ~~have~~ been translated into French.
have	17. Half the peach pies ~~has~~ been eaten.
is	18. The number of books translated into French ~~are~~ small.
doesn't	19. She ~~don't~~ want to go to the workshop.
were	20. There ~~was~~ several people waiting for the tickets.
are	21. The latest figures on yesterday's sale ~~is~~ available.
is	22. Each report and letter ~~are~~ on my desk.
has	23. A report on the accounts ~~have~~ been completed.
greet	24. In Spain men who are close friends often ~~greets~~ each other with a hug.
has	25. Each of the countries ~~have~~ distinct cultural characteristics.
packs	26. Neither he nor his assistant ~~pack~~ the items carefully.
C	27. Here's the software you ordered.
would	28. "We'd like to meet you when you visit Chicago," he wrote.
leave	29. Michelle and her aide, who drive in from Kenansville, ~~leaves~~ early.

does	**30.** He usually ~~do~~ his work carefully.
doesn't	**31.** The new copier ~~don't~~ work well.
works	**32.** My cousin, as well as a number of my aunts and uncles, ~~work~~ here.
_is_	**33.** Neither the attorney nor the paralegal ~~are~~ able to be here.
_go_	**34.** There ~~goes~~ the new models down the runway.
has	**35.** Another batch of envelopes ~~have~~ arrived incorrectly addressed.

Please check your answers on page 399.

IF I WERE A MILLIONAIRE

We've all indulged in fantasies about what we would do if we were someone else or if conditions were enormously different from our present reality. The English language provides a special verb form for the unreal—that is, to express ideas contrary to reality. Use this special form—were—principally when the subject follows *if* or *wish*. Use *were* regardless of what the subject is if the statement is contrary to reality.

Was/were is the only subjunctive form in which significant non-Standard usages occur.

We ordinarily use *was*, *is*, or other forms ending with *s* when the subject is a singular noun or a singular pronoun. Continue to use *was* or *is* if the statement is true or **might** be true (except, of course, with *you* or *I*).

ORDINARY OR MIGHT BE TRUE	CONTRARY TO REALITY
I was not at home Monday.	If *I were* you, I would stay at home on Mondays. [I can't be you.]
Everyone is going to the meeting.	If *everyone were* to go to the meeting, who would mind the store? [Everyone *won't* go . . .]
He was staring at the princess, but he is not a prince.	Peter wishes that he *were* a prince. [His parents are not king and queen.]
I was a millionaire last year.	If I *were* a millionaire, I would buy a 22 carat diamond for you at Tiffany's.

If Lincoln (was/**were**) alive, he would be pleased with the unity of our nation.

Make necessary corrections. If the sentence is already correct, write C in the blank.

were	**1.** If you ~~was~~ to read items 1–10 carefully, you would find 7 errors.
_C_	**2.** If I were not sure of how to get there, I wouldn't give you directions.
were	**3.** If he ~~was~~ faster at the keyboard, we would offer him the job.
_C_	**4.** Because she was ill so often last year, she couldn't complete her work.
were	**5.** I wish that I ~~was~~ your secretary instead of your husband.
_C_	**6.** He was a millionaire who spent his money to help the poor.

were	**7.** I wish that I ~~was~~ a millionaire.
was, wasn't	**8.** When I ~~were~~ your assistant, your office ~~weren't~~ in this building.
were	**9.** Cinderella wishes that she ~~was~~ a princess.
were	**10.** We wish the apple tree ~~is~~ healthy, but it never will be.

Answers to this Replay are on page 399, and your Pop Quiz is on page 345.

Word Power

Arithmetic Problem

2 plus 2 _is_ 4, but 2 and 2 _are_ 4, and 2 times 2 _is_ 4.

I don't know why; that's just the way it is . . . are?

Read 41

A SWARM OF BEES

A swarm is a collection of bees. If you completed Read 34, "A Gaggle of Geese," you made pronouns agree with **collective** nouns. _Gaggle_ and _swarm_ are Read titles because they're interesting words, but you don't hear them much in business communication. Please review pronoun agreement with collective nouns in Read 34 now. Additional collective nouns are club, herd, staff, management, family, name of a business or government organization, class, faculty, company, committee, crowd, and jury. A collective noun is like a single package containing several items. When you refer to the entire "package," the collective noun is thought of as a single unit. If the package is broken, you consider the items separately—resulting in ordinary plural nouns.

> If a collective noun acting as a single unit is a subject, use a singular verb—the one ending with _s_—for the present tense or for a helping verb.

NO The **faculty are** meeting in Room 406.

YES The **faculty is** meeting in Room 406. [_Faculty_ is a collective noun acting as a unit and, therefore, singular.]

The **members** of the faculty _are_ meeting in Room 406. [A plural verb is correct with the plural subject _members_; _of the faculty_ is a prepositional phrase and cannot include a subject.]

The **teachers are** meeting in Room 406. [_Teachers_ is a plural subject.]

NO **Macy's have** many employees. [Organizations are always singular.]

YES **Macy's has** many employees.

> A collective noun is a plural subject if the members act as separate individuals or disagree; in that case don't use the _s_ form of the verb.

NO The **faculty disagrees** about the new grading policy.

YES The **faculty disagree** about the new grading policy.

NO	The **jury goes** home every weekend. [Twelve people go off in different directions to different homes. They don't act as one.]
YES	The **jury go** home every weekend. [Collective noun is plural, thus requiring the plural verb *go*. Singular would be *goes*.]

But It Doesn't Sound Right!?

When a correct expression doesn't sound right to you, don't replace it with a grammar error, but don't leave it as is. Instead change the wording so that it is both grammatically correct **and** sounds right.

BETTER	The **jurors go** home every weekend. [Subject is now *jurors*, a plural noun requiring the plural verb form—go.]

Writing for Your Career

Collective noun principles are more important in business and professional writing than in speech. If in doubt about whether a collective noun is singular or plural, just rephrase the sentence.

International Note

If you studied British English outside the United States, you'll find that collective noun principles of British English differ from those of American English.

Replay
41

If subject and verb don't agree, make the needed correction only; don't change the rest of the sentence. Write *C* for correct in two blanks. Review Read and Replay 34 now so that you'll find two pronoun errors below.

take	1. My family takes separate vacations each year.
C	2. At noon today my family piles into the truck, leaves the old homestead, and heads for Anchorage, Alaska.
has its	3. Wong & Lopez, Inc., ~~have their~~ offices at 10 Park Avenue.
needs	4. Hiteki Corp. also need^s new headquarters.
were	5. Several groups ~~was~~ invited to the meeting.
were	6. For years J. C. Penney stores ~~was~~ called The Golden Rule Stores.
favors	7. The Senate favor^s new tax laws.
C	8. The company was warned by the fire chief to clear the aisles.
is/its	9. The Social Services Department ~~are~~ (is) submitting ~~their~~ (its) applications.
works	10. This class work^s quietly.

See page 399 for solutions.

CHECKPOINT

In the 1600s mathematician and grammarian John Willis was disturbed to discover that *shall* and *will* had about the same meaning. He worked out a plan to give them different meanings, and defenseless students have suffered ever since.

If you studied grammar in the past, you may wonder how you made it through a verb chapter without *will/shall–would/should*. Well, the simple truth is that our ever-changing language makes this instruction unnecessary. Some people who use British English still distinguish between these pairs of words. Other equally careful writers and speakers in business, politics, and the media now ignore these rules dating back to Mr. Willis. Only currently required verb principles are included in this chapter.

Repeat the correct forms aloud until they sound natural. For example, perhaps you noted a principle you've been careless about following before. Practice saying (to yourself) "If I *were* you" several times a day. Then, try to sound ignorant and uneducated and say (to yourself) "If I *was* you."

DANGER

The three most serious common verb errors follow. If you make errors like these, hurry to acquire the habit of using Standard English verbs.

1. Using the present verb form when the past is needed ☹

   Yesterday Ms. Wong talk to the business class about being flexible on the job. [should be *talked*]

2. Not adding *s* to a verb when it is needed—or adding *s* when it should not be there ☹

   Marg Taylor always explain the business English principles carefully. [For present tense, use *explains*; for past, use *explained*]

   Professors Weigand and Wardell teaches in Washington. [should be *teach* or *taught*, depending on the meaning intended]

3. Using the simple past with a helping verb or the past participle without a helping verb ☹

   He had wore that same shirt yesterday. [should be *worn*]

   I know he done a good job because I seen it. [Use *did* and *saw*, or use helping verbs before *done* and *seen*]

NOTES:

SPECIAL ASSIGNMENT

A. Compose sentences with any singular subjects except *you* or *I*. Use the present tense form of the verbs.

EXAMPLE   **beat**   **She beats** her husband regularly at tennis.   Sample Responses

1. go   _____ She goes to the tennis club at 8 a.m. _____

2. drive   _____ Professor Jagel drives her there. _____

3. see   _____ Mr. Agresta sees everything. _____

B. Compose sentences using the *ing* form of the verbs that follow the examples; use them as **verbs, not adjectives or nouns.**

EXAMPLES   **dance**   YES My husband is dancing. [*Dancing* is a verb.]

NO My hobby is dancing. [*Dancing* is a noun.]

NO My husband attends dancing school. [*Dancing* is an adjective.]

1. run   _____ Our horse is running very fast. _____

2. rise   _____ The sun is rising now. _____

3. raise   _____ Have you been raising your hand in class? _____

C. Compose sentences using the verbs that follow the example to show past tense **without a helping verb.**

EXAMPLE   **talk**   He **talked** on the phone for an hour.

1. win   _____ My horse won the race. _____

2. sing   _____ Then I sang, "Yankee Doodle Went to Town." _____

3. quit   _____ After that I quit the choir. _____

D. Compose sentences using the verbs that follow the example to show past tense **with a helping verb.**

EXAMPLE   **drink**   Victoria **had drunk** the last glass of punch before the clock struck 12.

1. hide   _____ The stepsisters have hidden her shoe. _____

2. pay   _____ George had paid the usher to lead her astray. _____

3. be   _____ The unhappy prince has been in the palace all night. _____

E. Compose sentences showing the present tense of these verbs with a plural subject. **Be sure to use the words as verbs.**

EXAMPLE   **give**   **They give** shoe polish with all shoe purchases.

1. cost   _____ New shoes cost a great deal. _____

2. ship   _____ We ship shoes to shoe shops every day. _____

3. produce   _____ Mr. and Mrs. Larson produce vegetables on their farm. _____

Complete this assignment by _____ . (date)

F. Solve an Action-Word Puzzle

¹C	A	U	G	²H	T		³G		⁴T		⁵S	U	N	G
H				A			O		O					
O		⁶L	O	S	T			⁷R	A	⁸W			⁹T
S								E		A			U
E		¹⁰R	U	N			¹¹U	¹²P		¹³S	E	E	N
N		U						U					I
		N			¹⁴P	U	T						N
			¹⁵B							¹⁶R	¹⁷A	N	G
		¹⁸N	E	E	D	E	¹⁹D		²⁰D		G	
²¹H	E					O		O		²²B	E	E	²³N
A			²⁴I			N		E				O
²⁵B	E	²⁶A	T	S		E		²⁷S	E	A	L	E	D
I		R										
T		²⁸E	²⁹D	S		³⁰S	O		³¹F	E	³²L	T
S			U			A		³³A			E	
	³⁴B	A	G		³⁵S	W	U	M			³⁶D	I	D

Across

1. Simple past and past participle of *catch*
5. Past participle of *sing*
6. Simple past and past participle of *lose*
7. Adjective meaning *uncooked*
10. Past participle of *run*
11. Antonym of *down*
13. Past participle of *see*
14. Simple past and past participle of *put*
16. Simple past of *ring*
18. Simple past and past participle of *need*
21. Singular masculine subject pronoun
22. Past participle of *be*
25. Form of *beat* with singular masculine pronoun subject (present tense)
27. Past tense of *seal*
28. Two guys named Ed
30. Coordinate conjunction too informal for business writing
31. Simple past tense of *feel*
34. They do it to your groceries at the register
35. Past participle of *swim*
36. Simple past tense of *do*

Down

1. Past participle of *choose*
2. Form of have for use with *he, she,* or *it*
3. Present tense of *went*
4. Simple past tense of *tear*
8. Simple past tense of *be*
9. Present participle of *tune*
10. Present tense of *run*
12. Past participle of *put*
15. "To _____ or not to _____," wrote Shakespeare
17. To grow older
19. Past participle of *do*
20. Present form of *do* with a singular noun subject
21. Doing Recaps and Replays are good ones
23. Verb meaning to show agreement silently
24. Present tense of *be* with *she* as subject
26. Present tense of *be* with *you* as the subject
29. Past participle of *dig*
30. Simple past of *see*
32. Past tense of *lead*
33. Present tense of *be* with *I* as the subject

PROOFREADING FOR CAREERS

A. Please save the English language by undoing these atrocities to verbs:

1. It ~~don't~~ seem right. _____doesn't_____
2. If you had ~~wrote~~, he would ~~of~~ answered. _____written, have_____
3. He ~~done~~ the work. _____did_____
4. We ~~seen~~ her yesterday. _____saw_____
5. He ~~ain't goin'~~. _____isn't going_____

B. Correct spelling, typing, capitalization, pronoun, and verb errors in this letter, which isn't really from a certified financial planner. Do not make changes other than correcting specific errors just mentioned. Glide your pen or pointing finger below each line as you proofread and correct the letter. Never let written material leave your desk without careful proofreading.

Dear Mr. and Mrs. Shue:

From a mutuel aquaintance I herd of your recent marriage. I want to be among those whom wish you many years of happiness and success.

All of us knows that young married people needs to build there homes on a firme foundation if they are to have the best chances for future success. That is why I offer are services at this time to explain how you can have an insurance program to cover you're needs but fit you're budget.

Please telephone me at (206) 448-7737. I look forward to speaking with you about our various insurance plans. You will have absolutely no expense or obligation.

Sincerly,

Eric A. Smith, CFP

How did you do? Excellent _____ Good _____ Poor _____

PRACTICE QUIZ

www.prenhall.com/smith

Refer to your dictionary when in doubt. Circle the correct answer.

1. Ms. Hixon explained that many fine people (live/lived) in Torrance now.
2. The people of this country (believe/believes) it can be done.
3. If I (was/were) not certain how to get there, I wouldn't give you directions.
4. Mr. Sedirko has (chose/chosen) three people to take deposition notes.
5. The film industry (employ/employs) many extras for walk-on parts.
6. At present 3,800 extras (is/are/be) on the roll of this bureau.
7. If I (were/was) you, I would take advantage of this opportunity.
8. Robert Ball, as well as all his aides, (deserve/deserves) a raise.

9. Neither the engineer nor the designer (is/are) here.
10. The costs of salaries and travel expenses (have/has) been determined.
11. Ms. Timm told me that I (did/done) the right thing in Monterey.
12. I wish that the new plans for San Antonio College (was/were) ready.
13. You have (broke/broken) one of the rules.
14. Neither the banker nor the lawyer (is/are/be) in Alabama.
15. Each girl and boy (is/are/be) doing well.
16. One-third of the pies (have/has) been sold.
17. Accuracy in figures (mark/marks) the expert accountant.
18. She is the only one of the attorneys who (do/does) the research.
19. They (sunk/sank) the ship.
20. The Bureau of Mines (is/are) now preparing to transfer its offices.
21. Rosenberg & McNeil, Inc., (insist/insists) on prompt shipments.
22. The Hartford City Council (were/was) disappointed in the results.
23. The committee (have/has) been unable to agree on the agenda.
24. The senior class (arrive/arrives) an hour before graduation.

THE GLOBAL MARKETPLACE—SPAIN AND SPANISH-SPEAKING COUNTRIES

Spanish	Spell	Say	Currency
GREETING	Buenas Dias	BWAYN.os.DEE.is	Spain: peseta (pta)
DEPARTING	Adios	a.dee.OS	
PLEASE	Por favor	por faVOR	
THANK YOU	Gracias	GRAH.see.is	
I DON'T UNDERSTAND	No comprendo	Nō comprĕndō	

Random House Unabridged Dictionary

Words That Describe

Describe with Adjectives and Adverbs

After Completing Chapter 8, You Will

> Use adjectives and adverbs skillfully.

"The adjective is the banana peel of the parts of speech."

—*Clifton Fadiman*

See your *Instructor's Resource Kit* for Chapter 8 Pretest and for adjective-adverb exercises to introduce this chapter.

Adjectives and adverbs are more fun than pronouns and verbs. Imagine how dull our language would be without adjectives and adverbs like *generous, happily, stingy, cheerfully, prudish, confidently, weaker, meaner, strictest, domineering, shabby, purple*, and *most comfortable*. Now imagine how vague our language would be without adjectives like *several, those, fifth, often, sometimes*, or even the little word *the*. Adjectives and adverbs are "modifiers"; that is, they tell about, describe, tell how many, or add special meaning to other words. They add precision, character, liveliness, and color to our language.

Communication is difficult at best; we often misunderstand one another. Skillful communicators draw from a wealth of adjectives and adverbs, using those that most effectively communicate the shade of meaning desired. In Chapter 8, you will review Standard English principles for adjectives and adverbs.

Adjectives tell something about—or *modify*—nouns and pronouns. The four kinds of adjectives are **pointers, articles, describing**, and **limiting**.

Adverbs tell something about—or *modify*—verbs, adjectives, and other adverbs. Many adverbs are formed by adding **ly** to adjectives; for example, *peaceful* is an adjective and *peacefully* is an adverb. Other adverbs, however, are totally different from adjectives—words such as *well, often*, or *sometimes*.

Writing for Your Career

The adverb VERY is probably the most overused adverb. If an intensifying word is used too much, it loses its power. William Allen White, famous writer of the early 1900s, was editor of the *Emporia Gazette*, a well-written Kansas newspaper. To discourage the staff from overuse of *very*, he sent them this memorandum:

If you feel you must write "very," write "damn."

Since the copy desk had instructions to delete reporters' profanity, the writing style of the newspaper was improved: "It was a very fine victory" was written, "It was a damn fine victory," and was printed, "It was a fine victory."

Read
42

POINTERS FOR POINTERS

Here are the only pointers—or tips—you need for *pointing* adjectives:

> When *this, that, these,* or *those* precede a noun, they are called pointing adjectives; otherwise they are pronouns.

This soup is mine. [*This* is a pointing adjective; *soup* is a noun.]
This is mine. [*This* is a pronoun.]

> Use *this* and *that* as adjectives pointing to singular nouns.

This kind of spice is too hot for **that type** of soup.

> Use *these* and *those* as adjectives pointing to plural nouns.

NO   **Those kind** of spices are hot.

YES   **Those kinds** of spices are hot.

NO   Will **these type** of homes sell in those neighborhoods?

YES   Will **these types** of homes sell in those neighborhoods?

Writing for Your Career

If a singular pointing adjective makes sense, avoid the plural construction. Instead of these or those kinds, types, or sorts, use this or that kind, type, or sort—or omit *kind, type,* or *sort*. For example, this kind of book OR this book INSTEAD OF these kinds of books.

INCORRECT	Those kind of roofs are fireproof.
OK BUT AWKWARD	Those kinds of roofs are fireproof.
IMPROVED	That kind of roof is fireproof. OR Those roofs are fireproof.

Avoid unnecessary *a* or *an* after kind of, type of, or sort of.

| OK BUT CAN BE IMPROVED | This kind of a house . . . |
| IMPROVED | This kind of house . . . |

Never use *them* as a pointing adjective; it is a pronoun only.

| NO | I plan to give them boys a million dollars. |
| YES | I plan to give those (or these) boys a million dollars.   OR   I plan to give them a million dollars. [*Them* is correct as a pronoun.] |

NEVER use the following expressions: *this here, that there,* and *them there.*

You probably don't write those expressions. If, however, you might **say** them, ask a friend or an instructor to help you break the habit.

NO	This here is my office.
YES	This is my office.
NO	That there phone is out of order.
YES	That phone is out of order.
NO	Them there books should be stored in the basement.
YES	Those books should be stored in the basement.

Replay
42

Correct the following sentences.

EXAMPLE: These kind $\overset{s}{\wedge}$of dogs can be vicious. **Possible Answers**

1. Professor Stagnaro ordered ~~those~~ $\overset{that}{\wedge}$ kind of PC.

2. ~~Them~~ $\underset{\wedge}{\text{Those or These}}$ books should be returned to the library.

3. If you are not careful, these type^s of errors will occur frequently.

4. Those sort^s of books are extremely interesting.

5. I don't like those type^s of people. or . . . that type of person.

6. He wants to buy this ~~here~~ book.

7. Please give that ~~there~~ calculator to Michelle Miller of Milwaukee.

8. Ask ~~them there~~ ^those people to do the work themselves.

9. That ~~there~~ is the way the cookie crumbles.

10. That kind of ~~an~~ advertisement doesn't attract our customers.

Answers are on page 399.

Read 43 is short and easy for students whose everyday speech includes *an*. For those using only *a* in normal speech, Read 43 requires major effort to correct this non-Standard usage—e.g., I ate *a* apple.

THREE LITTLE WORDS

Another kind of adjective is the *article*. **The**, **a**, and **an** are the only three articles. **The** is a "definite" article because it makes the noun following it *definite*—that is, *specific*. **A** and **an** are "indefinite" articles. Notice the difference in meaning between "the book" and "a book" or between "the apple" and "an apple."

If English is your second language, you may be unsure of when to use articles and when to omit them. Look at these two sentences:

We are going to school today.   **BUT**   We are going to *the* office today.

Why don't we use *the* before *school* in the first sentence? Why *do* we use *the* before *office* in the second sentence? These choices "sound right" to a native of an English-speaking country, but no simple rule governs them. Differences even occur between British English and American English:

BRITISH—She is in hospital.   **BUT**   **AMERICAN**—She is in **the** hospital.

If you're learning English now, the solution is to read English a great deal; listen to the radio, TV, and films; and ask co-workers, teachers, and friends to correct you. Your brain will gradually develop a sense of what "sounds right."

One use of articles, however, can be a problem even for native-born speakers of English—the two little words *a* and *an*. Some people use *a* almost exclusively and rarely use *an* when it's needed. Non-Standard use of *a* and *an* is noticeable in workplace communication. Since useful rules do determine the use of *a* or *an*, correcting this usage is easier than learning where and when to use articles. Not everyone needs to study the *a/an* rules. Many people automatically use these words correctly. The following Recap is a pretest to determine whether you need to study all, some, or none of the rules that follow.

Recap

Write *a* or *an* in the blank before each word. Choose what comes to you immediately. Don't try to figure out the right answer.

EXAMPLE _a_ $10 bill

1. _an_ addition
2. _a_ carrot
3. _an_ egg
4. _an_ apple
5. _a_ giant
6. _an_ honor
7. _a_ heater
8. _an_ $11 gift

9. _a_ hand
10. _a_ one-day sale
11. _a_ manager
12. _an_ onion
13. _an_ owl
14. _an_ uncle
15. _an_ Englishman
16. _an_ heir

17. _an_ island
18. _a_ European
19. _an_ IBM office
20. _a_ CIA report
21. _a_ UN member
22. _a_ 2 percent tax
23. _an_ X-ray
24. _an_ unknown admirer

Check your answers on page 399.

If your answers for the preceding Recap are correct, just go directly to the next Replay. If some of your answers are incorrect, study "Managing *A* and *AN*," and refer to the information as a reference when you're writing.

MANAGING *A* AND *AN*

> If a word following an indefinite article begins with a vowel sound (the sound of *a, e, i, o,* or *u*) use <u>an</u>. Otherwise use <u>a</u>.

Use of *a* or *an* depends on the **sound** of the word following *a* or *an*, not necessarily the written letter with which the word begins.

> **an a**pple   **an e**gg roll   **an i**llness   **an o**wner   **an u**ncle   **an** honor
> **a 10** percent raise [*t* sound]   **a h**ot rod   **a b**ookkeeper   **a p**uzzle
> **a u**niform

> Some words that begin with a vowel actually begin with a consonant *sound*; they require *a*, not *an*.

For example, the *u* in *union* sounds like the consonant *y* in *you*: **a u**nion, **a uni**form, **a eu**logy,* and so on. Also the *o* in *one* sounds just like the consonant *w* in *winner*: **a** one-cent stamp.

> *H* beginning a word is sometimes silent. For such words, use *an*.

The first sound of *honor* is *o*, a vowel sound: **an** honor, **an** honest person, **an** herb (pronounced "urb") garden. If you pronounce the *h* in herb—as is common in British English—say and write **a** *herb*.

---

*What does *eulogy* mean and how do you pronounce it? If in doubt, look it up now.

> **When using a letter of the alphabet alone or as part of an abbreviation, the letter is preceded by *a* or *an*, depending on the *sound*.**

USE *AN* an A, an E, an F, an H, an I, an L, an M, an N, an O, an R, an S, an X

USE *A* a B, a C, a D, a G, a J, a K, a P, a Q, a T, a U, a V, a W, a Y, a Z

- He needs **an FBI*** report for **a CIA*** officer.
- Write a **T** or an **F** in each blank.

> **Abbreviations pronounced as words instead of individual letters are called *acronyms*, such as NASA,* pronounced *nas'uh***

He was **a** NASA* employee and **an** NAACP member.

WORD TO THE WISE

Use **a** before a consonant **sound.**
Use **an** before a vowel **sound.**

Replay 43

Write *a* or *an* in each blank.

EXAMPLE __A__ pessimist sees the difficulty in __an__ opportunity. __An__ optimist sees the opportunity in __a__ difficulty.

1. Jan Hinkle was given __a__ one-day unpaid leave of absence and __an__ eight-day paid vacation.

2. __A__ union member ate __an__ onion during __an__ 80-day world tour.

3. It is __an__ unusual combination to be __a__ CFP and __an__ MD.

4. They left for __a__ meeting in Pasadena about __an__ hour ago.

5. __A__ thesaurus is __an__ invaluable tool for writers.

6. __An__ X-ray was needed to determine whether there was __an__ injury.

7. __A__ UNICEF** representative and __a__ NATO** representative were seated next to each other at __a__ UN** meeting.

8. __A__ *yes* or __a__ *no* in Japanese doesn't always mean what __an__ American thinks it means—because of aspects of Japanese culture differing from American culture.

9. Each time he received __a__ B on __a__ report card, he was given __an__ $11 gift.

10. He walked down __a__ hall to get __a__ history book for __an__ honest man.

International Business To some Japanese and other Asian people, saying "no" may be considered bad manners. Therefore, a "yes" response could mean something like " I understand what you want, and I'll think about it."

---

*If you're unsure of what any of the abbreviations and the acronym (NASA) mean, look them up now either in your dictionary or in Appendix D, Abbreviations.
**Pronounce acronyms UNICEF and NATO as *unisef* and *naytoe*. Pronounce the abbreviation UN as two separate letters.

11. <u>An</u> heir expected to inherit <u>a</u> one-million-dollar home.

12. His sister, <u>an</u> heiress, was not <u>an</u> honorable woman.

13. <u>An</u> uncle of mine planted <u>an</u> herb garden. (say *herb* as *erb*)

14. After receiving <u>an</u> AA degree, he earned <u>a</u> BA in sociology.

15. <u>A</u> European executive was <u>a</u> CEO in <u>an</u> American firm.

Please check your answers on pages 399–400.

I DON'T WANT NO BROCCOLI

George Bush, Sr., the 41st president of the United States and father of President George W. Bush, said his mother made him eat broccoli when he was a boy. He joked that as long as he's president, nobody could make him eat it. Of course, he didn't speak in double negatives, but here's how the dialogue might sound if he had.

CHILD   I **don't** want **no** broccoli!

PARENT   If you don't want **no** broccoli, you must want **some** broccoli, so eat up.

When two negative words express one negative idea, the result is a double negative. The idea of a double negative equaling a positive was thought up by an 18th-century British bishop, Robert Lowth. In his book, *Short Introduction to English Grammar*, he wrote he would "lay down the rules" and "judge every form and construction." He based the double negative rule on classical Latin, although the structure of English is different from that of Latin. He didn't apply the Latin principle correctly, since double negatives *are* used in Latin for emphasis. Two negative words to express a single negative idea does not really affect your listener's or reader's ability to understand. You know the child doesn't want any broccoli. If your employer asks you what time it is and you reply, "I don't got no watch," she will certainly understand what you mean.

BUT

She will be reluctant to involve you in written or oral communications with the public. In fact, using double negatives usually prevents applicants from being hired for better jobs. It isn't important to identify which negative words are adverbs and which are adjectives. Just remember that two negatives shouldn't be combined to express one negative idea. Here are examples of negative words not to be combined:

no	none	doesn't	aren't	never	shouldn't
not	nowhere	won't	wouldn't	scarcely	neither
nobody	can't	couldn't	don't	haven't	hardly

When two of the preceding words are combined, the result is one of the most serious and embarrassing errors in the English language. This really happened: A student showed me a punctuation quiz before turning it in. "I didn't put no comma there, Ms. Smith," he said, pointing to one of the sentences. Trying to save him from the horrors of a life of double negatives, I said, "I didn't put *any*

See the Instructor's Resource Kit for practice in avoiding double negatives, which you might need to distribute selectively to students "in need."

*Some **nouns** change to **adjectives** by adding endings like ish, y, ify, ous, al, or ful: girl**ish**, curl**y**, beauti**ful**, marvel**ous**, commun**al**, tact**ful**. Other nouns become adjectives without changing the word; function in the sentence makes the difference: He attends **college** (noun). He plays **college** football (adjective).*

*Past tense and past participles are not only verbs but may also be adjectives; e.g., I prefer **broiled** chicken rather than **fried** eggs.*

---

comma there, Joe." He replied, "Oh, good, you didn't put none there neither," thus increasing his original double negative to a triple.

If you suspect you use double (or even triple) negatives, kick the habit by asking an instructor or a friend to tell you privately when you say or write them.

Replay 44

Correct these sentences so that they conform to Standard English. Four sentences have verb errors. Write C next to the two sentences that are already correct.

_____ 1. I ~~scarcely never~~ do that.
           _{rarely or hardly ever}

_____ 2. I never said ~~nothing~~ about it to ~~nobody~~.
           anything   anybody

_____ 3. Don't put ~~none~~ over there as it might spill.
           any

_____ 4. He won't eat ~~no~~ more pizza if he goes to Afghanistan.
           any

___C___ 5. If you have a negative attitude, you won't succeed.

_____ 6. He ~~don't~~ know ~~nothing~~ about chemistry.
     doesn't   anything

_____ 7. Taylor won't go ~~nowhere~~ with me on Saturday.
           anywhere

_____ 8. You don't pass ~~no~~ serving dishes during a meal in China.
           any

___C___ 9. Instead you reach for food with your chopsticks.

_____ 10. She hasn't gone ~~nowhere~~ yet.
           anywhere

_____ 11. Dan ~~ain't never~~ in when you want to see him.
     isn't ever

_____ 12. You ~~couldn't~~ hardly expect Mr. Blank to join that organization.
     could

_____ 13. You can't win ~~no~~ friends that way.
           any

_____ 14. Nobody ~~doesn't like~~ Sara Lee Cheesecake. _{or Everybody likes}
     dislikes

_____ 15. I ~~ain't~~ never ~~gonna~~ use ~~no more~~ double negatives.
   _{I will never use double negatives again.}

See page 400 for answers. Take the Pop Quiz on page 347.

Word Power

The *principal* in colonial America was the "principal" teacher in a school with two or more teachers. The principal kept the records and saw that the schoolhouse was kept clean. *Dean* was originally a title for a superior over ten monks. The word is from the Latin *decanus*, which was derived from the Latin word for ten (*decem*).

Read 45

GOOD, GOODER, GOODEST??

Short, Lively Activity
Write on the board or dictate an easy-to-describe noun; e.g., *music, language, company.* Ask students for adjectives to describe the noun(s). Use a different color for the adjectives.

When with young children, you hear sentences like "Mine is the goodest of all." Very soon, children learn to say "best." They don't need a rule. It just happens naturally from imitating their elders. However, certain adjective and adverb principles are learned only with specific instruction.

RECOGNIZING ADJECTIVES REMINDER

Adjectives modify—that is, describe, limit, or point to—nouns or pronouns. They answer questions like What kind? Which? or How Many?

> He is **strong.** [adjective *strong* describes pronoun *he*]
>
> He is **a strong** man. [*a* and *strong* are adjectives describing noun *man*]

COMPARISONS

Descriptive adjectives, which let you show comparisons, come in three degrees—**positive, comparative,** and **superlative.** The comparative and superlative degrees are used for comparisons by adding *er/est, more/most,* or *less/least.*

Positive Degree—for One

> Positive degree adjectives describe, or modify, a noun or pronoun without making a comparison.

> old   young   happy   valuable   modern   affluent

> The queen's jewels are **old.** [*old* modifies the noun *jewels*]
>
> They are **valuable.** [*valuable* modifies the pronoun *they*]

Comparative Degree—for Two

> Comparative degree adjectives compare two nouns or pronouns. To make an adjective comparative, either add *er* to the end of the word or use *more* or *less* before the adjective.

> older   younger   happier   more valuable   less modern   more affluent

> Jean is the **younger** of the two sisters, but Marian's life is **happier.** [Two sisters are compared.]
>
> That building is **older** than the one on Vine, but the plumbing is **more modern.** [Two buildings and the plumbing are compared.]

Superlative Degree—for Three or More

Ask for comparative and superlative forms of various adjectives, such as: *many, angry, ridiculous, capable, new, lovely, sudden, serious, soon.*

> A superlative degree adjective compares three or more nouns or pronouns. To make an adjective superlative, add *est* to the end of the word, or use *most* or *least* before the adjective.

oldest	youngest	happiest	smartest
most valuable	least modern	most affluent	most careful

Melvin is the **youngest** of the three sons, and he is also the **most affluent.**

It was the **happiest** day of my life when you gave me your **most valuable** diamond. [You had three or more diamonds.]

Recap

Underline the correct answer.

Who is (<u>wisest</u>/wiser), the judge, the minister, or the professor?
Who is (wisest/<u>wiser</u>), the judge or the minister?

Check your answers on page 400.

If English is your first language, you probably easily distinguish between words that end in *er/est* and those words that require *more/most* or *less/least*. If in doubt or if English is your second language, the following principles help:

One-Syllable Words

Add *er* or *est:*   big, bigger, biggest   smart, smarter, smartest

Two-Syllable Words—Ending With Y

If the word ends with *y*, such as *easy*, change the *y* to *i* and add *er* or *est*—*easier, easiest.*

Two- or More-Syllable Words—Not Ending With Y

For words with two or more syllables that don't end with *y*, such as *careful* or *beautiful*, use *more/most* or *less/least*; for example, more or most careful, less or least beautiful, more/most or less/least important.

Avoid Double Comparatives and Superlatives

NEVER use *more, most, less,* or *least* before a modifier that ends with *er* or *est.*

HORRIBLE	He is the **most laziest** child I've ever seen. [double superlative]
CORRECT	He is the **laziest** child I've ever seen.
TERRIBLE	I am **more carefuler** than anyone else. [double comparative]
CORRECT	I am **more careful** than anyone else.
ATROCIOUS	Mine is **more better** than hers.
CORRECT	Mine is **better** than hers.

IRREGULAR ADJECTIVES

Several adjectives have comparative and superlative forms that differ from those just reviewed. In the dictionary, the entry word for an adjective is usually in the positive degree. Next to it you will find the irregular forms for that adjective. For example, look up the adjective *bad* right now. Next to it, instead of *badder* and *baddest* you find *worse*, which is the comparative form, and then the superlative, *worst*, in that order*. Some other irregular adjectives are *far, ill, good,* and *many.*

---

*In newer college dictionaries, *badder* and *baddest* are somewhere in the "bad" entry with the usage label *slang* along with the definition.

Avoid the comparative *more* or the superlative *most* before words like *unique* or *perfect*. Since UNIQUE means *one of a kind*, it isn't logical to say something is *more* unique or *the most* unique. "Unique" is an example of an "absolute" adjective. In the same way, *perfect* means no imperfections; therefore *more* or *the most* perfect do not make sense.

Replay
45

A. Look up these irregular adjectives in the dictionary and fill in the blanks. You'll find the comparative and superlative forms next to the positive, which is the entry word.

EXAMPLE   many   _____more_____   _____most_____

Although the distinction between *farther* and *further* is easy, it is disputed by some experts. Just recommend *farther/est* for physical distance and *further/est* for *additional*. (See "Grammar for the Expert" in your *Instructor's Resource Kit*.)

	COMPARATIVE		SUPERLATIVE	
1. far	farther	or further	farthest	or furthest
2. bad	worse		worst	
3. little	littler	or less	littlest	or least
4. much	more		most	
5. good	better		best	

B. Correct the adjective errors in these sentences. Write *C* beside the only correct sentence.

EXAMPLE   This lot is the ~~most~~ *more* valuable of the two we saw today.

_____   1. The ~~most~~ safest investments are in blue-chip companies.

_____   2. Of the two reports, his is the ~~worst~~ *worse*.

___C___   3. He is the younger of the two brothers.

_____   4. He is the ~~older~~ *oldest* of the three brothers.

_____   5. When the figures of the two accountants were compared, the controller found Mr. Higgins' work to be ~~best~~ *better*.

_____   6. This new alloy is ~~more~~ heavier than any other metal.

_____   7. This file contains ~~recenter~~ *more recent* information than that one.

_____   8. If you use our detergent and brand X, which one will give you the ~~brightest~~ *brighter* wash?

_____   9. He's ~~more~~ friendlier than the other sales manager.

_____   10. When you examine the two diagrams, you discover that the one on the right is ~~biggest~~ *bigger*.

Answers are on page 400. Please take the Pop Quiz on page 348.

TO *LY* OR NOT TO *LY*, THAT IS THE QUESTION

Class Responses Ask for adjective and adverb forms of the noun *convenience* (convenient, conveniently). What are some adverbs that can describe *convenient* and/or *conveniently?* (e.g., convenient: *more, most, so, not, never, always, very, especially, exceptionally, extremely, very, more, most*)

That's not exactly the way Shakespeare wrote it, but it rhymes with the original anyway. We add *ly* to many adjectives to form adverbs. For example, adding *ly* to the adjective *occasional* results in the adverb *occasionally*. The information that follows enables you to avoid some common adverb and adjective errors. The first step is to recognize which are adjectives and which are adverbs.

IS IT AN ADJECTIVE OR AN ADVERB?

Most (but not all) words ending in *ly* are adverbs.

ADVERBS	happily	busily	attractively	cheaply	carefully
ADJECTIVES	happy	busy	attractive	cheap	careful
OTHER ADVERBS	(not ending with ly) always, never, often, seldom, very				

Some adjectives end in *ly*; don't mistake them for adverbs.

ADJECTIVES curly, godly, friendly [If the *ly* word describes a noun or pronoun—curly hair, friendly man—you know it's an adjective.]

Student Alert Which is it? Mark feels bad. OR Mark feels badly. Which dog needs a bath—the one who smells **bad** or the one who smells **badly**?

Sometimes the same word is either an adjective or an adverb.

He is a **fast** worker. [*Fast* is an adjective describing the noun *worker*.]

He works **fast.** [*Fast* is an adverb describing the verb *works*.]

WHICH SHOULD IT BE—ADJECTIVE OR ADVERB?

Use adverbs (not adjectives) to modify verbs, adjectives, or other adverbs.

NO He arrived on time and **worked quiet.** [The adjective *quiet* is incorrect because an adjective cannot describe a verb—*worked*.]

YES He arrived on time and **worked quietly.** [The adverb *quietly* tells **how** he *worked*.]

Recap

Circle the correct choice.

This engine runs (smooth/(smoothly)).

_____

See the answer on page 400.

To make *ly* adverbs comparative (for two) or superlative (for three or more), use *more/less* or *most/least* before them.

ADJECTIVE	ADVERB	COMPARATIVE ADVERB	SUPERLATIVE ADVERB
efficient	efficiently	more/less efficiently	most/least efficiently
beautiful	beautifully	more/less beautifully	most/least beautifully
polite	politely	more/less politely	most/least politely

NO   Of all the custodians, the new man works the **most efficient.** [*Most efficient* is an adjective and is not to be used to describe a verb: *works.*]

YES   Of all the custodians, the new man works the **most efficiently.** [*Most efficiently* is an adverb correctly modifying a verb: *works.*]

NO   We obtain parking permits **easier** than do the other stores. [*Easier* is an adjective incorrectly modifying a verb: *obtain.*]

YES   We obtain parking permits **more easily** than do the other stores. [The adverb *more easily* correctly modifies the verb *obtain.*]

Recap

Using an easy-to-describe verb, such as **play, kiss,** or **work,** ask for adverbs for that verb. Examples are *hard, fast, sloppily, often, slowly, rapidly.* Point out that *fast* and *hard* are adjectives as well as adverbs; e.g., **hard work** is an adjective followed by a **noun,** but **work hard** is a verb followed by an **adverb.** Then ask for adjectives to modify *play, kiss, work,* e.g., *boring* play, *sloppy* kiss, *endless* work; next ask for adverbs to follow those words; e.g., play *carefully,* kiss *lovingly,* work *safely.*

Circle the correct choice in item 1, and write the answers in items 2 and 3.

1. He writes (clearer/(more clearly)) than his assistant.
2. What is the verb in item 1? _____writes_____
3. What part of speech is the circled answer to item 1? _____adverb_____

Answers are on page 400.

> **A modifier after a *being verb* must be an adjective modifying the subject.**

The most common being verbs are forms of *be* such as *is, am, are, was, were, been.*

> She is intelligent. [The adjective *intelligent* modifies *she,* the subject of the **being** verb.]

In addition to forms of *to be,* verbs of the senses are often (but not always) "being verbs." Typical verbs of the senses are *appear, become, seem, look, taste, sound, smell, feel.* To determine whether a verb of the senses is a being or an action verb, decide whether it refers to action or state of being in that sentence.

> She **looks** carefully for the eraser. [*Looks* is an action verb referring to the action of using her eyes to look.]
>
> She **looks** intelligent. [*Looks* is a being verb because she is not doing the action of looking; *looks* refers to her appearance.]
>
> I **tasted** the potatoes. [Action verb because tasting is action.]
>
> The potatoes **tasted** delicious. [Being verb; potatoes are not doing the action of tasting.]

NO   I feel **badly** today. [Adverb *badly* changes *feel* to an action verb.] The sentence then means I do a bad job of "feeling"; for example, I probably can't feel the difference between wood and velvet.]

YES   I feel **bad** today. [*Bad* describes how the subject (which is *I*) feels.]

YES   Susan Chin seems **friendly** and her hair is **curly.** [Adjectives *friendly* and *curly* follow being verbs *seems* and *is.*]

Remember that *ly* words—such as *curly* and *friendly*—aren't necessarily adverbs. In this case, they are adjectives describing the nouns Susan and hair.

Replay 46

Correct the adjective and adverb errors in these sentences, and write C beside the three correct sentences.

EXAMPLE   She works quick. (ly ^)

_____ 1. The fumes from the refinery smell badly today.

_____ 2. Businesspeople should write clear (ly ^) and correct (ly ^).

_____ 3. His sister feels sadly about her loss.

_____ 4. Be sure to do the problems careful (ly ^).

_____ 5. The doctor wrote legible (y ^).

___C___ 6. The pie is excellent today.

_____ 7. The manager should think deep (ly ^) about that subject.

_____ 8. I hope you will treat him fair (ly ^).

_____ 9. This one works as efficient (ly ^) as the new one.

_____ 10. Ms. Teller dances the most graceful (ly ^) of all the dancers.

_____ 11. Our assistant feels badly about the mistake.

___C___ 12. The vegetables taste more delicious than ever.

_____ 13. That which we call a rose,

        By any other name would smell as sweetly—William Shakespeare

_____ 14. They work more quiet (ly ^) today than usual.

_____ 15. Some days they work ~~quieter~~ (more quietly) than other days.

_____ 16. The Infinity runs ~~smoother~~ (more smoothly) than the other cars.

_____ 17. You did satisfactory (ily ^) on all the tests.

_____ 18. She appears more calmly (or calmer) than her sister.

___C___ 19. Of the two Las Vegas hotels, the Luxor has the higher rating.

_____ 20. The music on this boat sounds ~~more~~ loudly (er ^) today.

_____ 21. He feels ~~badder~~ (worse) today than he did yesterday.

_____ 22. The engine in the truck runs (more ^) quieter (ly ^) than the one in the car.

_____ 23. She appears the (most ^) capablest of all the candidates.

_____ 24. Compared with Mr. Beligusi, Mr. Rosenberg presented his case (more ^) conciser (ly ^).

_____ 25. Roberta is the oldest (r ^) of my two children.

_____

Answers are on page 400.

Read
47

MORE SHAKESPEARE/COMPARISONS

Shakespeare compared a woman to a summer's day in these romantic lines:

Shall I compare thee to a summer's day?
Thou art more lovely and more temperate:
Rough winds do shake the darling buds of May.
And summer's lease hath all too short a date.
But thy eternal summer shall not fade.

Though less romantic, we move along to more adjective and adverb comparisons.

FEELING GOOD

The suit fits *well* (adv.) and looks *good* (adj.).

The irregular adjectives **good** and **bad** and the irregular adverbs **well** and **badly** call for extra attention when using them to make comparisons:

	POSITIVE	COMPARATIVE (2)	SUPERLATIVE (3 OR MORE)
Adjective	good	better	best
	bad	worse	worst
Adverb	well	better	best
	badly	worse	worst

> To choose between good or well, decide whether you need an adjective or an adverb. If an adjective is required, choose *good*. If an adverb is required, choose *well*.

Professor Stagnaro wrote a **good** report. [The adjective *good* describes the noun *report*.]

YES The report looks **good.** [The adjective *good* describes the noun *report*; *looks* is a being verb.] It looks **good.** [*Good* describes the pronoun subject *it*; *looks* is a being verb.]

NO He plays the drums **good.** [*Good*, an adjective, incorrectly describes *plays*, which is an action verb; do not use an adjective to describe an action verb.]

YES He plays the drums **well.** [The adverb *well* correctly describes the action verb *plays*.]

YES The drums sound **good.** [*Sound* is a being verb; *good* is an adjective correctly describing the noun subject *drums*.]

NO She **knows** English **good.**

YES She **knows** English **well**. [The adverb *well* tells **how** she **knows**; do not use an adjective to describe a verb; therefore, *good* is incorrect.]

> When referring to health or state of being, either *good* or *well* is correct with the being verb *feel*.

YES He feels **good.** OR He feels **well.** She feels **bad.**

> When being verbs other than *feel* refer to health, use *well*, which becomes an adjective on the subject of health.

YES   They are all **well.** He is not **well.** She seems **well.** [*Good* would refer to their behavior.]

> If health is not the subject, use an adjective after a being verb.

YES   The sauce smells **good.** [This is not about health. *Smells*, a being verb, requires an adjective—in this case, *good*—to modify the subject *sauce*.]

> Use an adverb—never an adjective—to modify an action verb.

Use an adverb to modify the action verb *write*.

They write **badly.** They write **poorly.** They write **neatly.**

 Recap   Circle the correct choice.

1. Greta writes (good/well/either good or well).
2. Do you get along (good/well/either good or well) with people?
3. I hope he feels (good/well/either good or well) today.
4. He spoke (good/well/either good or well) of Ms. Sorenson.
5. Professor Sanneh feels (bad/badly) about losing her motorcycle.

Check your answers on page 400.

> Comparative and superlative forms are easy for *good* or *well:* The comparative is *better*, and the superlative is *best.* Both the following sentences are correct.

YES   Our widgets are **better** than theirs. [The comparative **adjective** *better* modifies the noun *widgets*.]

YES   Our widgets are designed **better** than theirs. [The comparative **adverb** *better** modifies the verb *are designed*.]

Word Power

Memo from Grandma

Good, better, best!
Never let it rest.
Till your good is better,
and your better is best.

Better is both an adjective and an adverb—the comparative form of *good* and *well;* see dictionary.

Real as an adverb is incorrect. Use a true adverb such as *really, very, extremely, exceptionally,* or *truly*; e.g., I'm *truly* sorry. He's *exceptionally* good at that. Use *real* as an adjective meaning "genuine."

Being real smart is not really smart. Speaking English good is not speaking English well.

REALLY SURE?

> Since "real" and "sure" are adjectives, do not use them to describe *other* adjectives. Adjectives can describe nouns or pronouns only.

Only **adverbs** can describe adjectives. Here's a quick test for the adjectives *real* and *sure:*

> If you can substitute **very,** you need an adverb.

NO   She is **real smart.** [The adjective *real* cannot correctly modify the adjective *smart*; switch to an adverb like *really* or *extremely*.]

YES   She is extremely smart.

NO   That report is **sure good.** [The adjective *sure* cannot correctly describe the adjective *good*; replace it with an adverb such as *especially, certainly, exceptionally*.]

YES   That report is exceptionally good.

Change the incorrectly used adjectives to adverbs.

1. I'm ~~sure~~ *very or extremely* happy you decided to buy a new water cooler.
2. We're ~~real~~ *really or especially* disappointed about losing the account.

Check answers on page 400.

Writing for Your Career

Excessive describing words interfere with the effectiveness of your writing. When you see a sentence in context, decide whether to include or eliminate the *really, surely, extremely, especially, very,* or other "intensifying" word. Particularly avoid overuse of *very.* A sentence such as "The presentation is good" sometimes carries more "punch" than "The presentation is really (or very or extraordinarily) good."

What adjective has the five vowels in alphabetic order? HINTS: 1. The first letter is *f* and last letter *s*. 2. This adjective has nine letters; if you add *ly*, it becomes an adverb. 3. It means something like humorous. ANSWER: facetious

> Sometimes adverbs are acceptable without the *ly* in specialized language such as advertisements and road signs.

Buy direct.   Go slow!

This practice is unacceptable for most other types of business communication.

> *Way* and *where* may be parts of compound words, such as *anyway* or *nowhere.* They are often used as adverbs; NEVER add *s* to these words—either in writing or in speech.

NO   anyways, somewheres, everywheres; adding *s* to those words makes one appear uneducated

YES   anyway, anywhere, everywhere, somewhere, nowhere

These sentences provide practice for Read 47 as well as the preceding "Reads" in this chapter. If the sentence is correct, write C. Otherwise circle the adjective or adverb error, and write the correct form in the blank.

EXAMPLE ___C___   The sauce smells good.

really, very	**1.** They are (real) unhappy about the declining profit.
surely, certainly	**2.** We (sure) wish you would participate in the conference.
logically	**3.** Tony Carter writes more (logical) than anyone else here.
better	**4.** Of the two trumpets, the new one sounds (best.)
better	**5.** Ms. Teller speaks (weller) than Mr. Keller.
more widely	**6.** These PCs are (wider) used than the others.
better	**7.** His cooking is (more) better than his brother's.
really well	**8.** I did (real good) in the grammar part.
most poorly	**9.** Of all our stores this one is managed the (poorest.)
C	**10.** The chemicals don't smell as bad today as they did yesterday.
differently	**11.** This office is furnished (different) from the others.
bad	**12.** He is feeling (badly) today.
well, an	**13.** She thinks she did (good) on (a) interview.
kinds, an	**14.** These (kind) of books are not suitable for (a) officer.
C	**15.** The view today is more beautiful than ever before.

Match answers with those on pages 400–401. Then take the Pop Quiz on page 348.

WORD TO THE WISE

Spelling Help Is Here: The *ily* ending always has one *l*—as in adverbs *easily, busily, happily*. The *ally* ending always has two *l*'s—as in adverbs *accidentally, occasionally, officially*.

*or simply delete 1. real   2. sure   8. real.

CHECKPOINT

Place a check in the blank after you review the information and understand it.

ADJECTIVES

_____ **Adjectives** modify, or tell something about, nouns and pronouns.

_____ **Pointing adjectives:** this, that, these, those.
Singular: this kind, **that** type **Plural: these** or **those** kinds or types.

_____ **Articles:** a, an, the. Use *a* before a word or letter beginning with a **consonant sound** and *an* before a word beginning with a **vowel sound**.

_____ **Limiting adjectives:** These limit in the sense of quantity: **42** hats, **several** coats, **few** children, **some** applesauce. Limiting adjectives, explained in Read 19, are usually used correctly and therefore not studied in this section.

_____ **Describing adjectives:** These describe a noun or pronoun:
   adj   n   adj   n   adj   n   pro   adj
blue eyes, black hair, good manners; he is responsible

USING ADJECTIVES ILLOGICALLY: Have students noticed the many ads offering **true facts** and **free gifts?** Those expressions virtually define redundancy! Can a **fact be false?** Can a **gift not be free?**

ADVERBS

_____ **Adverbs** modify, or tell something about, verbs, adjectives, or other adverbs:
eat **carefully** [*Eat* is a **verb** modified by the adverb *carefully.*]
very good [*Good* is an **adjective** modified by the adverb *very.*]
so softly [*Softly* is an **adverb** modified by the adverb *so.*]

_____ **Add ly** Many adverbs, but not all, are adjectives to which *ly* has been added: accidental**ly,** happi**ly,** cheerful**ly;** not all adverbs end with *ly*—for example, *well, often, always.*

THREE DEGREES OF ADJECTIVES AND ADVERBS

_____ **Positive Degree** modifies without making a comparison.
That building is **safe.** [The adjective *safe* modifies *building.*]
The plane is **extremely** safe. [The adverb *extremely* modifies the adjective *safe.*]

_____ The **Comparative Degree** ends in *er* or is preceded by **more** or **less.** Use the comparative degree to compare two only.
This building is **safer** than that one.
He writes **more neatly** than his brother. [The adverb *more neatly* modifies the verb *writes.* The adjective *neater* would be incorrect since an adjective cannot modify a verb—*writes.*]

_____ The **Superlative Degree** ends in *est* or is preceded by **most** or **least.** Use the superlative degree to compare three or more.
This investment is the **most secure** of all those I've made.
Of the three senators, Senator Clinton worked the **hardest.**

WANDERING MODIFIERS: Adjectives and adverbs should be as close to the words they're modifying as possible—to avoid confusing the reader and bringing ridicule to the writer.

ADJECTIVE AND ADVERB TIPS

_____ Negative words are either adjectives or adverbs, but it isn't important to identify them as such. Just be sure to use only one negative word to express one negative idea.
NO He doesn't want **nothing** to do with them. (two negative words: doesn't, nothing)
YES He doesn't want **anything** to do with them.

_____ A describing word following a being verb is an adjective; it modifies the *subject* of the being verb.

Professor Kostner **feels bad** about your low grade. [subject, being verb, adjective]

_____ To modify an action verb, use an adverb.

She plays the saxophone **loudly**. [The adverb *loudly* modifies action verb *plays*; *loud*, an adjective, is incorrect for describing verb *plays*.]

_____ Using an adjective or an adverb where the other is needed is a common error in writing: An adjective modifies a noun or pronoun, and an adverb modifies a verb, an adverb, or an adjective.

NO He is **higher qualified** than the other applicants. [The adjective *higher* incorrectly modifies *qualified*, which is also an adjective.]

YES He is **more highly qualified** than the other applicants. [The adverb *more highly* correctly modifies the adjective *qualified*.]

_____ **Good/Well** Use *good* as an adjective and *well* as an adverb. The comparative for good or well is *better*. The superlative for good or well is *best*.

The steak smells **good.** [*Good*, an adjective, describes the subject *steak*; *smells* is a being verb.]

He writes *well* (not good). Use an adverb (*well*) to modify a verb (*writes*).

Either *good* or *well* is correct after a being verb that refers to health or state of being.

I feel *good*. OR I feel *well*. BUT He feels *bad* (not *badly*) about that.

_____ **Sure** and **Real** *Sure* and *real* are adjectives; don't use them to modify other adjectives. Instead use adverbs *surely, really*, or some other adverb—or omit altogether, which is often the best choice.

NO I'm sure happy that you invited me. I'm real happy that you invited me.

YES I'm surely happy. . . . or I'm extremely happy. . . . or I'm happy. . . .

_____ Never add *s* to *anyway, somewhere*, etc.

Avoid wandering modifiers: Walking down the street in Fort Lauderdale, a banana peel suddenly appeared in front of me.

WORD TO THE WISE

Circle the sentence that refers to mending a sock.

It's darned good.   OR   ⟨It's darned well.⟩

Circle the sentence about the dog that will do poorly in police work or hunting.

The dog smells bad.   ⟨The dog smells badly.⟩

See page 401 for answers.

SPECIAL ASSIGNMENT

Write a short report (about 15 to 20 typed lines) comparing two co-workers who differ from each other in some way. Choose a specific difference to write about—such as appearance, personality, character, intelligence, or ability. Introduce both people and your subject in the opening sentence. Use concise and correct language.

Start immediately with a **topic sentence** to get the reader's interest. A topic sentence gives the main idea (the topic) of the paragraph.

THESE ARE INEFFECTIVE OPENINGS

I am going to write about. . . . [We know you're going to write; just get started.]

This is a comparison of two people I know who differ from each other. [Who cares?? Tell me who they are and what they do so that I'll want to know more about them. Be specific.]

THESE ARE EFFECTIVE OPENINGS

Amy Lopez and Josh Stern are successful department managers at Magnasoft, Inc.; however, their leadership styles differ considerably.

Although Professor X and Professor Y are both excellent instructors, their teaching styles are quite different.

Latisha is a perfect example of how a supervisor should dress for the office, while Carmen is just the opposite.

BODY

Continue with sentences supporting your opening statement—or topic sentence. Describe the people with interesting, accurate, and precise words. Anything that doesn't support the topic sentence does not belong in the report.

CONCLUSION

Conclude with a summarizing sentence. Avoid introducing new information in the closing.

Submit your report on _____. (date)

Writing for Your Career

Instead of writing less or least before an adverb or adjective, you can sometimes choose an antonym (opposite). For example, you could replace *less easily, less easy,* or *least easy* with *harder* or *hardest* or *more/most difficult.*

PROOFREADING FOR CAREERS

Find and correct spelling, capitalization, and grammar errors as well as abbreviations, numbers, and typographical errors. Avoid unnecessary abbreviations. If you keyboard the letter, use your spelling checker **before** your final proofreading. The checker will probably not pick up wrong words that are spelled correctly; for example, *be/bee* are both spelled right. See "Mini Reference Manual" beginning on page 370 for number, abbreviation, and capitalization information. Check the entire letter, not just the body. Except to correct errors, do not change wording. The punctuation is already correct. Use a pen or pencil as a pointer and mark errors as you go along.

IDAHO POTATO GROWERS EXCHANGE
COMMONTATER DRIVE
BOISE, IDAHO 83700

(208) 462–3200   FAX (208) 462–3232

March 6, 2005

Ms. Maryann Lamb
Editor, Health Magazine
1300 5th St., WA
Wenatchee, ~~Wash~~. 98801

Dear Ms. ~~Maryann~~ [Lamb]:

In response to your Feb. 27 letter, we are pleased to share the following facts about potatos with you: Potatos ~~is~~ [are] composed of 78 percent water, about 18 per cent carbohydrate, and about 2 per cent protein. ~~Their~~ [They're] a good source of iron and vitamines.

It is an errogeneous idea that potatoes ~~is~~ [are] fatening; nutritionists recognize that all foods eaten in excess are fattening. Actualy potatoes aren't ~~no~~ [any] more fattening then most items in a typical american diet. Potatoes alone ~~is more~~ [are] lower in calories per pound than bread and many other foods. When ~~fryed~~ [fried] or served with butter and sour cream, however, the total calorie intake is high.

Potatoes ~~has~~ [have] a high satiety value and gives a full feeling that checks overeating. Sodium content is so lowly that the American Heart Association reccomends potatoes for low-salt diets; in addition, they taste ~~well~~ [good].

The enclosed pamflet was ~~wrote~~ [written] by Nancy Borden, who lives in Southwick, Idaho, an area known for ~~their~~ [its] potato industry. Also enclosed are ~~too~~ [two] recipes, but the Potatoes Granada recipe would be ~~best~~ [better] for your publication.

Please let us know if we can be of further help to you.

Very truely yours,

Charlene W. Martindale
Consumer Relations

lr

enclosures

After checking with the corrected version of this letter, evaluate your skill. Excellent____ Good___ Fair___ Other___

PRACTICE QUIZ

www.prenhall.com/smith

Take the Practice Quiz as though it were a real test. In the blank write the letter that identifies the correct answer. Use the dictionary when in doubt.

EXAMPLE _____b_____ (a) A (b) An eager history professor assigned 200 pages.

_____a_____ **1.** She is the (a) less (b) least efficient of the two clerks.

_____a_____ **2.** (a) A (b) An union official would probably

_____b_____ **3.** refuse (a) a (b) an hourly wage.

_____b_____ **4.** These (a) type (b) types of properties are advertised

_____c_____ **5.** in (a) this here (b) that there (c) either this or that newspaper.

_____b_____ **6.** She doesn't want to lose (a) a (b) an $11 commission.

_____b_____ **7.** If he doesn't understand the work, he will get (a) a (b) an F on the test.

_____a_____ **8.** (a) A (b) An European businessperson would probably

_____b_____ **9.** consider this award to be (a) a (b) an honor.

_____a_____ **10.** Since both brands are good, order the (a) less (b) least expensive one.

_____a_____ **11.** Which is (a) easier (b) easyer (c) easiest (d) easyest for you to do, a graph or a chart?

_____b_____ **12.** I (a) can't (b) can hardly believe what I saw.

_____c_____ **13.** Although Mr. Shue and Ms. Farr are both skilled at keyboarding, Mr. Shue is (a) more faster (b) most faster (c) faster (d) fastest.

_____b_____ **14.** The class was asked to sit (a) quiet (b) quietly and read.

_____e_____ **15.** I hope you (a) won't do no more (b) won't do any more (c) will not do no more (d) will not do any more (e) either b or d work on that project.

_____b_____ **16.** This is the (a) newer (b) newest (c) most new of the six computers in our office.

_____b_____ **17.** Mr. Young spoke at Southwestern College (a) brief (b) briefly and to the point.

_____a_____ **18.** The sewage treatment plants don't smell as (a) bad (b) badly today.

_____a_____ **19.** When we evaluated our three facilities, we found this one is run (a) most poorly (b) most poorest (c) more poorly.

_____a_____ **20.** He seems (a) worse (b) more badly (c) badder this week.

_____b_____ **21.** Of the four letter styles, full block seems (a) better (b) best (c) more better for our correspondence.

_____b_____ **22.** The file is missing, but we know it must be (a) somewheres (b) somewhere (c) either a or b (d) neither a nor b in this office.

_____c_____ **23.** My sister is in the hospital but feels (a) good (b) well (c) either good or well today.

_____a_____ **24.** The software-design team worked (a) really well (b) really good (c) either good or well (d) real well (e) real good together.

_____d_____ **25.** We would (a) sure (b) really (c) certainly (d) b or c like to meet him.

THE GLOBAL MARKETPLACE—DENMARK

Danish	Spell	Say	Currency
GREETING	Goodag	Gu.DAYH	Dansk Krone (DKK)
DEPARTING	Farvel	Far.VEL	
PLEASE	Versavenlig	Ver.sah.VEN lee	
THANK YOU	Tak	Tak	
I DON'T UNDERSTAND	Jeg forstaar ikke	Yeg forstar ikee	

Courtesy of Alice Larsen Smith

The Taming of the Apostrophe

9

Avoid Apostrophe Catastrophe

> "No one worth possessing can be quite possessed."
> —Sara Teasdale

After Completing Chapter 9, You Will

> Place apostrophes where they belong and not where they don't belong.

The shrew (a scolding, nagging woman) in Shakespeare's *Taming of the Shrew* didn't know "her place." She didn't know how to behave as a dutiful wife until her husband "tamed" her—now, of course, an unacceptable concept. **Our** concern, however, is taming **apostrophes** (the punctuation mark showing *possession*). Untamed apostrophes appear in the wrong places and are missing from the right places. The result is unclear or distracting writing.

Use this opportunity to learn correct use of apostrophes! Since this little mark is all you need think about in Chapter 9, you'll become an "apostrophe expert," which means putting them exactly where they belong and never where they don't belong. Best of all, you'll never have an apostrophe catastrophe!

Word Power

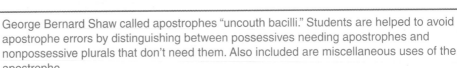

WHICH BUTLER WOULD BE FIRED FOR EXTREME RUDENESS? No.1 or No. 2? ____1____

1. The butler stood at the doorway and called the guests names.
2. The butler stood at the doorway and called the guests' names.

George Bernard Shaw called apostrophes "uncouth bacilli." Students are helped to avoid apostrophe errors by distinguishing between possessives needing apostrophes and nonpossessive plurals that don't need them. Also included are miscellaneous uses of the apostrophe.

Chapter 9's apostrophes are kept far from Chapter 5's plural-noun practice. You could stress that 's does not make a noun plural, except for those discussed in Read 50, "More Apostrophes."

Read 48

THE UBIQUITOUS "S"

Chapter 9 helps students speaking and writing in dialects in which *s* isn't pronounced in plurals or possessives, as well as students whose primary language doesn't include possessive forms (e.g., Spanish, French, Italian). To translate *Jose's house* into Spanish, we say *la casa de José.*

However, native English speakers also make apostrophe errors. Discouraged by misplaced apostrophes, some British companies instruct employees to omit all apostrophes. Since this isn't true in the United States or Canada, students need to learn correct apostrophe usage.

It's here again. But then that's what *ubiquitous* means—seeming to be everywhere at the same time. The letter *s* begins more words than any other letter in the alphabet and is the second most frequently used consonant (*t* is the first).

In Chapter 5 you worked with the ubiquitous **s** while it played the original numbers game—**plurals.** Because the **s** also makes nouns **possessive,** be careful to avoid confusing plurals with possessives.

> For possessive nouns, use an apostrophe (') and an *s*. BUT Do not use an apostrophe in just plain plural nouns that are not possessive.

Conscientious students, aware of the need for apostrophes, sometimes sprinkle them like raindrops wherever an **s** happens to end a word. In this chapter, your apostrophes will be tamed, and you will *never* be an apostrophe sprinkler. However, you *will* use apostrophes where needed—in possessive nouns and contractions.

DISTINGUISH BETWEEN PLURALS AND POSSESSIVES
Possessive Nouns

> Possessive nouns show the relationship between one noun and another noun. The first noun shows who or what possesses; the second shows who or what is possessed. The relationship is made clear by the use of an *s* and an apostrophe in the first noun. (Don't add an apostrophe to the second noun just because it happens to ends with *s*.)

Various possessive relationships for nouns are shown below:

PERSONAL   Latisha's brother, the auditor's friend, teachers' salaries

OWNERSHIP   women's hats, girls' pearls, boy's toys

PLACE OF ORIGIN   Springfield's population, Tennessee's weather

AUTHORSHIP   representatives' speeches, chair's report

TYPE OR KIND   children's clothes, soldiers' uniforms

TIME   two years' delay, a day's vacation

Sometimes the apostrophe is before the *s*, and other times it is after.

A possessive noun can replace a prepositional phrase.

Prepositional phrases (word groups beginning with a preposition) may create the same meaning as the possessive expression, but possessives are usually more concise.

> brother **of Latisha,** friend **of the auditor,** salaries **of the teachers**
> hats **for women,** pearls purchased **by girls,** toys belonging **to children**
> population **of Springfield,** weather **in Tennessee,** wife **of Charles**
> speeches given **by senators,** report **from the director,** paintings **by Monet**
> uniforms **for soldiers,** vacation **of a day**

Writing for Your Career

When a possessive sounds natural, use it instead of a prepositional phrase. If the result seems awkward or changes the meaning, use a prepositional phrase. For example, **the interior of the house** sounds better than **the house's interior.** In "The Star Spangled Banner," however, **the dawn's early light** is better than **the early light of the dawn.**

The old rule about not making inanimate objects possessive has too many exceptions (many of which appear in this chapter's examples) to be valid. Possessives enable us to communicate more concisely. Deciding whether to use a possessive depends on the *sentence's rhythm,* a concept perceived even by students with weak English skills. (See the Writing for Your Career box above.)

The following sentences show how to replace prepositional phrases with possessive nouns for more concise or clearer writing. Notice it's always the **first** of the two nouns—the possessor (or owner)—that gets the apostrophe.

PREPOSITIONAL PHRASES (WITHOUT APOSTROPHES)	POSSESSIVE NOUNS (WITH APOSTROPHES)
The records prepared **by the accountants** were taken to the **office of the secretary.**	**The accountants' records** were taken to **the secretary's office.**
Is the **population of Nevada** smaller than the **population of Arizona?**	Is **Nevada's population** smaller than **Arizona's?**
Clothes for children are on this floor, and **clothes for infants** are on the fourth floor.	**Children's clothes** are on this floor, and **infants' clothes** are on the fourth floor.
A delay of two hours would be disastrous.	**Two hours' delay** would be disastrous.

In the sentences above about Nevada and Arizona, "population" is understood but not stated after "Arizona's," thus avoiding needless repetition. To find out whether to use an apostrophe, reverse the order of the nouns and put "of" between them: **Twileens brother—brother of Twileen**. If reversing the nouns and inserting "of" delivers the intended meaning, then the first noun needs an apostrophe: Twileen**'s** brother.

Recap

Circle the possessor and the possessed. In the blank write the "understood" word for item 2. Use insert mark (^) to show location of "understood" word.

1. "Five minutes' planning might save an hour's work on New Year's Day," said George's mother to Twileen's father.

2. The purchase of Alaska was called "Seward's Folly," but history proved the foolishness was his critics' rather than Secretary of State William Seward's ^foolishness _____ .

Check your answers on page 401.

Plural Nouns

> **Just because a plural noun ends in *s* doesn't mean it needs an apostrophe. Use the apostrophe only for a possessive noun.**

Here are some sentences with plural nouns ending in *s*, but no possessive connection is shown. Apostrophes would be incorrect:

Plural Nouns, nonpossessive: The **novels** of recent **writers** show the social **changes** of our **times**.

The **brothers** are **partners** in software consulting **businesses.**

The **Browns** each play a few **instruments** in several **orchestras.**

The **Joneses** wear straw **hats** on **holidays.**

Your **records** indicate the **Lees** made several **errors.**

The **Nguyens** own **factories** in both small **towns.**

WORD TO THE WISE

Study the difference in meaning between the following two sentences. What part of speech is *work* in each sentence? 1. ___noun___ 2. ___verb___

1. We would like you to see our students' work.

2. We would like you to see our students work.

Check your answers on page 401.

Replay 48

Some nouns in the following sentences are possessive, but the apostrophes are left out. Underline the noun that is "possessed" or "owned." Then insert the apostrophe in the possessive noun. Write *C* for correct if a sentence doesn't require an apostrophe.

EXAMPLE   The artist^'s <u>books</u> were left in Mr. Fox^'s <u>office</u>.

___C___   1. The Byrneses have sent four altos to try out for the operas.

_____   2. Our editor^'s <u>stories</u> please his readers greatly.

Extra Practice— possessive nouns: We toured **Amsterdam**'s night spots. The **bride**'s flowers were pink **carnations** from her **sister**'s garden. We ate two **dollars**' worth of fudge.

_____C_____

_____C_____

_____

_____

_____

_____

_____

_____

_____

_____C_____

_____

_____

_____

_____

3. His brothers-in-law manage the offices.

4. His brother-in-law's manager has been transferred to Guam.

5. The Schwartzes own property in the swamp lands of Brazil.

6. The attorneys' offices are in new buildings.

7. South Dakota's resources are listed in the back pages of two *Almanacs.*

8. Men's and women's clothes are on sale in all our stores today.

9. Have you shipped Ms. Lopez's orders yet?

10. One of the film industry's most talented directors, Steven Spielberg, lectures at UCLA.

11. The crew's strength was spent in useless maneuvers. (one crew)

12. California's gold mines were less profitable than its orange groves.

13. These vineyards supply more than 75 percent of this nation's wine and raisins.

14. Mattel and Lego are among the world's largest toy manufacturers.

15. The former Claremont Men's College is now simply Claremont College.

16. Oral communications on the job include making introductions, giving directions, and greeting visitors.

17. Tom's new book was used twice before the errors were found.

18. Several hours' work took more than three days to do.

19. Important to James Cash Penney's success was how he treated employees.

20. Barbie's worldwide fame is probably Mattel's greatest success.

Please check your answers on page 401.

Read 49

BEFORE OR AFTER?

Several hundred years ago possession was expressed by using the pronoun "his" after the first noun. Instead of saying *the clerk's desk*, they said *the clerk his desk.* If you say this old-fashioned possessive form fast, you hardly hear the first two letters of "his." Therefore, people of that day, when spelling was much more individualized than it is today, began spelling the expression *the clerk s desk.* Since the apostrophe had previously been used to show omission of one or more letters, writers began to use this mark to show that the first two letters had been left off "his." That's how we ended up with the modern form, *the clerk's desk.*

If more than one clerk shared the desk, the original wording would have been *the clerks their desk.* This, in turn, was shortened to *the clerks' desk*—with the apostrophe after the *s* to show that "their" was left out after the plural word *clerks.*

This method still works for determining whether to place the apostrophe before or after the *s:* Once you've decided the noun is possessive, see whether it's singular or plural. If it's singular, put the apostrophe where *his* would have been in the 1700s. If it's plural, insert the apostrophe where *their* would have fit.

Although the *his* and *their* trick still works, here are three modern rules to help you form possessives correctly.

MAKING SINGULAR NOUNS POSSESSIVE

> Add *'s* to make a SINGULAR noun possessive.

The toothbrushes are on the **dentist's** desk. (one dentist)

We do not believe the **witness's** testimony. (one witness)

Please send the **technician's** reports to the lab. (one technician)

The **city's** need for extra funds has not been met. (one city)

Mr. Hawkes's office is at Kent. (one Mr. Hawkes)

Exception

> If adding apostrophe *s* to a *singular* noun makes the word hard to say, just add an apostrophe. Choose this exception only for a singular *proper* noun with two or more syllables and ending with an *s* sound.

President **Adams'** wife was one of the earliest **feminists**.

Mr. Stettinius' wife was not active in Washington's social life.

Joyce Simmons' office is on the main floor.

Moses' journey is described in the *Bible*.

Steinmetz' discoveries made him famous.

MAKING COMPOUND SINGULAR NOUNS POSSESSIVE

> If a compound singular noun (made up of two or more words) is possessive, add an apostrophe and *s* to the end.

We responded to the **commander in chief's** salute.

All words, whether individual words or compound nouns, form possessives at the end, not somewhere in the middle:

The reporters were amused by the **editor in chief's** remarks.

My **brother-in-law's** appetite amazes me.

Insert six apostrophes in items 1 and 2.

1. Mr. Smiths and Ms. Perkins assistants will tour New York Citys tallest buildings during a weeks vacation. (Her name is Pat Perkins.)

2. My son-in-laws business is as successful as my sons. (one son)

Check answers on page 401.

MAKING PLURAL NOUNS ENDING IN *S* POSSESSIVE

Add only an apostrophe to make a PLURAL noun that ends in *s* possessive.

One-Word Possessives

Several **weeks'** work, **uncles'** wives, **Fritzes'** assistant (plural possessors)

The **dentists'** desks are on wheels. (more than one dentist)

The **witnesses'** statements are false. (more than one witness)

Send the **technicians'** report to the lab. (more than one technician)

Three **days'** work is needed to complete the job. (more than one day)

The **Hawkeses'** office is in Rock Springs. (more than one person named Hawkes)

Exceptions

There are none! The exception rule shown under the "Singular Nouns" heading is for **singular** possessives only—*not* for plurals.

MAKING PLURAL NOUNS THAT DON'T END IN *S* POSSESSIVE

The **salespeople's** commissions (plural possessor)

If a plural possessive noun does not end in *s*, add *'s* to make it possessive.

Notice that the underlined nouns are plural but don't end in *s*.

<u>Men</u>'s suits are very colorful this year.

The trustee controls the <u>children</u>'s assets.

The <u>alumni</u>'s contributions to the <u>Women</u>'s Fund were small.

Compound Plural Possessives

Some compound words become **plural** by adding *s* to the first word—like *sons-in-law*. This plural word does not end in *s* but in *w*.

If a plural compound noun is possessive, first spell the plural. Then add the *'s* to the end of the compound noun.

Newspaper reporters listened attentively to five **editors in chief's** speeches. [The *s* after editor makes the compound expression plural. The *'s* after *chief* makes it possessive.]

My **brothers-in-law's** appetites amaze me. [plural possessive]

WORD TO THE WISE

Before placing an apostrophe, decide whether the **possessive** noun is singular or plural. Then make certain this noun is correctly and completely spelled before making it possessive.

The ladies' purses—plural possessive noun

The lady's purses—singular possessive noun [one lady with more than one purse]

It doesn't matter whether the second noun—the possessed—is singular or plural. Just consider the first—the possessor—and decide whether it's singular or plural.

Recap

Please insert *s*, *es*, and apostrophes where needed. Correct item 1 so that you can tell I have more than one brother; then make Martinez plural. Item 2 needs three apostrophes. In item 3 two of his sons' wives are business partners.

1. The meeting was held at my brother^{s'}office, and both the Martinez^{es} attended.

2. The old saying, "Women^{'s}work is never done," is harmful to women^{'s} and men^{'s}roles in modern society.

3. His daughter^s-in-law^{'s}business is bankrupt.

Check your answers on page 401.

SPECIAL CASE

> **If the name of an organization includes a possessive form, just see how the organization writes it, and do the same.**

Vons (a supermarket chain) has no apostrophe.

Macy's (the department store chain) has an apostrophe.

No Exceptions

> **Men's, women's, children's, man's, woman's, and child's (all possessive nouns) are always written with an ' before the *s*. No exceptions!**

Writing for Your Career

If a possessive sounds clumsy, reword the sentence to avoid the need for a possessive form.

Instead of	My **brothers-in-law's** huge appetites amaze me. [correct but clumsy]
Rephrase	I am amazed that my **brothers-in-law** have such huge appetites.
or	I am amazed that my **husband's brothers** have such huge appetites.

Replay 49

A. Write the singular possessive, the plural, and the plural possessive.

SINGULAR	SINGULAR POSSESSIVE	PLURAL	PLURAL POSSESSIVE
EXAMPLE lawyer	lawyer's	lawyers	lawyers'
1. representative	representative's	representatives	representatives'
2. week	week's	weeks	weeks'
3. witness	witness's	witnesses	witnesses'

4. James	James's	Jameses	Jameses'
5. country	country's	countries	countries'
6. Filipino	Filipino's	Filipinos	Filipinos'
7. man	man's	men	men's
8. Asian	Asian's	Asians	Asians'
9. wife	wife's	wives	wives'
10. father-in-law	father-in-law's	fathers-in-law	fathers-in-law's
11. congresswoman	congresswoman's	congresswomen	congresswomen's
12. family	family's	families	families'
13. Webster	Webster's	Websters	Websters'
14. hour	hour's	hours	hours'
15. Wolf	Wolf's	Wolfs	Wolfs'
16. wolf	wolf's	wolves	wolves'
17. organization	organization's	organizations	organizations'
18. boss	boss's	bosses	bosses'
19. woman	woman's	women	women's
20. child	child's	children	children's

WORD TO THE WISE

Regular nouns become plural by adding *s* or *es*, not by adding *'s*.

No   The Smith's and the Jones's are server's at Trendy's.

Yes   The Smiths and the Joneses are servers at Trendy's

B.   Insert an apostrophe and an *s* or just an *s* where needed, or write C for correct. Show clearly whether an apostrophe is before or after the *s*. Make any necessary spelling changes to nouns. Do not change verb forms. Read for sense before correcting:

EXAMPLE   Mr. Williams' book is the manager's choice.

_____C_____   1. The Columbuses never dreamed Chris would become so famous.

_____   2. Two years' interest is due on the note.

_____   3. Health is a person's most valuable possession.

_____   4. Men's fashions change almost as quickly as women's.

_____   5. Be prepared to come at a minute's notice.

_____   6. Mr. Childress's signature was needed two days ago.

_____   7. The store was having a sale on ladies' coats.

_____   8. The Goldsteins of West Palm Beach will join us in two days

_____ 9. Brunswick population has increased during the past five year.

_____ 10. Mr. Jenkins desk is to your right.

_____ 11. Several coach reports included details about their player health.

_____ 12. We studied Keats poetry in our literature class.

_____ 13. Montreal and Quebec are in Mr. Hendrix territory.

_____ 14. Three Marx brothers film were shown on TV.

_____ 15. The butler stood at the doorway and called all the guests name.

Check answers on pages 401–402.

Read 50

MORE APOSTROPHES

CONTRACTIONS

> Contractions are words shortened by removing one or more letters, and replacing them with apostrophes.

Writing for Your Career

Successful written communication for the workplace has a natural and conversational style. One technique to achieve naturalness is using contractions. Do not, however, contract *would* as in *I'd* or *we'd*, etc., although these contractions are fine for conversation.

Here are some words frequently contracted in business writing. Look at them carefully, and remember that the apostrophe belongs where the missing letter or letters would have been.

is not = isn't	are not = aren't	was not = wasn't
were not = weren't	have not = haven't	would not = wouldn't
should not = shouldn't	could not = couldn't	will not = won't
cannot = can't	has not = hasn't	do not = don't
does not = doesn't	of the clock = o'clock	I have = I've
you have = you've	we have = we've	I shall or will = I'll
we shall or will = we'll	you will = you'll	he is = he's
I am = I'm	we are = we're	you are = you're
that is = that's	they are = they're	it is = it's
what is = what's	who is = who's	she is = she's

POSSESSIVES OR CONTRACTIONS

> *One*, *body*, and *thing* pronouns often end with *'s* and may be used as either possessives or contractions, depending on the meaning desired.

anyone's	someone's	no one's	everyone's
anything's	something's	nothing's	everything's
anybody's	somebody's	nobody's	everybody's

CONTRACTION   The apostrophe represents the missing *i* in the word *is*:

Everybody's going to that party. *No one's* at home. *Who's there?*

POSSESSIVE   Now the apostrophe makes the following words possessive:

Everybody's coats are in the closet. *No one's* home is available for the party, but *someone's* here now.

The preceding possessives and contractions are correct and appropriate in conversation and in most business writing. However, in legal or other formal documents, avoid contractions. Possessives are always appropriate. Use other types of contractions such as *nat'l* for *national* or *sec'y* for *secretary* only in e-mails, informal notes, memos, or tables where saving space is important.

Plural Abbreviations

> An apostrophe is unnecessary to form the plural of capital letter abbreviations, and they are often written without periods—unless a misunderstanding might result.

All CODs should be sent to my office.

There are two YWCAs in Toledo. (See Appendix D for abbreviations often used in the workplace.)

> Add *'s* to make lowercase abbreviations plural if they might be misread without an apostrophe.

NO   Please be sure to dot your is.

YES   Please be sure to dot your i's.

NO   They asked Joey to put on his pjs.

YES   They asked Joey to put on his pj's.

NO   Too many *etc.s* usually mean the writer isn't sure of the facts.

YES   Too many *etc.'s* usually mean the writer isn't sure of the facts.

> Plural abbreviations without apostrophes are suitable for such documents as specifications or invoices. However, within letters or reports, spell out "quantity words."

5 yds., 6 gals., 7 ctns.

Thank you for shipping five gallons of Ubet Chocolate Syrup in seven cartons.

We'll use it to make 250 quarts of chocolate egg cream drinks for the party.

> **Apostrophes are unnecessary for stating academic grades unless required for consistency.**

Her college transcript showed two Fs in math classes.

Three students earned A's in biology but F's in anatomy. [*As* without an apostrophe might cause confusion. The apostrophe is used in F's for consistency.]

Plural Numbers and Words

> **Do not add an apostrophe to form the plural of numbers or words.**

The temperature in New Brunswick is in the 70s.

The young child wrote 3s backwards.

Please omit all *therefores*. (plural of word)

POSSESSIVE ABBREVIATIONS

> **Use an apostrophe in possessive abbreviations, just as with any other noun.**

The AMA's position is clear. [singular possessive]

Our R.N.s' uniforms are yellow. [plural possessive; make *R.N.* plural by adding *s*; then make it possessive by adding an apostrophe to a plural noun ending in *s*.]

MISCELLANEOUS

> **As a symbol, the apostrophe has several meanings.**

FEET AND MINUTES 4′ means either 4 feet or 4 minutes depending on the context.

CENTURY Although ′06 means 2006, avoid this style in business writing, except sometimes for class graduation years or for decades.

The Class of ′95 will hold its next reunion at the Fritz Karlton.

In the 20th century more cultural changes occurred during the ′60s than in any other decade.

Ordinarily use the full number for the century.

In 2010 our company will celebrate 25 years in business.

QUOTATION WITHIN A QUOTATION The candidate said, "It was Abraham Lincoln who spoke of 'government of the people, by the people, and for the people'; and that is also my credo."

Apostrophes may be used to make words plural as in "The suitor was discouraged by all her no's and not enough yes's," if the words would be difficult to read without the apostrophes.

Replay 50

Insert apostrophes (or numbers) where needed. Write *C* beside the correct sentences. Draw a line through incorrectly used apostrophes.

EXAMPLE He's not aware that Winston Churchill was in his 80s in ′54. [19]

_____ **1.** I couldnt meet you at five oclock.

_____ **2.** Please be sure to dot your is and cross your ts.

_C___ **3.** MBAs are given preference when we recruit mid- or top-level management.

_C___ **4.** They experienced many ups and downs before achieving their astounding success.

_____ **5.** We believe the next CFOs convention will be held in ‚'07. [2007]

_____ **6.** Don‚t use too many *ands* and *buts* in your writing.

_C___ **7.** Several M.D.s and R.N.s usually have lunch here.

_____ **8.** In tables or charts, it‚s all right to use abbreviations like yd's., ft., or amt's.

_____ **9.** During the 19th century era known as "The Gay '90s," worker‚s suffered while the wealthy held lavish parties.

_C___ **10.** Three CPAs have offices in this building.

_____ **11.** If you cross your Is, they‚ll look like sloppy t‚s.

_____ **12.** "Won‚t you please marry me?" asked George for the 22nd time.

_____ **13.** Only five As were recorded.

_____ **14.** In the early 1900s, women‚s dresses were long.

_____ **15.** Couldn‚t you make your 9s look less like 7s?

Answers are on page 402. Take the Pop Quiz on page 349.

CHECKPOINT

Tame apostrophes, like pets, obey the rules and are in the right place at the right time. After checking off Chapter 9 principles summarized below, your apostrophes will be ready for company—any company you work for. Place a check beside each principle you're sure of. If your conscience will not permit the check mark, review the Read explanations.

_____ Use apostrophes in singular and plural possessive nouns.

Maria's office and the managers' offices are always open.

_____ If a plural noun is not possessive, don't use an apostrophe.

The brokers work from 6 a.m. to 3 p.m.—the New York Stock Exchange hours.

_____ A possessive noun ends with *s* and precedes another noun or "an understood" noun.

Charley's aunt is eccentric.

Of all the aunts here, Charley's is the most eccentric. [*aunt* is understood after *Charley's*]

_____ To test for possessives, reverse the order of the two nouns and insert **of** between them.

aunt **of** *Charley* is *Charley's aunt*

_____ Make a singular noun possessive by adding *'s*.

Shinji's job pays well.

_____ If adding *'s* to a **singular proper noun with two or more syllables** makes the word hard to pronounce, add an apostrophe only.

Mr. Watkins' home is near mine. [instead of *Watkins's* home]

BUT Morris's home is far away. [*Morris's* doesn't sound awkward.]

_____ A plural noun that ends in *s* becomes possessive by adding an apostrophe only.

Ladies' shoe styles sometimes result in injuries to their feet.

_____ A plural noun that does **not** end in *s* becomes possessive by adding 's.

Please don't buy children's toys that encourage violence.

_____ Insert an apostrophe within a contraction at the exact place where letters are left out.

A toy gun **isn't** a good gift for a child. [apostrophe where *o* would be]

_____ To form the plural of a lowercase abbreviation or of a lowercase single letter, use an apostrophe.

Mind your p's and q's, and don't use too many e.g.'s or i.e.'s.

_____ Avoid apostrophes for plurals of numbers, words, capitalized letters, or capitalized abbreviations (unless confusion might result).

The VCRs are in the large meeting hall for the use of the PhDs.

This quarter he received two A's, two B's, and three 100's on his exams. (To avoid confusion, add apostrophes.)

_____ The apostrophe symbol is sometimes used for feet, minutes, and centuries. You can also use the apostrophe as single quotation marks to enclose a quotation inside another quote.

_____ Now is a good time to review possessive **pronouns** in Read 31. Review the difference between possessive pronouns and contractions containing pronouns; for example, *you're/your, it's/its, they're/their, who's/whose.*

 SPECIAL ASSIGNMENT

A. In Replay 48 you underlined the words that are "possessed" or "owned." List the possessors below and insert the apostrophes. If no apostrophe is required in a sentence, just write "none" in the blank.

EXAMPLE   See Replay 48 Example. artist's,   Fox's

1. _____ none _____
2. _____ editor's _____
3. _____ none _____
4. _____ brother-in-law's _____
5. _____ none _____
6. _____ attorneys' _____
7. _____ South Dakota's _____
8. _____ Men's   women's _____
9. _____ Lopez's or Lopez' _____
10. _____ industry's _____

11. _____ crew's _____
12. _____ California's _____
13. _____ nation's _____
14. _____ world's _____
15. _____ Men's or Mens (whichever the college chooses) _____
16. _____ none _____
17. _____ Tom's _____
18. _____ hours' _____
19. _____ Penney's _____
20. _____ Barbie's   Mattel's _____

B. Write sentences in which you shorten these phrases by using a possessive noun or a contraction.

EXAMPLE   son of Mr. Ames   _Mr. Ames's son is the auditor_. Possible answers

1. books of George _____ George's books are overdue at the library. _____
2. wife of Mr. Adams _____ Mr. Adams' wife is my dermatologist. _____
3. vacation of a week _____ A week's vacation is insufficient for a grand tour of Europe. _____
4. home of the Adamses _____ The Adamses' home is near everyone's offices. _____
5. store of my sisters _____ My sisters' store is on Fifth Avenue. _____
6. name of the server _____ A restaurant server's name is not "Honey" or "Garcon." _____
7. problems of the members _____ We tried to solve the members' problems. _____
8. commissions of the salespeople _____ The salespeople's commissions are computed daily. _____
9. work of two years _____ After two years' work, the plans for the subway were canceled. _____
10. studio of my mother-in-law _____ My mother-in-law's studio is in Paris. _____
11. words of Moses _____ Charlton Heston quoted Moses' words. _____
12. streets of Dallas _____ Dallas's streets or Dallas' streets are safe. _____
13. report of the auditor _____ Please give me the auditor's report. _____
14. notice of ten minutes _____ Ten minutes' notice is not enough. _____
15. expense accounts of the supervisors _____ Supervisors' expense accounts are audited. _____

C. Correct the following sentence:

Student's who put apostrophe's into plain plural's will receive shock's when they get grade's on the examination's in a few day's. Remove _____ all _____ apostrophes from this sentence.

_____

Please complete this assignment by _____. (date)

PROOFREADING FOR CAREERS

Add eight apostrophes to this short essay, and correct all other errors.

WHISTLER'S MOTHER

In the worlds most famous museum, the louvre in Paris, hangs a painting by Americas celebrated artist, James McNeill Whistler. This paintings formal title is "An Arrangement in Gray and Black," but it is better known by the simple name "Whistlers Mother."

Studies have been made to explain this portraits almost universal appeal, but what criterias can an art critic use to judge a painting? Critics are not like scientists. They cannot set up controlled experiments in which a number of stimulus are shot into subjects and data collected on the subjects reactions. No, an art critic relys on inner emotions and sensitivity when analyzing a painting. Analyses of a painting is very personal.

When you visit Paris and look at "Whistler's Mother," what will you see? Will you, like most of us, be left wondering about the source of this portrait's greatness?

After verifying your corrections, choose an adjective to evaluate your proofreading and English skill today: _____ Please check the appropriate blanks. Do you need to review: apostrophes in Chapter 9? _____, singulars and plurals in Chapter 5? _____, capitalization in Chapter 5? _____, using the dictionary in Chapter 3? _____ I don't need to review anything! _____ Really? _____

Word Power

Doormats in Suburbia

A stroll through a middle-class suburb reveals incorrect apostrophes on doormats and mailboxes and even in elaborate wrought-iron signs. Circle the unneeded apostrophes and correct the names identifying these homeowners:

The Smith's   The Gomez's   The Anderson's   The Fox'es

To form the plural of nonpossessive proper nouns, merely add s or es:

The Smiths   The Gomezes   The Andersons   The Foxes

Practice Quiz

www.prenhall.com/smith

Insert apostrophes and s's exactly where they belong, delete unneeded apostrophes, and fix incorrect plurals and contractions. Write C for correct to the left of sentences needing no change.

EXAMPLE   The architect's report includes the data regarding the workers' cafeteria. [one architect]

_____   1. The dean's view is that your attitude needs drastic changes. [one dean]

_____   2. Hiring 1,000 employees for the merger with Fisher-Price will add to Mattel's 22,000-member workforce.

_____   3. Many Americans take their mothers out to restaurants on Mother's Day.

_____   4. Yesterday's techniques cannot succeed in today's marketplace.

_____   5. Intel Corp. has two weeks to accept or reject our company's offer.

_____   6. My sister's-in-law's bookkeeper completed three years of college. [Two of my brothers' wives are partners and share a bookkeeper.]

_____   7. Yamada and Jones is one of the city's finest law firms.

_____   8. Emily Jones's reputation as a criminal lawyer is excellent.

_____   9. The MIS personnel have the CEO's attention in this company.

_____   10. You're going to hire that mechanic, aren't you?

___C___   11. Ship 5,000 yds. of cement to the PTA's president.

_____   12. During the 1990's I bought several VCRs for the convention facilities.

202   Chapter 9 | The Taming of the Apostrophe

_____C_____ **13.** Within the last 50 years, many famous pieces of computer history have been discarded in landfills across the world.

_____C_____ **14.** These losses are painful for historians because these items would have been important artifacts in the future.

_____ **15.** Some of these items were as revolutionary as Gutenbergs printing press and James Watts steam engine. [These inventors names are spelled Gutenberg and Watt.]

_____C_____ **16.** During the World War II era, British Colossus computers helped crack the Nazi war codes.

_____ **17.** Your going to hire the electrical engineer, arent you?

_____ **18.** Xeroxes Palo Alto Research Center developed the Alto, the first computer to use a mouse.

_____C_____ **19.** Although Charlie's aunt was eccentric, her antics have entertained countless theater-goer's since the late 1800s.

_____ **20.** Womens roles have changed enormously since comedies like *Charlie's Aunt* were written.

_____ **21.** Mens roles have also changed, and its not unusual for successful, well-educated men to be the primary caretakers for their children.

_____ **22.** A baby's first cry may be called it's inaugural ball.

_____ **23.** Show us someone who habitually oversleeps, and well show you cause for alarm. [Dear Student: It's OK to groan after reading items 22 and 23. Leila]

_____ **24.** Be aware for your international business dealings that public praise for Latino's and Asian's is often embarrassing because modesty is an important cultural value.

_____ **25.** In Arab countries, it's considered inappropriate for wives to accept gift's from other men.

_____ **26.** Bussiness's consistently rank communication skills as a top requirement for ~~they're~~ their employee's.

_____ **27.** Shakespeare's play *A Midsummer Night's Dream* is about fairies, magic, king's, and queen's.

THE GLOBAL MARKETPLACE—GERMANY

German	Spell	Say	Currency
Greeting	guten tag	GOOT.n.tahg	deutsche mark (DM)
Departing	auf Wiedersehen	of VEED.er.san	
Please	bitte	BEET.eh	
Thank You	danken	DONK.n	
I Don't Understand	Ich verstehe nicht.	Ekh verstay nisht.	

American Heritage Dictionary

Secret Life of a Sentence Revealed

10

Say No to Blunders and Gaffes

"Vigorous writing is concise. A sentence should contain no unnecessary words . . . for the same reasons that a drawing should have no unnecessary lines. This requires not that the writer make all sentences short, but that every word *tell*."

—*William Strunk, Jr.*

After Completing Chapter 10, You Will

> Identify and correct fragments, run-ons, and comma splices.

> Construct complete and correct sentences.

> Say NO to sentence blunders and gaffes.

See *Instructor's Resource Kit* for Chapter 10 Pretest.

By completing Chapters 1–10, students develop a command of putting words together correctly in sentences. Chapter 11 provides more advanced understanding of sentence structure techniques, which continue in greater depth throughout the rest of the text, particularly in Chapter 13. Chapter 14 culminates the development of successful English for career skills.

When we combine the following letters in the order shown, we can pronounce a word:

S-P-E-L-L

We can't, however, be sure whether SPELL refers to magic or to arranging letters to form a word, which may also be magic. Therefore, to increase our understanding of words, we arrange them in groups:

> **A when not word spell how to you sure are.**

The preceding word group begins with a capital letter and ends with a period, but you know it isn't a sentence. You don't receive a sensible message from how the words are arranged.

> **When you are not sure how to spell a word.**

Ah! Now you've received a sensible message—but you have a feeling of incompleteness: you sense something is missing. Although you understand the word group, you wait for completion.

> **When you are not sure how to spell a word, use the dictionary.**

At last the familiar signals—capital letter and period—enclose a complete thought that you understand. Expressing ideas and information clearly is essential to successful business writing, and complete sentences are the first step toward that goal. Incomplete sentences are called **fragments.**

FRAGMENT When you are not sure how to spell a word.

The opposite of a fragment is a run-on. **Run-ons** occur when a writer forgets to stop. The writer goes on and on and doesn't insert a period or a connecting word where needed.

RUN-ON The location of vending machines at Yavapai College is important they must be placed in areas with large amounts of foot traffic.

Sometimes writers stop between two complete thoughts, but not firmly enough. This means they are guilty of the **comma splice**—a run-on with a comma but no connecting word between the complete thoughts.

COMMA SPLICE The location of vending machines is important, they must be placed in areas with large amounts of foot traffic.

Fragments, run-ons, or comma splices may keep the reader from understanding the message in a letter or report. They may also cause the reader to wonder about the writer's education.

You'll be confident of your ability to avoid fragments, run-ons, and comma splices after completing this chapter.

Read 51

BASIC NEEDS OF A SENTENCE: IDENTITY, ACTION, AND INDEPENDENCE

Just as you and I have basic needs, so too does a sentence. If our needs are not met, we feel incomplete and fragmented. When a sentence's needs are not met, it is incomplete and called a **fragment. A complete sentence** reveals three characteristics:

- Identity—who or what
- Action—doing, having, being, helping
- Independence—able to stand alone

IDENTITY—THE SUBJECT

> A sentence includes at least one word to identify who or what the sentence is about. This part of the sentence is the *subject* and is always a noun or a pronoun.

The following subjects in bold type identify **who** or **what** the sentence is about.

> **Mr. Escobar** is our newest administrative assistant. [who]
>
> **He** received the promotion after a year as a data entry clerk. [who]
>
> **It's** a good job. [what]
>
> **Learning** a second language takes time and effort. [what]
>
> **To leave** now would be a mistake. [what]
>
> A **book** can help you learn English, but **you** also need to listen to English-speaking radio, TV, and films. [what and who]
>
> Do **you** want to improve your pronunciation and vocabulary? [who]

Understood Subjects

Sometimes a subject is a "missing person." Look at these examples:

> Have lunch at the company café today! [The understood subject is *you*.]
>
> Let's have lunch at the company café today. [The understood subject is *you*.]
>
> Leave the file on my desk. [The understood subject is *you*.]

In commands—sentences that tell somebody what to do—*you* is understood as the subject even though the word isn't used. Often commands are expressed more courteously by adding "please," but the subject is still **you.**

> Please have lunch at the company café today.
>
> Please have lunch with me at the company café today.
>
> Please leave the file on my desk.

ACTION, HAVING, BEING, OR HELPING—THE VERB

> A sentence must have at least one verb—the word that tells what the subject *does* or *has* (action or having) or *is* (being). A helping verb sometimes precedes the action, having, or being verb. If needed, briefly review Chapter 7 now.

A sentence may have more than one verb, as in items 4 and 7 that follow. Sometimes a verb consists of more than one word; that is, a helping verb and a main verb, as in items 5 and 8 that follow. The verbs are in bold type:

1. We **saw** the new building on January 6. [action verb]
2. Ms. Hirsch from Ergonomics, Inc., **is** the electrician. [being verb]
3. You **have** the right of way. [action (having) verb]
4. The crime **was** embezzlement, and the punishment **fits** the crime. [being and action verbs]
5. They **will** both **speak** at the annual meeting in Omaha. [helping + action verb]
6. Many successful companies **operate** with honesty and social responsibility. [action verb]

7. Please **show** me the building and **describe** the facilities. [2 action verbs]
8. **Does** he **know** Eric Smith **is** Seattle's best CFP? [helping + action verb and being verb]

Here are the verbs from the preceding examples. In the blanks, write the subject(s).

1. _____We_____ saw
2. ____Ms. Hirsch____ is
3. _____You_____ have
4. _____crime_____ was
   ____punishment____ fits

5. _____They_____ will speak
6. ____companies____ operate
7. (understood you) show
   (understood you) describe
8. _____he_____ does know
   ____Eric Smith____ is

Answers are on page 402.

Answers are on page 402.

RECOGNIZING SUBJECTS AND VERBS

Just because a word is a noun or a pronoun doesn't mean it's the subject. For example, the object of a preposition is always a noun or a pronoun but is *not* a subject. To identify a subject, first find the verb. If the noun or pronoun does, is, or has what the verb states, it is the subject.

Read the following sentence:

Successful multicultural communication is vital to the nation's productivity.
What is the subject? ___communication___ Which is the verb? ___is___
What is the prepositional phrase? ___to the nation's productivity___

See answers on page 402.

INDEPENDENCE

The third (and final) requirement of a complete sentence is independence.

> **A sentence needs to be able to "stand alone"—that is, it must be independent to be complete.**

A word group with a subject and verb is called a **clause.** A clause may be either dependent or independent. A clause starting with a dependent conjunction is a **dependent clause.** Every sentence requires at least one **independent clause:** a word group that doesn't begin with a dependent conjunction and that has a subject and verb.

An independent clause that begins with a capital letter and ends with a period, question mark, or exclamation mark is a complete sentence. If, however, a capital letter begins a **dependent** clause and a closing punctuation mark ends it, the result is a **fragment.**

RECOGNIZING DEPENDENT CLAUSES

You recognize a **dependent** clause by its first word—a dependent conjunction. Look over the dependent conjunctions listed on page 00 and the additional ones shown below. You don't need to memorize them:

after	before	since	until	which
although	even though	so that	when	while
as	if	that	where	who(m)
because	provided	unless	whether	why

SENTENCE   Prices will continue to fall.

FRAGMENT   While prices will continue to fall. [dependent conjunction *while*]

SENTENCE   PCs are relatively cheap.

FRAGMENT   Since PCs are relatively cheap. [dependent conjunction *since*]

A. The example sentences numbered 1–8 on pages (206–207) contain six prepositional phrases. The first one is in EXAMPLE below. Write the five others in the blanks. Write NONE in the blank for the three sentences with no prepositional phrases.

EXAMPLE

1. on January 6.

2. from Ergonomics, Inc

3. of way

4. NONE

5. at the annual meeting / in Omaha

6. with honesty and social responsibility

7. NONE

8. NONE

B. Subjects tell who or what the sentence is about. Verbs tell what the subject does, is, or has. Write the subject(s) in the left blank and the verb(s) in the right blank. One sentence has the "understood subject" *you.* Another has two subjects and two verbs.

	SUBJECTS	VERBS
EXAMPLE The new secretary is in the office now.	secretary	is
1. Steam Gene could have shot down the Messerschmitt.	Steam Gene	could have shot
2. You must be computer literate to get a good job.	You	must be
3. The winner sees an answer for every problem.	winner	sees
4. The loser sees a problem in every answer.	loser	sees
5. The books were put on the shelf.	books	were put
6. What you earn depends on what you learn.	you/you	earn learn
7. The new members of the team have uniforms.	members	have
8. Three visitors arrived yesterday.	visitors	arrived
9. Please give me the ledger.	(You)	give
10. Professor Friede found five errors.	Professor Friede	found

WORD TO THE WISE

Notice that "of the team" in B, No. 7 above is a prepositional phrase. The object in this phrase is "team." An object of a preposition can't be a subject.

C. A clause is a word group with a subject and verb. Each item below is a clause. Six clauses are independent and six are dependent:

(a) In the blanks write *I* for **independent** or *D* for **dependent.**

(b) Circle the word that makes the clause dependent. Capitalize the first letter of the next word.

(c) Change the independent clauses to sentences with capital letters and closing punctuation.

EXAMPLE ___D___ (since) the new assistant didn't know the meaning of *chronological.*

___I___ 1. she couldn't arrange the reports in chronological order.

___D___ 2. (because) many organizations have similar problems with employees.

___D___ 3. (although) an accountant needs excellent communication skills.

___I___ 4. in the workplace you may converse with many people.

___D___ 5. (who) They have limited English skill.

___I___ 6. don't laugh at someone's pronunciation or grammar error.

___D___ 7. (when) you transport something by car.

___I___ 8. it's called a shipment.

_____ D _____	**9.** (If) however, you transport it by ship⊙	
_____ I _____	**10.** it's called cargo⊙	
_____ I _____	**11.** isn't that strange ?	
_____ D _____	**12.** (after) you've learned a few phrases in other languages⊙	

The answers are on pages 402–403.

FRAGMENTS, ANYONE?

A fragment is an incomplete word group beginning with a capital letter and ending with a period. Avoid fragments in most business writing.

One type of fragment is a dependent clause that begins with a capital letter and ends with a period, question mark, or exclamation mark.

Recap

Change these fragments to sentences as follows: Cross out the dependent conjunction, begin the next word with a capital letter, and insert a closing punctuation mark at the end.

1. ~~when~~ she dines at the college cafe⊙
2. ~~where~~ the beans taste like caviar⊙

See answers on page 403.

Another way to correct fragments is to add an independent clause before or after the dependent clause. Then you have a complete sentence. The dependent clause is in bold letters in the following examples; the independent clause is not bold.

(a) **Since prices will continue to fall,** competition will get stronger.

Competition will get stronger **because prices will continue to fall.**

(b) **Although software is now easier to use,** well-trained people still command high salaries.

(c) Well-trained people still command high salaries **although software is now easier to use.**

You'll become a comma expert in Chapter 11. In the meantime, use a comma between clauses if the dependent clause precedes the independent—as in sentences (a) and (b) but not (c) above.

Recap

Convert the fragment to a sentence by adding an independent clause of your choice. Replace a period with a comma where required.

1. When she dines at the college cafe, <u>the beans taste like caviar. (answers will vary)</u>

2. Rewrite No. 1 so that the independent clause you used above opens the sentence and the dependent clause follows. <u>The beans taste like caviar when she dines at the college cafe. (answers will vary)</u>

Check your answers on page 403.

Two other types of fragments follow:

NO   Mr. Henry while jogging around the block. He sprained his ankle. [fragment followed by complete sentence]

YES   Mr. Henry, while jogging around the block, sprained his ankle. **or** Mr. Henry sprained his ankle. He was jogging around the block when it happened.

NO   Mr. Henry's office, which is closed today. It will reopen next week. [fragment followed by complete sentence]

YES   Mr. Henry's office, which is closed today, will reopen next week. **or** Mr. Henry's office is closed today. It will reopen next week.

Some people mistakenly believe it is "businesslike" to omit a subject. A word group without a subject that begins with a capital letter and ends with a period is a fragment.

FRAGMENT   Hope to hear from you soon. **or** Hoping to hear from you soon. [These are OK for postcards or informal e-mails.]

SENTENCE   We hope to hear from you soon. **or** We are hoping to hear from you soon. [The latter is correct but needlessly wordy.]

Omitting the subject is correct only in a command or request with *you* as the understood subject.

SENTENCE   Write soon. **or** Please write soon. [*you* understood]

Writing for Your Career

Do not omit subjects to write sentences more concisely. Concise writing is important—but not if the result is poor grammar or sentence construction.

Replay 52

A. Write **F** in the blanks next to the fragments and **S** next to the sentences.

EXAMPLES   __F__   Whether it can be done.

          __S__   We do not know.

__F__   **1.** If you do a good job on the Mendocino project.

__S__   **2.** You will get a salary increase after the first of the month.

__F__   **3.** Although most corporations use the services of an auditor to examine the books.

__S__   **4.** Some errors may never be found.

__S__   **5.** Don't ever think.

__F__   **6.** That you know it all.

__F__   **7.** Because sexual harassment is illegal and immoral.

__S__   **8.** Twileen filed a complaint.

__F__   **9.** When Mr. Lopez became general manager.

__S__   **10.** He asked the staff to greet Spanish-speaking customers in Spanish.

B. Use the ten items in Part A to write five sentences. Combine each fragment (dependent clause) with one of the sentences (independent clauses).

EXAMPLE   <u>We do not know whether it can be done.</u> (See Examples above—Part A)

1. If you do a good job on the Mendocino project, you will get a salary increase after the first of the month.

2. Some errors may never be found although most corporations use the services of an auditor to examine the books.

3. Don't ever think that you know it all.

4. Because sexual harassment is illegal and immoral, Twileen filed a complaint.

5. When Mr. Lopez became general manager, he asked the staff to greet Spanish-speaking customers in Spanish.

C. Write **F** for fragment or **S** for sentence in the blanks below.

EXAMPLE   __F__   A stock certificate, which is a valuable document.

__F__   **1.** The man whom we met yesterday in Fullerton.

__F__   **2.** Our Human Resources Department on the third floor.

__F__   **3.** Twileen, believing that she was right.

__S__   **4.** Mr. I. M. Shady also felt that he was right.

__S__   **5.** Mr. Yee continued to argue the point.

__F__   **6.** George, having been on vacation last week.

__S__   **7.** The staff was on vacation last week.

__F__   **8.** The team that won all the games in Oklahoma City last year.

_S_   **9.** That team won all the games in Philadelphia last year.

_S_   **10.** The Bridgeport team that won all the games last year is losing today.

     D.   Five items in Part C are fragments. Write the item number for each fragment in the blank; then change the fragment to a sentence. Make up necessary information to complete the sentence.

1	The man whom we met yesterday in Fullerton is the treasurer of the Cambridge Corporation. Answers will vary.
2	Our Human Resources Department is on the third floor.
3	Twileen, believing she was right, took the matter to the Human Resources Department.
6	George, having been on vacation last week, was shocked by the news.
8	The team that won all the games in Oklahoma City last year will be hard to beat.

Please check your answers on pages 403–404.

CAPITAL PUNISHMENT

RIDDLE   What kind of sentence requires capital punishment? Do you want a hint? Here's a sentence that needs capital punishment:

> List everything you want to achieve in life, be creative and adventuresome.

ANSWER—COMMA SPLICE   One way to correct a comma splice is with a period followed by a *capital*. Now here's what happens after "**capital** punishment":

> List everything you want to achieve in life. **Be** creative and adventuresome.

Point out that length doesn't determine whether a word group is a run-on, comma splice, or correct.

A **comma splice** occurs when you join two or more independent clauses with a comma but without a coordinate conjunction. If the independent clauses just run together with neither conjunction nor punctuation, we call the error a **run-on.**

> One way to avoid run-ons and comma splices is to separate independent clauses with a period and capital letter, resulting in two sentences.

RUN-ON   Use gestures to reinforce your message don't cross your arms.

COMMA SPLICE   Use gestures to reinforce your message, don't cross your arms.

CORRECT   Use gestures to reinforce your message. Don't cross your arms.

RUN-ON   The vigorous handshake common in American business is inappropriate in France it is seen as rude behavior.

COMMA SPLICE   The vigorous handshake common in American business is inappropriate in France, it is seen as rude behavior.

CORRECT   The vigorous handshake common in American business is inappropriate in France. It is seen as rude behavior.

RUN-ON   Business is part of our society there is no escape.

COMMA SPLICE	Business is part of our society, there is no escape.
CORRECT	Business is part of our society. There is no escape.
RUN-ON	American Telephone and Telegraph Company has equipment worth billions of dollars it is one of the giants of American business.
COMMA SPLICE	American Telephone and Telegraph Company has equipment worth billions of dollars, it is one of the giants of American business.
CORRECT	American Telephone and Telegraph Company has equipment worth billions of dollars. It is one of the giants of American business.
RUN-ON	Arnold said that the recession is over salaries and prices are rising.
COMMA SPLICE	Arnold said that the recession is over, salaries and prices are rising.
CORRECT	Arnold said that the recession is over. Salaries and prices are rising.

WORD TO THE WISE

A run-on and a comma splice may have the same wording. **The comma splice, however, has a comma between independent clauses;** the run-on does not.

Replay 53

Spot and correct run-ons and comma splices. Write **C** for correct, **R** for run-on, and **CS** for comma splice. Use a period and capital letter to correct the run-ons and comma splices. Don't place a period after a dependent clause; if you do, you create a fragment.

EXAMPLES __CS__	Sticks and stones will break my bones, words will never harm me.
__C__	Sticks and stones will break my bones. Words will never harm me.
__C__	Although sticks and stones will break my bones, words will never harm me.

__R__   **1.** The winner always has a program the loser always has an excuse.

__C__   **2.** The English translation of a sign in the window of a travel agency in Spain reads, "Go Away."

__CS__   **3.** These statistics deal only with symptoms they do not reveal the fundamental economic problems.

__C__   **4.** Geriatrics, the study of dealing with problems of the aged, is a relatively new science.

__R__   **5.** Geriatrics is the study of dealing with problems of the aged it is a relatively new science.

__R__   **6.** This tax return is inaccurate ask your accountant to prepare a new one.

__C__   **7.** As your tax return is inaccurate, ask your accountant to prepare a new one.

__CS__   **8.** Many students are attracted to a business career they like the challenge and rewards it offers.

_C_ 9. Because of the challenge and rewards it offers, many students are attracted to a business career.

_C_ 10. Each year *Fortune* magazine publishes a list of the top five hundred American corporations.

_C_ 11. Although many small companies have manufactured automobiles in the United States, only General Motors, Ford, Chrysler, and American Motors have survived and grown.

_CS_ 12. Many of today's corporations started in colonial times; they became far bigger than anyone expected.

_CS_ 13. Most large corporations are multinational; they do business in many foreign countries.

_CS_ 14. A good example of such a corporation is IBM; it has an office in almost every major city in the world.

_C_ 15. A good example of such a corporation is IBM, which has an office in almost every major city in the world.

_C_ 16. Our in-service training includes instruction in Excel and in PowerPoint.

_C_ 17. The job requires someone who can operate folding machines, drills, and stitchers.

_C_ 18. Entry-level positions in the manufacturing technology field often require a knowledge of die making and tool design.

_C_ 19. After you have made your list of goals, prioritize them in order of importance.

_C_ 20. Set a realistic, but ambitious, timetable for each of your goals.

See answers on page 404. Then review Reads and Replays 51–53 before taking the Pop Quiz on page 350.

Read
54

THREE IMPORTANT CONNECTIONS

Some ideas should be connected (or joined) and others separated. In Read and Replay 53, you **separated** independent clauses correctly. In Read and Replay 54, you **connect** them correctly. When writing business documents, you must decide quickly whether to separate or connect clauses. If independent clauses are closely related and not too long, connecting is often (not always) preferable to separation. Three correct ways to connect independent clauses are with a semicolon, a comma and a coordinate conjunction, or just a coordinate conjunction. The correct way to separate independent clauses is with a period followed by a capital letter for the first word of the next sentence.

CONNECTING WITH A SEMICOLON

Closely related independent clauses may be connected with a semicolon.

Get organized; handle each piece of paper only once.

CONNECTING WITH A COMMA AND COORDINATE CONJUNCTION

Closely related independent clauses may be connected with a comma followed by a coordinate conjunction: *and, but, or, nor, for,* and sometimes *so* or *yet.**

You cannot control what anyone else does, **but** you can take charge of your own life.

CONNECTING WITH A COORDINATE CONJUNCTION ONLY

If the independent clauses total no more than ten or eleven words and you join them with *and* or *or,* a comma isn't needed.

Get organized **and** handle each piece of paper only once.

Efficiency helps, **but** skill is also required. [Because *but* is the conjunction, add the comma.]

REVIEW THREE IMPORTANT CONNECTIONS

Here we connect the clauses instead of separating them.

RUN-ON | Business is part of our society there is no escape.

COMMA SPLICE | Business is part of our society, there is no escape.

CORRECT | Business is part of our society; there is no escape. [Connect with a semicolon.]

OR | Business is part of our society and there is no escape. [Connect with a coordinate conjunction. A comma isn't needed since *and* is joining short clauses.]

RUN-ON | American Telephone and Telegraph has equipment worth billions of dollars it is one of the giants of American business.

COMMA SPLICE | American Telephone and Telegraph has equipment worth billions of dollars, it is one of the giants of American business.

CORRECT | American Telephone and Telegraph has equipment worth billions of dollars; it is one of the giants of American business. [Connect with a semicolon.]

OR | American Telephone and Telegraph has equipment worth billions of dollars, and it is one of the giants of American business. [Connect with a comma and a coordinate conjunction.]

RUN-ON Don't fill a business letter with long sentences or with words of many syllables it doesn't impress anyone.

COMMA SPLICE Don't fill a business letter with long sentences or with words of many syllables, it doesn't impress anyone.

Learning this rule—no comma before *and* or *or* in short compound sentences—helps students avoid calling such constructions run-ons because of no comma.

For is a coordinate conjunction when it means because; otherwise, it is usually a preposition. Use of *so* and *yet* as joiners are explained later.

Correct the preceding run-on and comma splice in three different ways:

1. Connect with a comma and a coordinate conjunction. Don't fill a business letter with long sentences or with words of many syllables, for it doesn't impress anyone.

2. Connect with a semicolon. Don't fill a business letter with long sentences or with words of many syllables; it doesn't impress anyone.

3. Separate with a period and capital letter. Don't fill a business letter with long sentences or with words of many syllables. It doesn't impress anyone.

Solutions are on page 404.

THE TRANSITION TRAP

Transitions help readers cross over from one idea to the next closely related idea. Here are some common transitions:

also	hence	moreover	then
consequently	however	nevertheless	therefore
for example	in addition	otherwise	thus
furthermore	in fact	that is	yet

> **When a transition word connects _independent_ clauses, insert a semicolon or a period—not a comma—before the transition word. A comma would result in a comma splice and would mean you've been caught in the _transition trap_.**

The examples show how to fix comma splices by correctly separating or connecting the independent clauses. The transition words are in bold type.

COMMA SPLICE   About 50 percent of our employees are engaged in the distribution of goods and services, **however,** about 20 percent are in production.

RUN-ON   About 50 percent of our employees are engaged in the distribution of goods and services however, about 20 percent are in production.

CORRECT   About 50 percent of our employees are engaged in the distribution of goods and services; **however,** about 20 percent are in production. OR About 50 percent of our employees are engaged in the distribution of goods and services. **However,** about 20 percent are in production.

COMMA SPLICE   In the United States eye contact is extremely important, **in fact,** Americans don't trust someone who won't look them in the eye.

CORRECT   In the United States eye contact is extremely important; **in fact,** Americans don't trust someone who won't look them in the eye.

In the United States eye contact is extremely important. **In fact,** Americans don't trust someone who won't look them in the eye.

COMMA SPLICE   Order the new computers, **then** you can find out about training.

CORRECT   Order the new computers; **then** you can find out about training.

Order the new computers, and **then** you can find out about training. [coordinate conjunction makes comma correct]

Order the new computers. **Then** you can find out about training.

> **Use commas before and after a transitional adverb that does *not* connect independent clauses.**

NO   We; therefore, will be pleased to help you. [Test for independent clauses by imagining a period after *we* and a capital *T* for *therefore*; then decide whether two sentences would result.]

YES   We, therefore, will be pleased to help you. [*Therefore* doesn't join independent clauses but merely separates a subject and a verb.]

SO AND YET

So as a coordinate conjunction, while not wrong, is too informal for most business writing; it's fine in conversation, memos, and informal notes. When **yet** joins independent clauses, use either a comma or a semicolon. If in doubt, use the comma for a short sentence and the semicolon for a longer sentence. A period **separating** the two independent clauses is also correct.

CORRECT
1. Average business letter cost is increasing, **yet** the percentage of increase is declining.

2. Average business letter cost is increasing; **yet** the percentage of increase is declining.

3. Average business letter cost is increasing. **Yet** the percentage of increase is declining.

If a sentence seems too long, the period or the semicolon is better than the comma, but any one of the three is correct in the above-shown examples.

WORD TO THE WISE

A comma **after** a transition expression has nothing to do with whether a word group is a comma splice or run-on. Do use a comma **after** a transition of more than one syllable—except the short word *also*. Do not use a comma after one-syllable transitions like *yet*, *thus*, and *then*. (Details of correct punctuation are in Chapters 11 and 12.)

Replay
54

A.   Write **C** beside the two correct sentences, **R** beside the run-on, and **CS** for comma splices. Insert semicolons to correct the run-on and comma splices.

EXAMPLE ___CS___   A stock does not have a fixed worth, it's only as valuable as people think it is.

___CS___   1. Teenagers are very fashion-conscious, however, they closely follow the dictates of their friends.

___CS___   2. Consumers have the last word, that is, if they buy a fashion, retailers profit and continue to stock the item.

___R___   3. Some people buy because they want to be distinctive; they want to be recognized as leaders.

___C___   4. Many businesspeople wear the latest fashions because they want to make a good impression on their colleagues and clients.

<u>   R   </u> **5.** Greet your clients by name; welcome them with a friendly smile and a handshake.
                                              ∧

<u>   C   </u> **6.** Greet your clients by name and welcome them with a smile.

<u>   CS   </u> **7.** Greet your clients by name, then welcome them with a friendly smile and a handshake.
                                      ∧

     B. Write **C** for correct, **R** for run-on, and **CS** for comma splice. Then make corrections with a comma and one of the seven coordinating conjunctions (and, but, or, nor, for, and sometimes yet or so); or insert an appropriate dependent conjunction.

EXAMPLE <u> R </u> You can vote in person by attending a corporation's annual meeting **,or** you can vote by using an absentee ballot.

<u> R </u> **1.** Professor Wayne Moore was able to repair the motor quickly, *since* he is an expert mechanic with years of experience. ∧

<u> C </u> **2.** He didn't notice any of the errors, nor would he have corrected them if he had.

<u> R </u> **3.** The Personnel Department is on the third floor, *but* the interviewers are out to lunch now. ∧

<u> R </u> **4.** The new equipment is being shipped to you at once, *and* you should receive it by the end of the week. ∧

<u> R </u> **5.** Possibly Joyce can attend the conference, *but* maybe Michael can go in her place. ∧

<u> CS </u> **6.** Sticks and stones will break my bones, *but* words will never harm me. ∧

<u> C </u> **7.** Goods are delivered either to the receiving area of a department store or to the central warehouse of a chain.

<u> C </u> **8.** Experience is an expensive school, but fools will learn in no other.—Benjamin Franklin

     C. Answer the following questions.

1. List six commonly used transition expressions:

<u>however</u>   <u>therefore</u>   <u>for example</u>   <u>also</u>   <u>yet</u>   <u>then</u>

2. If a transition expression joins **independent** clauses, use a semicolon or a period before the transition. (a) true (b) false

3. When a transition is between a **dependent** and an **independent** clause, use a semicolon before the transition. (a) true (b) false

     D. Replace a comma with a semicolon before a transition expression that joins **independent clauses.** Write **C** beside the three correct items. Show needed corrections for the others.

<u> C </u> **1.** He will not, however, take the blueprints to the laboratory until the end of next week.

<u>   </u> **2.** He will complete the blueprints today; then he'll send them to the laboratory. ∧

<u>   </u> **3.** The average person spends almost 80 percent of awake time in communication; however, communication efficiency for the untrained is about 25 to 30 percent. ∧

<u>   </u> **4.** High achievers do mental rehearsals; that is, they visualize themselves performing tasks successfully. ∧

_C_   **5.** High achievers can picture themselves doing something well; for example, they might visualize themselves keyboarding rapidly and accurately.

_____   **6.** Your mind never sleeps; consequently, visualization is effective just before falling asleep since your mind probably continues to work on the desired achievement as you sleep.

_C_   **7.** Mr. Higgins was right, however, in thinking a poor flower girl could be changed into "my fair lady."

Please check your answers on pages 404–405.

Read
55

Students with good language skills profit from practice on when to subordinate, when to coordinate, and when to separate into two sentences. However, students must first clearly grasp concepts of fragment, comma splice, and run-on.

MAKING MORE CONNECTIONS

You can create a correct sentence by connecting the independent clause or clauses to a dependent clause (a clause that begins with a dependent conjunction).

> **When a dependent clause is connected to an independent clause, the result is a complete sentence. [Review dependent conjunctions in the page 208 box.]**

A run-on or comma splice means two (or more) independent clauses joined without a coordinate conjunction or semicolon. One way to correct such a sentence is to make one clause dependent by beginning it with a dependent conjunction. Voilà, you no longer have a run-on or comma splice! Usually the dependent conjunction slightly changes the emphasis or meaning of the original sentence and might help express a particular idea.

A sentence needs at least one independent clause; it may be either before or after the dependent clause(s), whichever sounds better for that sentence. In the following examples you see how to correct a run-on or comma splice by making one clause dependent. The dependent conjunction is in bold type:

RUN-ON   He thought he would be sick something was wrong with the potato salad.

COMMA SPLICE   He thought he would be sick, something was wrong with the potato salad.

CORRECT   He thought he would be sick **because** something was wrong with the potato salad.

Because something was wrong with the potato salad, he thought he would be sick.

RUN-ON   The picnic was fun the ants thought so too.

COMMA SPLICE   The picnic was fun, the ants thought so too.

CORRECT   The picnic was fun **although** the ants thought so too.

Although the picnic was fun, the ants thought so too.

Two closely related ideas are, in a way, like people deciding whether to join or separate. Outsiders can't decide for them. When writing for business, *you* decide whether to join or separate closely related ideas. Writing practice, reading well-written business communications, and good judgment enable you to decide quickly.

Skillful business writers often use a dependent clause to de-emphasize an idea they don't want the reader to focus on and use an independent clause to *emphasize* another idea.

┌──────── Dependent Clause ────────┐ ┌──────── Independent Clause ────────┐

Although we don't give refunds on earrings, we'll be happy to exchange them for any other jewelry in the store.

With this writing technique, you emphasize the positive and make the negative less important.

Re items 2 and 15: *which* (No. 2) and *that* (No. 15) function here as relative pronouns **and** dependent conjunctions. *Which* and *that* are subject pronouns and dependent conjunctions at the same time in these sentences. Explanations about "relative pronouns and restrictive clauses" do not help any but the more advanced students. These sentences show various ways to construct a dependent clause and help if a student asks about a similar sentence found elsewhere.

Replay 55

A. Write **R** for run-on, **CS** for comma splice, or **C** for correct sentence. Underline the dependent conjunctions. Insert needed punctuation and capitalization.

EXAMPLE ___C___ <u>When</u> you buy more shares of stock, you increase your voting power at the corporation's annual meeting.

___R___ **1.** He knows the definition of *vitamin.* it's what to do <u>when</u> guests come to your door.

___C___ **2.** He knows the definition of *vitamin*, <u>which</u> is what to do <u>when</u> guests come to your door.

___R___ **3.** The highest achievers are passionately committed to their work they are not workaholics.

___C___ **4.** <u>Although</u> the highest achievers are passionately committed to their work, they are not workaholics.

___C___ **5.** The highest achievers are passionately committed to their work, <u>although</u> they are not workaholics.

___CS___ **6.** High achievers take more short vacations <u>than</u> the average person, they often get new ideas for their work during <u>these</u> vacations.

___R___ **7.** High achievers take more short vacations <u>than</u> the average person; they often get new ideas for their work during <u>these</u> vacations.

___C___ **8.** Workaholics work long hours <u>because</u> they fear losing the job or not impressing the boss favorably.

___C___ **9.** <u>Because</u> high achievers feel a strong sense of commitment to their work, they work long hours.

___C___ **10.** High achievers define skills needed for their career and then go out to get them.

___CS___ **11.** Statements about high achievers are from Charles Garfield's book *Peak Performers,* Mr. Garfield studied hundreds of top achievers.

_____C_____ **12.** Always part from employers on good terms, for you may need a reference some day.

_____R_____ **13.** During an interview, don't forget to ask *your* questions; also remember to thank the interviewer for the appointment. ∧

_____CS_____ **14.** Listen attentively during an interview; make eye contact with the interviewer. ∧

_____C_____ **15.** Personnel directors are skilled at asking questions <u>that</u> cannot be answered by a simple "yes" or "no."

_____C_____ **16.** The winters are long and cold in Indianapolis, but this gives IVY Tech students more time to study.

_____CS_____ **17.** Love competence in the performance of your tasks; begin now.—Lao Tzu ∧

_____C_____ **18.** A rolling stone gathers no moss, <u>yet</u> it does get a certain smoothness from its rolling.

_____CS_____ **19.** Mix a little foolishness with your serious plans; it is lovely to be silly at the right moment.—Horace (Roman poet and philosopher, but he wrote ∧ it correctly.)

_____C_____ **20.** Mix a little foolishness with your serious plans, <u>as</u> it is lovely to be silly at the right moment.

      B. Correct each comma splice by adding a dependent conjunction that makes one clause dependent. Write **C** beside the one sentence that is already correct.

When
EXAMPLE ∧A company finds its high stock price is discouraging new investors, it might initiate a stock split.

_____C_____ **1.** The future, Twileen, is that time you'll wish you had done what you're not doing now.

_____ **2.** He places big orders with us, since George gives him a special discount. ∧

_____ **3.** Although ∧The plant was operating on a 24-hour basis, management refused to adopt a three-shift schedule.

_____ **4.** When ∧Professor Brown explained that studying business communication at Wright Business School is enjoyable, the new student wouldn't believe it.

_____ **5.** Then Veronica said we need the latest dictionary for Chapter 3, because ∧language changes constantly.

Check your answers on page 405. Next do the Pop Quiz on page 352.

FULL STOPS AHEAD

Oscar Wilde, a witty British playwright of the late 1800s, wrote, "The English have really everything in common with the Americans except of course language." One of many examples of the differences between British and American English is the word for the mark that ends most sentences. In the United States, we say *period*; in Britain it's a *full stop*.

Although Replay 56's business letter is worded appropriately, commas are not only where they belong but also where a period or "full stop" is needed. Draw a line through the incorrect commas, and insert periods in the blank space just above the commas. As you read this letter and other examples of well-written business letters throughout the text, notice the friendly, easy-to-understand style. Only the salutation (greeting), body (message), and closing information are included, not the date or address.

Following the name and title of the letter's originator, a postscript (PS) appears. Postscripts should be used rarely—only to emphasize something in the letter or add a personal message. If after you print a letter, you discover you forgot something, insert what's needed where it belongs, and reprint the document.

Replay 56

At the end of each sentence, please cross out the comma, insert a period/full stop in a contrasting color, and capitalize the first letter of the new sentence. Do not delete, add, or change any words. If you close the sentences at the correct places, you will have no fragments, comma splices, or run-ons.

Dear Professor Head: Salutation

Thank you for the time and courtesy you extended to our representative, Laura Mann, at your college last month, she enjoyed her visit with you.

At Laura's request we have sent you the new edition of *Mathematics of*   Body *Business*, this was sent to you several weeks ago, and you should have it by now, we do hope you'll look it over carefully, in addition, your name has been placed on our mailing list to receive an examination copy of *Business Math: Practical Applications*, a new edition of this book by Cleaves, Hobbs, and Dudenhef is expected off the press sometime next month.

We'll send your copy just as soon as it is available, if there is any way we can be of help to you, Professor Head, please let us know, best wishes for a happy holiday season.

Cordially,   Complimentary close

Hal Balmer   Handwritten signature

Hal Balmer   Writer's name and title
Vice President

lrs   (initials of assistant who keyboarded the letter)

PS Don't write comma splices or run-ons, they are hard to read.   Postscript

See page 405 to check your responses. Then take the Pop Quiz on page 352.

Be sure you have checked all Recap and Replay answers before proceeding to Checkpoint.

Place a check next to each item below when you're sure you understand it.

_____ A subject is a noun or pronoun that tells who or what a sentence is about.

_____ A verb tells what the subject does, is, or has.

_____ A prepositional phrase begins with a preposition but has no subject and verb.

Examples: near the old red schoolhouse   during the past year   to Janice

_____ A clause is a word group with a subject–verb combination. A clause beginning with a dependent conjunction is a dependent clause. A clause *not* beginning with a dependent conjunction is an independent clause.

We did enjoy the show [independent clause]

because the acoustics were perfect [dependent clause since the first word is a dependent conjunction]

_____ An independent clause may be used as a sentence:

We did enjoy the show.   The acoustics were perfect.

_____ A fragment is an incomplete idea masquerading as a sentence since it begins with a capital letter and ends with a period. A fragment might be missing a subject or verb, or it could be a dependent clause.

When we received your June 15 letter. [dependent clause]

Received your June 15 letter. [subject missing]

Your records during the past year. [fragment; verb missing]

_____ A run-on is two or more independent clauses with neither punctuation nor coordinate conjunction between the clauses.

Yes, we did enjoy the show the acoustics were perfect.

_____ A comma splice has a comma between the independent clauses of a run-on.

Yes, we did enjoy the show, the acoustics were perfect.

_____ Identification of fragments, run-ons, and comma splices is not based on length, but on structure.

Birds sing they also fly. [run-on]

Birds sing, they also fly. [comma splice]

Although most large and colorful birds not only sing but also fly. [fragment]

_____ Run-ons and comma splices are unacceptable in workplace writing; fragments are usually unacceptable—unless in a very informal note.

A run-on or comma splice may be corrected with a semicolon (;) between the independent clauses or with a comma and a coordinate conjunction (and, but, or, nor, for, or yet) between the independent clauses. The comma may be omitted if the sentence is short and the conjunction is *and* or *or*.

We did enjoy the show; the acoustics were perfect.

We did enjoy the show; however, the acoustics were not good.

We did enjoy the show, **for** the acoustics were perfect.

We did enjoy the show **and** the acoustics were perfect.

_____ A comma splice or run-on may be corrected by separating the independent clauses with a period followed by a capital letter.

We did enjoy the show. The acoustics were perfect.

_____ A comma splice or run-on may be corrected by making one of the clauses dependent.

We did enjoy the show **because** the acoustics were perfect.

Because the acoustics were perfect, we did enjoy the show.

Writing for Your Career

Experienced writers who clearly understand sentence structure sometimes use fragments to create certain effects, particularly in advertising material. After you're confident of your business writing expertise, you may intentionally use a fragment. Until then, keep fragments out of your writing.

SPECIAL ASSIGNMENT

A. In conversation, fragments are frequently used and are acceptable. Imagine this conversation:

INSTRUCTOR   What did you study in Chapter 10?

STUDENT   The difference between sentences and fragments.

If you were to write your reply, you would insert a subject and a verb. This would change the acceptable conversational fragment above to an acceptable written sentence. Write the acceptable **written** reply in the blank. We studied the difference between sentences and fragments.

Special Assignments B and C begin on page 227 and can be done after you complete the Practice Quiz and check your answers.

PROOFREADING FOR CAREERS

Use your spell checker, proofread from the screen, and make corrections before printing. Proofread again from printed pages. It's not good enough to rely on "sensing" when you've made a typographical error. Sometimes errors are not "felt" while keyboarding. Proofread for "sense," not just spelling and correctness.

When someone has keyboarded for you, do not submit, sign, or okay the document until you proofread it yourself. Don't rely on anyone else's accuracy or skill when a document is your responsibility. Everyone makes errors, but successful people find and correct their own before anyone else finds them. If your proofreading isn't too good, try this technique: After proofreading a document slowly and carefully, reread it **starting with the last word** and continue until you reread the first word.

Using your dictionary, correct the following article for spelling, typographical, or sentence construction errors. Change punctuation **only to correct a run-on, comma splice, fragment, or apostrophe error.** Keep the wording as is. Your instructor will tell you whether to use proofreader's marks or to keyboard the article correctly as a short report—double-spaced, first line of each paragraph indented, and title centered in all capital letters.

CORPRATE ORGANIZATION: A VIEW FORM THE TOP

The owners of a corporation are called stockholders or shareholders, each stockholder has one vote for each share of stock he or she owns. At the annual meeting, The stockholders elect a board of directors, Which usually has between 7 and 17 members. Because this bord (board) represents the stockholders, it elects the high-ranking corporate officers, or at least the president. The officers are responsible for the day-by-day operation the of corporation.

The board of directors, the officers, and the stockholders, Are required by law to meet at least once each year. The corporate secretary notifies the stockholders of the meeting by male, it is common practice to enclose a proxy form. With the notice of the meeting. Use of such a form allows stock holders to submit there votes in lieu of attending the meeting.

TOTAL ERRORS = 14 (Each sentence construction error counts as one.) Your score is (Circle one) Excellent, Good, Fair, Ugh.

Keyboard a corrected version of the CORPORATE ORGANIZATION article.

PRACTICE QUIZ

www.prenhall.com/smith

Take the Practice Quiz as though it were a real test. After the quiz is corrected, review the chapter if needed.

Part I Write **A** in the blank for correct sentences, **B** for fragments, **C** for run-ons, and **D** for comma splices.

A	**1.** About an hour later the CEO arrived.
C	**2.** Anthony is the new accountant in our office he is not a CPA.
D	**3.** Here's the truth, the difference between achieving and not achieving is goalsetting.
A	**4.** Wal-Mart's employees must recite and practice founder Sam Walton's "10-foot rule."
D	**5.** The "10-foot rule" is explained in item 22, it is good advice for anyone working in retail sales.
D	**6.** The carpenter is working very hard, nevertheless, we don't think she can finish by one.
A	**7.** While the carpenter is working hard, we don't think she can finish by one.
D	**8.** The carpenter is working hard, we don't think she can finish the job on time.
B	**9.** Although Macy's first efforts at the dry goods business in Boston failed miserably.
B	**10.** After trying his luck with six other doomed retailing ventures.
D	**11.** It is the weak who are cruel, gentleness can be expected only from the strong.
D	**12.** Leaders have different styles, they all need, however, to be skilled in handling people.
C	**13.** Meetings must proceed according to rules furthermore, business must be conducted fairly.
B	**14.** Because many informal meetings occur.

<u> A </u> **15.** Stay calm and don't blame or name-call.

<u> D </u> **16.** Service can come in many forms, for example, it can mean having the right merchandise at the right time.

<u> A </u> **17.** The word *business* originally meant the state of being busy with anything.

<u> C </u> **18.** Ms. Scott flew to Spain last year then she took a Mediterranean cruise.

<u> A </u> **19.** A characteristic of high achievers is that they believe they are responsible for most of what is good or not good in their lives.—Charles Garfield

<u> B </u> **20.** Mrs. Powell, carrying the coffee in one hand and the keys in the other.

Part II Correct these run-ons and comma splices by inserting a dependent conjunction in each sentence. Your corrected sentence will show an independent and a dependent clause. If the dependent clause is first, follow it with a comma.

EXAMPLE Setting goals may be scary, *although* it is necessary.

21. *When* You can do all the common things of life in an uncommon way, you will command the attention of the world.

22. *If* a customer comes within 10 feet, the employee stops what he or she is doing, speaks to the customer, and offers help—San Walton's 10-foot rule for Wal-mart employees.

23. The manager was finishing the report *while* her husband was busy baking a chocolate cake.

24. In France dinner guests send flowers *before* they arrive at the host's home.

25. *If* You point your forefinger to your chest in Japan, it means you want a bath.

SPECIAL ASSIGNMENTS

A. See page 225.

B. Select the fragments from Nos. 1–20 of the Practice Quiz. On a separate sheet of paper, write the number of the fragment, followed by a complete sentence to correct the fragment.

EXAMPLE 14. Many informal meetings occur. OR Because many informal meetings occur, the conference room must always be ready.

C. Now select the run-ons and comma splices from Nos. 1–20 of the Practice Quiz. Write the item number, and then correct each one by either separating or correctly connecting the independent parts.

EXAMPLE 2. Anthony is the new accountant in our office, but he is not a CPA.

Submit this assignment to your instructor on _____. (date)

THE GLOBAL MARKETPLACE—INDIA

Hindi	Spell	Say	Currency
GREETING	namaste	NAH.mas.TA	rupee (Re)
DEPARTURE	kkhuda haviz	KOO.da HA.feez	
PLEASE	krupa	crew.pa	
THANK YOU	Shukrya	shook.RI.ah	

Courtesy of Sean Chandra

The Pauses That Refresh

Use Commas, Exclamation Marks, Periods, Question Marks

After Completing Chapter 11, You Will

> Use commas and end-of-sentence punctuation according to principles established for business communication.

A period is to let the writer know he has finished his thought, and he should stop there if he will only take the hint.

—*Art Linkletter*

See the *Instructor's Resource Kit* for Chapter 11 Pretest.

Because commas are the most frequently used punctuation and so many comma principles need to be learned, we have almost an entire chapter for one little mark. Chapter 11 covers mainly comma principles but also end-of-sentence punctuation.

In the 15th century, Aldus Manutius the Elder invented logical punctuation, printed the first semicolon, invented italics, and developed the first standard punctuation system. (Imagine what reading would be like without Manutius's invention.)

Read
57

I suggest starting with the **nonrules**—the three most common **comma myths**: (1) Use a comma where it sounds like it's needed. (2) Use a comma to break up a long sentence. (3) Use a comma before *and*.

Extra examples are in red to the left of comma rules.

We provide child care, services for the elderly, training programs, and much more.

When you speak, brief pauses help listeners follow the ideas you express. If you don't pause at exactly the right places, listeners will often let you know they don't understand. Then you repeat or rephrase some words to be sure your listeners understand. Writing, however, needs to be understood on sight since readers cannot easily ask for a repeat or rephrasing. When you're writing, punctuation replaces the pauses.

Good writers have developed a system for placing these pausing symbols—or punctuation marks. Although punctuation through hearing (placing a mark wherever it seems there should be a pause) is helpful, it's too inexact by itself for workplace documents. When you apply the following established principles, you no longer rely on a kind of guesswork for punctuation. These logical rules help make your writing precise and clear. We start by reviewing the system for the most commonly used punctuation mark—the comma. Then you will practice correct use of end-of-sentence punctuation (period, exclamation mark, and question mark).

A WARNING

As you complete the Recaps and Replays, insert a comma only when the rule applies. If you don't think of the rule first, you'll continue to make whatever errors you made in the past. So practice applying the Chapter 11 rules in each Recap, Replay, and end-of-chapter exercise.

WINE, WOMEN, AND SONG

Johann Voss, an 18th-century poet who may have been a bit of a reprobate,* wrote:

Who does not love wine, women, and song
Remains a fool his whole life long.

"Wine, women, and song" is an example of a series.

> **Use commas to separate items in a series of three or more words, phrases, or clauses.**

Notice the commas between the items in the series below—as well as before *and* or *or* that precedes the last **word, phrase,** or **clause.** Business communication experts use that final comma, although literary writers and journalists often omit it.

WORD SERIES Do you prefer **Catalina, Hawaii,** or **Bermuda** for your honeymoon?

PHRASE SERIES Joe believes in government **of the people, for the people**, and **by the people.**

CLAUSE SERIES **Amish men wear black felt hats all the time, Sikh men always wear turbans,** and **Orthodox Jewish men usually wear skullcaps.** [three independent clauses]

We believe **that your idea was good, that you planned carefully,** and **that you had a good product.** [one independent clause, *We believe*, followed by three dependent clauses]

---

*If in doubt, look up *reprobate*, now before you forget.

The film portays anger and envy and fear and love.

If a conjunction precedes each item in the series, omit the commas.

NO COMMAS Engineers **and** physicists **and** chemists are working on the project.

Apply the series rule for commas in these sentences. If you are tempted to add a comma for any other reason, DON'T! Write **C** beside the only sentence that doesn't require a comma.

_____ 1. Business etiquette is now so important to career success that the market is booming with videos, Web sites, books, and personal advisers on this subject.

___C___ 2. Employees of CBS and St. Vincent Hospital and G & D Construction all attended the workshop given by cultural consultant Angi Ma Wong.

_____ 3. Buicks, Fords, and Hondas are generally less expensive than Jaguars, Lincolns, or Audis.

_____ 4. Taiwan, Hong Kong, Singapore, and Korea have achieved high levels of industrialization.

_____ 5. First we studied the annual report, then we verified its accuracy, and finally we bought the stock.

_____ 6. Major differences among cultures include language, religion, politics, and the role of women.

_____ 7. Find out what kind of world you want to live in, what you are good at, and what you need to work at to build that world.

_____ 8. When you attend a business lunch, dress appropriately, use good table manners, avoid ordering something difficult to eat, and order nonalcoholic beverages.

_____ 9. The job of a top manager is to set broad objectives for the company, formulate strategies to meet them, and decide among alternate possibilities.

_____ 10. Executive etiquette classes offer instruction in how to meet, greet, and eat with clients of varied nationalities and races and religions.

See pages 405–406 for answers.

A cool, alert agent chose an innovative tracking system.

FOR ADJECTIVES ONLY

Apply the "series rule" to _three_ or more items. For adjectives, however, use a comma between just _two_ or more that describe the same noun. Imagine _and_ between consecutive adjectives. If it makes sense, use a comma to replace _and_, which is understood but omitted.

English is a living language—ever-changing, accepting, rejecting, expanding. (*SPELL Members' Handbook*)

AND OMITTED TEST

They died during the long, severe winter. [*And* makes sense between adjectives *long* and *severe*; therefore, use a comma instead of *and*.]

They died during the long and severe winter. [*And* **is** already between *long* and *severe*; therefore, no comma is needed.]

Insert a comma in the sentence that needs it, and write **C** beside the correct sentence.

_____ 1. She is an intelligent‸loyal employee.

___C___ 2. She is an intelligent and loyal employee.

The answers are on page 406.

REVERSAL CLUE

Another clue helps decide whether to use a comma between consecutive adjectives: reverse the order of the adjectives. If they make sense either way, you need a comma.

She is a **loyal, intelligent** employee

She is an **intelligent, loyal** employee.

Since the sentence makes sense either way, use a comma between the adjectives.

NO COMMA BETWEEN ADJECTIVES

If *and* doesn't make sense between the adjectives, do not insert a comma between them. Also try the reversal test. If the adjectives don't make sense reversed, don't use a comma between them.

YES **Many elderly** people died during the long, severe winter.

AND-IMAGINED TEST *Many* and *elderly people* doesn't make sense; therefore don't use a comma between these adjectives.

REVERSAL TEST *Elderly many people*—reversal doesn't make sense, further proof not to use a comma between the adjectives.

Try both the ***and*-imagined** and **reversal** tests with the adjectives in bold type below. Insert a comma in one sentence, and write **C** beside the other.

___C___ 1. Our **new advertising** booklet was completed last week.

_____ 2. The **annual financial** report was prepared by a **highly paid‸famous** accountant.

Check your answers on page 406.

Replay 58

Apply only Reads 57 and 58 comma rules. If you are tempted to use a comma for another reason, DON'T. Write **C** beside the sentence that needs no comma.

EXAMPLE   That company needs strong‸ aggressive sales representatives.

_____   1. A dazed‸ demure debutante dated a disheveled‸ double-dealing dancer in December.

_____   2. They are about to begin a unique‸ exciting adventure.

_____   3. He asked why you came‸ what you wanted‸ and what you expect to do.

_____   4. Our showroom is in a small‸ elegant building in a new part of Milwaukee.

___C___   5. Eric Smith is an ethical and expert financial planner specializing in socially responsible investing.

_____   6. Carla will call on accounts in Texas City‸ Dallas‸ and Austin.

_____   7. John Sculley's marketing expertise in a profitable‸ highly competitive industry was what Apple needed at that time.

_____   8. Sculley was offered a million-dollar bonus‸ a million-dollar salary‸ and stock options.

_____   9. Use *nd*‸ *rd*‸ *st*‸ or *th* with the day of the month only when that day precedes the month or stands by itself; for example, the *4th of May* or the *4th*—but *May 4*.

_____   10. That country has a settled‸ well-educated population.

Answers are on page 406.

Read 59

INDEPENDENTS' DAY

No, the title isn't spelled wrong, but then neither is it July 4. This is the day to review **independent** clauses and join them with a comma—as you did in Chapter 10.

> **Use a comma before a coordinate conjunction that joins independent clauses.**

Coordinate conjunctions are *and, but, or, nor, for,* and sometimes *yet*. Although *so* is a coordinate conjunction and appropriate in conversation, avoid using it in business writing. Each independent clause has a subject and a verb:

> Joe faxed printouts of the e-mail messages, but he forgot to send the invoices. [two independent clauses joined by a comma and but]

> Joe faxed printouts of the e-mail messages but forgot to send the invoices. [**one** independent clause consisting of the subject *Joe* with its two verbs—*faxed* and *forgot*; NO comma, please]

More people speak Chinese than any other language, but English is used in more countries.

Typewriters were first patented in 1714 but did not become practical until the late 1860s. (No comma because an independent clause doesn't follow *but*.)

> *Either/or, neither/nor,* and *not only/but also* clauses don't *sound* independent. They *are* independent, however, if each word group has a subject and verb.

> **Either** he lives in Savannah, **or** he lives in a South Carolina suburb. [two independent clauses joined by *or* and a comma]

> **Either** he lives in Savannah **or** in some other Georgia city. [one independent clause with no reason for a comma]

Recap

Insert two commas and two apostrophes. Write **C** beside the sentence that is already correct.

_____ **1.** Everyone in the office is extremely pleased with Pam's work, and we especially appreciate her proofreading ability.

_____ **2.** I am pleased with Rosalyn's work and appreciate her skill.

_____ **3.** Not only does Professor Caroll like sports, but she also enjoys music and art.

___C___ **4.** Darcy not only likes sports but also music and art.

See answers on page 406.

> For a short sentence (no more than about ten words) omit the comma when *and* or *or* joins independent clauses.

If *but, nor, for,* or *yet* joins the independent clauses, however, do use a comma before it regardless of sentence length.

> He won **and** I lost.   Either he built it **or** she did.
> He won, **but** I lost.   He really won, **yet** I can't believe it.

Replay 59

Apply the comma rules you studied on the previous pages. If you are tempted to insert a comma for another reason, **resist!** Write **C** beside the five sentences that do not require commas.

EXAMPLE She bought her husband a microwave oven for Christmas, and he bought one for her also.

___C___ **1.** International students bring to classrooms not only a variety of cultures but also a wide range of learning experiences.

___C___ **2.** All the students are better prepared for today's international workplace if students share information about their countries and their culture.

___C___ **3.** Ms. Mack is a member of the prestigious Theta Alpha Delta honorary business education sorority and will join us at the state convention.

_____ **4.** Twileen wants to learn to tap dance, but will she do it well?

_____ 5. Jesse spoke harshly to Twileen today, for she still hasn't completed the research.

_____ 6. Neither one can do the job, nor is it fair to ask someone else to do it.

__C__ 7. Either make a down payment or pay the total now.

__C__ 8. Mr. Penney treated employees as he would want to be treated were the situations reversed.

_____ 9. Until the 1800s ice cream was only for the rich, but that changed after the ice cream freezer was invented.

_____ 10. Around this time insulated ice houses also became available, and even more people were able to enjoy ice cream.

Check your answers on page 406.

CURTAIN RAISERS

A curtain raiser is a short play before the principal performance. The "curtain raiser" for a sentence is an introduction preceding the main part of the sentence. This **introductory expression** is always **dependent;** that is, it's not a complete thought any more than the theatrical curtain raiser can be presented without the main show. The audience would want refunds.

Many, but not all, introductory expressions are followed by a comma to separate them from the main idea—or independent clause—of the sentence.

INTRODUCTORY EXPRESSIONS

> **Use a comma after these one-word or other very short introductory expressions:** *Yes, No, Well, Oh.*

Yes, we'll be glad to introduce Lester to Esther.

> **Use a comma after an introductory "direct address"; that is, the name of the person you're writing to.**

Major Messerschmitt, have you designed any new planes lately?

> **Use a comma after an introductory expression that includes a verb.**

If you have, please send the blueprints to Gene.

When you're ready to fly, please let me know.

If at first you don't succeed, you'll get a lot of advice.

Recap Insert commas after introductory expressions, apply the series rule in one sentence, and write **C** for two sentences that are already correct.

_____ 1. Since Twileen wants to be discrete, she meets George in out-of-the-way places such as Hernando's Hideaway, Rosie's Retreat, or Larry's Lair.

_____C_____ 2. Yesterday George planned to meet Twileen at their favorite rendezvous.

_____ 3. When George went to his car, he found a note on the windshield.

_____ 4. Being in a hurry to meet her, he didn't read the note and stuffed it into his pocket.

_____C_____ 5. He didn't make it because he drove too fast and rear-ended a police car.

_____ 6. Had he read the note, he would have taken his time getting to Hernando's Hideaway.

_____ 7. Twileen, we think you should drop George like a hot potato.

Check your answers on page 406.

After a salutation in a business letter, a colon is often used.

Place a comma after an introductory expression of five or more words— whether it has a verb form or not.

Because of the unusual circumstances, we urged Twileen to stop seeing Jesse.

Under the sponsorship of George and Jesse, a conference on psychosomatic medicine will be held on July 2.

Even though an introductory expression has fewer than five words and no verb form, use a comma if it is needed for clearness or if it is a transition expression.

Once inside, the man requested food.

In addition, a few parts suppliers make special deals.

For years the Penney stores were called _The Golden Rule Stores_.

Commas are not required after short "place" and "time" introductories— unless the comma is needed for clearness or emphasis.

Within a month George will know the results of this unfortunate office romance. ["time" introductory]

In Antarctica the winter nights will be long for George. ["place" introductory]

Replay 60

Insert commas based on the rules given so far. If tempted to use a comma for another reason, be strong—**DON'T!** Add two apostrophes and write **C** for four correct sentences.

EXAMPLE   Although George didn't know it, Twileen and Jesse often worked late on the same projects.

_____ 1. When George walked through the French doors of the exclusive Flintridge Inn, he spotted Twileen at a romantic corner table.

_____ 2. As he approached her, he saw that she looked nervous.

_____ 3. In this attractive, modern dining room, you always find a courteous staff at your service.

_____ 4. As we all know, real merit is hard to conceal.

_____	5.	Examples of English words taken from Native American languages are _igloo_, _kayak_, _moccasin_, _skunk_, and _persimmon_.
C	6.	You'll receive the pagers for the holiday sale if you order now.
_____	7.	George discreetly motioned to Twileen, and they escaped through a side door without being seen.
_____	8.	In 1969 the world's first Internet message was transmitted at UCLA.
_____	9.	Being an alert salesperson, he noticed the prospect's gesture of annoyance.
_____	10.	No, he hasn't called on us either this month or last month.
C	11.	Please e-mail the dinner chairperson explaining why we can't attend.
_____	12.	If it is to be, it is up to me.
C	13.	No one in the world has more courage than the person who stops after eating one peanut.
_____	14.	Jesse and Twileen think West Covina would be a good location for the aviary, but the client believes it should be in Cucamonga.
C	15.	In Chicago some people used to sunbathe on a cement beach near Lake Michigan.

After checking your answers on pages 406–407 and reviewing any you missed, take the Pop Quiz on page 353.

FOUR EASY COMMAS

STATES

He moved from Kentucky to Virginia last year.

> **Enclose in commas a state, province, or country that follows the name of a smaller unit, such as a city.**

Did you visit Paris, France, or Paris, Kentucky?

I lived in Springfield, Massachusetts, and taught business classes at Bay Path College in Longmeadow.

DATES

Return this form by April 1, 2008, in the envelope provided.

> **With the American-style date, enclose the year in commas when it follows the day. If a specific day isn't included, commas are unnecessary. Do not use commas with the international- or government-style dates.**

AMERICAN   The dinner meeting was held on May 5, 2004, in Prescott, Arizona, at Yavapai College.

NO DAY   The next meeting will take place in April 2005 in Frederick, Maryland.

INTERNATIONAL   I might see him in Amsterdam, Holland, at the sales meeting on 31 July 2008.

Use *th*, *st*, or *nd* if *of* precedes the month or if the month isn't named; e.g., the 10th of May or on the 10th.

Enclose in commas elements of a date or time that explains a preceding date.

At 9 a.m., Wednesday, January 3, the sale on women's shoes begins.

ABBREVIATIONS

Use commas to enclose a college or professional degree that follows a name.

Millie S. Perry, Ph.D., Brad Rosenberg, M.D., and Sean Chandra, L.L.M., will be the speakers at the Eclectic Convention in Visalia, Montana.

A person or company with a title such as Jr., Sr., II, III, Inc., or Ltd. (following the name) may set off this title with a comma before and after it.

Charles Davis, Jr., and Charles Davis, III, have worked for Rokia, Inc., for five years.

If the Davises and the company choose not to use commas between their names and the titles, the sentence looks different:

Charles Davis Jr. and Charles Davis III have worked for Rokia Inc. for five years. [Use either both commas or neither.]

NAMES

When using the name of the person you're writing to in "direct address," enclose the name in commas.

Because of your expertise, Avonel, we'll appreciate your leading the discussion. The following will explain, Dr. Hellwig, just how this system works.

Writing and Conversing for Your Career

Trained communicators tend to use "direct address," because it's courteous and because most people like to hear or read their name:

Thank you, Ms. Leslie, for showing me around Grossmont last Tuesday.

Overuse of the technique, however, appears insincere.

Insert commas where required. Write **C** next to the two sentences that don't need a comma. If you are tempted to add a comma but don't know why, **DON'T!**

EXAMPLE   Marietta, Georgia, is the home of Chattahoochee Technical Institute.

_____   **1.** On March 9, 2004, we shipped you five MultiSync XP17 monitors.

___C___   **2.** In April 2004 Professor Van Vooren became the vice president of the college.

_____   **3.** Steven Smith, Ph.D., will head the Industrial Arts Department.

_____   **4.** One of Billy Crystal's first "gigs" was in a Long Beach, New York, nightclub.

_____   **5.** The deposition was taken on Friday, January 4, 8:30 a.m., in my office.

_____   **6.** Irish Linens Ltd. opened a new shop, and Ms. O'Callahan is the general manager.

_____   **7.** Little Rock, Arkansas, was the scene of tragedy and strife in the 1960s.

_____   **8.** Goldfinger's Variety Stores, Inc., closed the Yonkers, New York, store in 1993.

_____   **9.** We visited Edinburgh, Scotland, on 3 May and London on 6 May 2004.

___C___   **10.** Parliament members interviewed me on 7 May 2003.

Please check your answers on page 407.

A COMMA MEDLEY

Mark Twain wrote about his typewriter, "I'm trying to get the hang of the newfangled machine."

A medley, which means a mixture, usually refers to a musical arrangement made up of a series of songs or short musical pieces. What follows, however, is a medley of comma-use principles.

NOTES ON QUOTES

> **Use commas to separate short quotations (not over two lines) from the rest of the sentence.**

George said, "Isn't that Stella Glitter signing autographs in the lobby?"

William Safire wrote in response to missing commas on a U.S. census form, "Cutting commas this way is no way for the government to cut costs."

When handwriting, some people place an apostrophe directly over an *s* because they don't know whether it goes before or after. Another kind of fence-sitter places quotation marks directly over a comma. The correct placement is easy to learn by reading the next punctuation tip.

> **Place a comma or period *before* the closing quotation mark, never after— no exceptions in American English.**

QUOTE AT END OF SENTENCE	Years later, Woolworth's former boss complained, "Every word I used in turning Woolworth down cost me a million dollars."
QUOTE AT BEGINNING OF SENTENCE	"Money is the root of all evil," he replied.
QUOTE AT MIDDLE OF SENTENCE	Jesse whispered, "I love you," and then fainted.
QUOTE THAT IS INTERRUPTED	"Money," Jesse explained, "is the root of all evil."

> **Do not add a comma when a quotation ends with a question mark or an exclamation mark.**

"Did you ever tell a lie?" asked Ms. Ripley.

"No, I never did and never would!" exclaimed Mr. Ripley.

NUMBERS

> **In American writing, a comma separates thousands, millions, and billions when the number refers to a quantity. It is becoming common, however, to omit the commas in four-digit numbers.**

Alertness is required in international business because in some countries a period is used, not a comma. Use the form preferred in the country you work or live in.

$2,000 or $2000

$1,321,000 (USA and some other countries)

$1.321.000 (some parts of the world)

> **Do not use commas in numbers that "identify," such as addresses, serial numbers, page numbers, and so on. However, do use dots (instead of parentheses and hyphens) in telephone numbers.**

page 1247   19721 Victory Boulevard   No. 23890   420.681.6666

ADDRESSES

> **When address parts are in sentence form, separate the parts with commas; however, never use a comma between the state and the ZIP.**

Please return the software to Computer Learning Center, 3600 Market Street, Philadelphia, PA 19104, before June 8.

The art director lives at 35 Wynford Heights, Don Mills, Ontario M3C 1k9, Canada.

Replay 62

A. The best way to remember "Notes on Quotes" is to compose sentences that apply them. Possible Answers

1. Compose a sentence that includes a quotation at the end. Jeff told the students, "Use last names in the workplace unless you're sure first names are appropriate."

2. Compose a sentence with a quotation at the beginning. "I've e-mailed the price list he requested," said Ms. Kato.

3. Compose a sentence with a quotation in the middle. <u>The instructor said,</u>
<u>"Beginning some sentences with *I* is all right," and then gave item 2 as an example.</u>

4. Compose a sentence with an interrupted quotation. <u>"Love," Jesse said, "is</u>
<u>often combined with heartache."</u>

5. Write a sentence beginning with a quotation that is a question or an
exclamation. <u>"I'm married!" shouted Jeff.</u>

B. Commas in the following sentences are correct. Fill in the blanks with the comma rules that were applied.

1. Please, Mr. Gilmore, ship the merchandise to our distributor: Pueblo Art Inc., 900 West Orman Avenue, Pueblo, Colorado 81004 <u>Use commas</u> <u>for direct address. Use commas to separate the parts of an address that are in sentence</u> <u>form. Do not use a comma between state and ZIP.</u>

2. Our home office is at 4800 Freshman Drive, but we have offices all over the world. <u>Use a comma before a coordinate conjunction that joins</u> <u>independent clauses.</u>

3. On Friday, May 27, 2005, a permit was issued allowing me to buy 5,000 yellow tennis rackets. <u>1. Enclose elements of dates within commas. 2. In</u> <u>the U.S. style date, enclose the year in commas when it follows the day. 3. A comma</u> <u>may be used in four-digit figures if the number refers to quantity.</u>

4. Fred, if you refer to men as *Mr.* in the workplace, then use the equivalent title *Ms.* for women. <u>1. Use a comma after direct address. 2. Use a</u> <u>comma after an introductory expression that has a verb.</u>

5. For successful careers, employees must be dependable, cooperative, self-disciplined, courteous to co-workers and clients, effective in speech and writing, and appropriately dressed. <u>1. When needed for clearness,</u> <u>use a comma after an introductory phrase. 2. Insert commas between items of a series.</u>

Check your answers on page 407.

"Nonessential" includes traditional grammar terms such as *apposition, nonrestrictive, conjunctive adverb, parenthetical*, etc. Those terms probably don't help most students and may discourage many. If students learn the concept with the words *"nonessential/essential,"* they will understand what they might read later in a reference manual with traditional grammar terms.

TO COMMA OR NOT TO COMMA

To decide whether commas are required, you sometimes need to know whether a word group is "nonessential" or "essential." A nonessential expression may help the reader understand the rest of the sentence or may simply give additional information. The "nonessential" word or words do not, however, change the main idea of the sentence.

ENCLOSING NONESSENTIAL WORD GROUPS

> Enclose in commas a nonessential word or word group that appears to interrupt or be added on to the main idea. "Nonessential" means providing additional information that might be useful but isn't essential for understanding the main idea of the sentence.

Extra Example: Managers, from supervisors to company presidents, are often envied by people who have never held such jobs.

Each nonessential expression that follows is in bold print and enclosed in commas. If you cover the nonessential words, the rest of the sentence keeps its original meaning.

Nonessential Words and Phrases

Anything, **little or big,** becomes an adventure, **however,** when the right one shares it with you.

Professor Ingram-Cotton is, **of course,** an expert in this field.

Dr. Ellis Jones, **an expert in this field,** will lead the other discussion.

A training DVD, **"Banking for Older Customers,"** is in the employee library.

Patricia Noor, **a Saddleback College professor,** is an exceptionally fine business English instructor.

Nonessential Clauses

If a sentence opens with a dependent clause, it is introductory and a comma follows. When, however, the dependent clause **follows** the independent, use a comma only if the dependent is nonessential, like "although we cannot stay long."

Ms. Lilly K. Hall, **who is a LAN expert,** will chair the Detroit conference.

This business plan, **which we have studied carefully,** is unsatisfactory and needs to be rewritten.

We'll meet you at 5 p.m., **although we cannot stay long.**

Recap

Insert commas around nonessential expressions unless they end the sentence, in which case insert just one comma. One apostrophe is also needed.

1. Home, where I learned to walk and talk, holds a special place in my heart.
2. The course is divided into two parts, "Job Content" and "Work Environment."
3. The job you want, therefore, may not be available when you want it.
4. Our new system, which was devised by the American Records Management Association, eliminates considerable waste motion.
5. The guide gave us a fascinating tour of the old courthouse, even though he was worried about his wife's injury.

See the answers on page 407.

ESSENTIAL WORD GROUPS

> **Do not enclose essential word groups in commas. "Essential" words mean words essential to the meaning of the main idea.**

Imagine the bold-print words omitted; then see how the meaning of the rest of the sentence changes:

The employee **who avoids the more unpleasant tasks of a job** isn't doing the job.

A person **who is a LAN expert** should lead the discussion. Compare this with the sentence about Ms. Lilly K. Hall under the previous heading, "Nonessential Clauses."

Someone **like you** should lead the discussion.

Bring your resume with you **when you apply for a job.**

A national organization **that has a research bureau** stands behind our product.

Any pedestrian **who doesn't obey the caution signals** might be fined.

A city **that can qualify as a depressed area** is eligible to receive federal aid.

The ones **who complain the loudest** are generally those who contribute the least.

My sister-in-law **Bea** lives in Niagara on the Lake. [As I have three sisters-in-law, her name is "essential."]

WORD TO THE WISE

Here are two correctly punctuated sentences with totally different meanings resulting from use or absence of commas. Circle the one in which Ron is believed to be generous. Then check the answer on page 408.

Brad thought Ron was very generous.

Brad, thought Ron, was very generous.

Cross out commas enclosing essential word groups, or write **C** for correct.

_____ C _____ 1. Our best video, *Dealing with Customers*, shows a simple method for dealing effectively with angry customers.

_____ 2. An Arizona mapmaker was fired, because he had no sense of Yuma.

_____ 3. A woman student, must not enter men's rooms, without a chaperone, approved by the principal, or her representative.—Oxford Intercollegiate Rules for Women, 1924.

Check your answers on page 408.

> **Most prepositional phrases are "essential" and shouldn't be enclosed in commas. In addition, do not use a comma to separate a subject from its verb.**

NO   We saw your sales representative, at the conference, on May 2.

YES   We saw your sales representative at the conference on May 2.

NO   The Board of Directors, is considering Internet sales, to increase income.

YES   The Board of Directors is considering Internet sales to increase income.

Concepts are grasped more readily if you confess that nonessentials/essentials are a gray area of punctuation. Sometimes whether or not commas are used determines whether the expression is read as essential or nonessential; for example, My brother who is a playwright will be the keynote speaker. To decide whether to enclose "who is a playwright" in commas, you must know whether I have more than one brother and emphasis is required for the fact that he's a playwright. Reassure students that tests include only expressions that are clearly one way or the other. If they understand the principle, they will apply it correctly in their own writing.

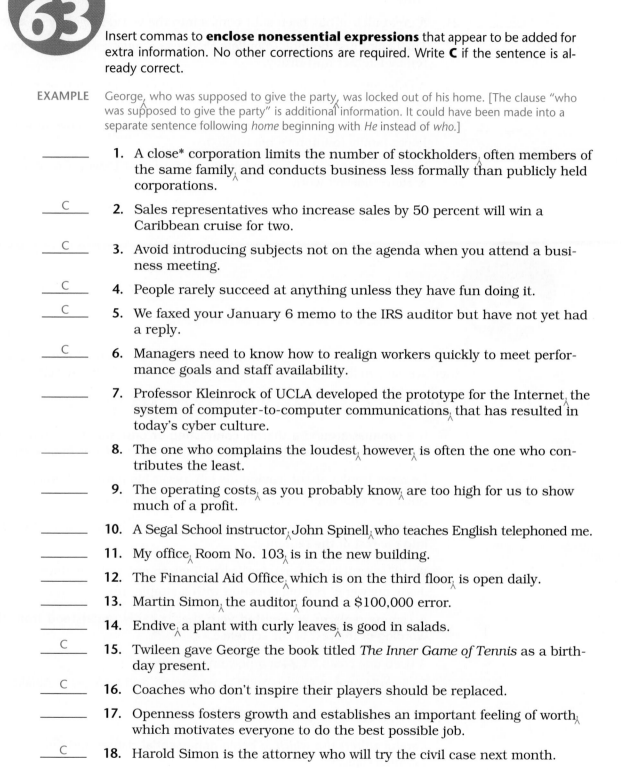

Replay 63

Insert commas to **enclose nonessential expressions** that appear to be added for extra information. No other corrections are required. Write **C** if the sentence is already correct.

EXAMPLE   George, who was supposed to give the party, was locked out of his home. [The clause "who was supposed to give the party" is additional information. It could have been made into a separate sentence following *home* beginning with *He* instead of *who*.]

_____   **1.** A close* corporation limits the number of stockholders, often members of the same family, and conducts business less formally than publicly held corporations.

___C___   **2.** Sales representatives who increase sales by 50 percent will win a Caribbean cruise for two.

___C___   **3.** Avoid introducing subjects not on the agenda when you attend a business meeting.

___C___   **4.** People rarely succeed at anything unless they have fun doing it.

___C___   **5.** We faxed your January 6 memo to the IRS auditor but have not yet had a reply.

___C___   **6.** Managers need to know how to realign workers quickly to meet performance goals and staff availability.

_____   **7.** Professor Kleinrock of UCLA developed the prototype for the Internet, the system of computer-to-computer communications, that has resulted in today's cyber culture.

_____   **8.** The one who complains the loudest, however, is often the one who contributes the least.

_____   **9.** The operating costs, as you probably know, are too high for us to show much of a profit.

_____   **10.** A Segal School instructor, John Spinell, who teaches English telephoned me.

_____   **11.** My office, Room No. 103, is in the new building.

_____   **12.** The Financial Aid Office, which is on the third floor, is open daily.

_____   **13.** Martin Simon, the auditor, found a $100,000 error.

_____   **14.** Endive, a plant with curly leaves, is good in salads.

___C___   **15.** Twileen gave George the book titled *The Inner Game of Tennis* as a birthday present.

___C___   **16.** Coaches who don't inspire their players should be replaced.

_____   **17.** Openness fosters growth and establishes an important feeling of worth, which motivates everyone to do the best possible job.

___C___   **18.** Harold Simon is the attorney who will try the civil case next month.

_____   **19.** You can buy the new instruction booklet, which gives complete information, for only $3.95.

---

*Also called *closed*. Close or closed corporations are regulated by the Securities and Exchange Commission (SEC), a government agency.

<u>C</u> **20.** Sandi uses English effectively and correctly and also speaks French fluently.

_____ **21.** A great idea, it has been said, comes into the world as gently as a dove.

_____ **22.** The book I just bought, *Complete Business Etiquette Handbook*, was written by Pachter and Brody.

_____ **23.** A redesigned Rolls Royce, the development of which is top secret, will be launched before too long.

<u>C</u> **24.** The man who just arrived from Taiwan was invited to move in with the Lopez family to improve his English.

_____ **25.** The new printer, the one we bought yesterday at Dewey, Cheatem, & Howe, doesn't work.

Please check your answers on page 408.

UNCOMMON COMMAS

Here are some comma rules that aren't required as often as the others. Since they are easy to learn, please add them to your comma repertoire.*

THREE OCCASIONALLY USED COMMA RULES

> **Use commas around a sharply contrasting or opposite expression. Such expressions are generally introduced by *not, never, seldom, but,* or *yet.***

He often thinks about leaving his job, never his parents' home.
She loves chocolate, yet breaks out when she eats it.

> **A comma is unnecessary before a word or words in quotations "woven" into the rest of the sentence.**

We all sang "Happy Birthday" when Twileen entered the office.
Jesse answered the big question with a simple "no."

> **A comma replaces an omitted verb that is easily understood from the wording of the rest of the sentence.**

A used one costs $1,245; a new one, $2,300.
Twileen is now in Cleveland in a new career and George, in Alaska.

WHICH AND *THAT* COMMAS

> **Careful writers use *which* to begin nonessential clauses and *that* to begin essential clauses.**

GOOD   Homes <u>that are beautifully decorated</u> sell faster. (essential clause)

Self-respect requires being treated like a human being, not a robot.

Mark Twain claimed to be the "first person in the world to apply the 'type machine' to literature."

Everything to the right of the @ sign is the domain name, **which is based upon the Domain Name System.** BUT . . . A meeting room **that is too comfortable** can make the participants sleepy.

---

*If you look it up, read all the meanings, not just the first; check pronunciation too. If you don't know any French, you'll be surprised at the pronunciation.

> However, if you can shorten a modifying clause or phrase with a modifying word, you improve the writing style through conciseness.

EVEN BETTER   Beautifully decorated homes sell faster.

GOOD   The office, <u>which has been rented by Mr. I. M. Rich</u>, is on the third floor.

EVEN BETTER   Mr. I. M. Rich rented the new third-floor office.

Replay 64

Insert commas where you can apply a Read 64 rule; otherwise, write **C**.

_____   **1.** He arrived at 10 p.m., not a.m.

_____   **2.** Twileen qualified for promotion to upper management; George, for transfer to Antarctica.

_____   **3.** Be sure to send for his brother, not his sister.

___C___   **4.** The system that was devised by Records Management Inc. is easy to learn.

_____   **5.** This system, which was devised by Records Management Inc., is easy to learn.

___C___   **6.** The system devised by Records Management Inc. is easy to learn.

_____   **7.** He works out at the gym every morning, never in the afternoon.

_____   **8.** In September we hired an administrative assistant; in October, a data entry clerk; and in November, a website manager.

___C___   **9.** My professor told me over and over again, "Learn touch keyboarding!"

___C___   **10.** In the old movie *Gone with the Wind*, Scarlett thinks she can do everything "tomorrow."

___C___   **11.** The No. 64 bus, which is usually late, was delayed for over an hour.

Check your answers on page 408. Then take the Pop Quiz on Reads 61–65, page 354.

Read 65

IN CONCLUSION . ! ?

We conclude this chapter with the end—the end of sentences. The three marks to end sentences are the period, the exclamation mark, and the question mark.

In an English grammar book written in 1582, the comma is described as "a small crooked point, which in writing followeth some small branch of the sentence, & in reading warneth us to rest there, & to help our breth a little."

Word Power

Periods are called *full stops* in British English. In American English the tendency is growing to say *dot* instead of *period*. When keyboarding, decide whether you want one or two spaces after end-of-sentence punctuation, but be consistent.

USE A PERIOD

Use a period after a statement, a command, a courteous request that is a statement but sounds like a question, words that stand for a sentence, or an indirect question.

We sent you the bill last week. [statement]

Pay your bill this week. [command]

Please pay your bill this week. [courteous request that is a statement]

Will (or Would) you please pay your bill. [courteous request that sounds like a question; a courteous request means action is desired, not a reply]*

Yes, of course. [words that stand for a sentence]

I asked whether you would pay your bill. [indirect question]

Use just one period after an abbreviation that ends a sentence.

Please ship the purple widgets c.o.d.

A new use of periods is to replace parentheses and hyphens in telephone numbers—although it's still all right to write phone numbers traditionally.

OK BUT GETTING OLD (800) 234-5678-9105

NEW LOOK 800.234.5678.9105

Word Power

Which one gets the job? No. 1 __1__ or No. 2 ____

1. He'll wear nothing that might discourage them from hiring him.
2. He'll wear nothing. That might discourage them from hiring him.

USE A QUESTION MARK

Use a question mark after a direct question—whether it's a sentence or words that stand for a sentence. A direct question calls for a reply.

Will you pay the bill this week or next?

Do you intend to pay this bill? If so, when?

Use a question mark after a sentence that might be considered presumptuous if punctuated as a courteous request.

Would you please handle my mail while I'm away? [question in a memo to your supervisor]

BUT

Would you please handle my mail while I'm away. [courteous request in a memo to your subordinate]

---

*When asking a customer to pay a bill, the seller wants money, not *yes* or *no.*

USE AN EXCLAMATION MARK

> **Use an exclamation mark to express strong feeling at the end of a sentence or words that stand for a sentence.**

Please send your check today!	He paid!	Run!
I can't believe it!	What great pizza!	How delicious!

An exclamation mark following a statement, command, or courteous request enables the reader to sense strong emotion or urgency. If the words are spoken instead of written, voice and facial expression transmit the strong feeling to the listener. Read these three correctly punctuated sentences aloud:

Will you order the pizza.   Will you order the pizza?   Will you order the pizza!

Writing for Your Career

An exclamation mark is often used in advertising copy and sales letters. Don't use it often in other workplace writing. With overuse, the exclamation mark loses effectiveness—like the parent who yells at the children frequently and finds they no longer respond to the yelling. Also avoid using an exclamation mark to knock the reader over the head with how wonderful, cute, or funny something is.

Replay 65

Add periods, question marks, and exclamation marks; correct comma splices with periods and capital letters.

1. I wonder whether he uses voice recognition software.
2. The pizza is good, but where's the pepperoni?
3. Would you please FedEx these items to us.
4. A winner says he fell, a loser says somebody pushed him.
5. Management makes important policies and decisions. we just carry them out.
6. Do you know what an FHA mortgage is?
7. Would Thursday be more convenient for you?
8. Buy UNEEDA now!
9. Will you please fax this report before you go to lunch.
10. Wonderful!

Check your answers on page 408. Then take the Pop Quiz for Reads 61–65, page 354.

Word Power

Which sentence has the exclamation mark in the right place?

Woman! Without her, man would be uncivilized. ___This one.___

Woman without her man would be uncivilized! _____

CHECKPOINT

Correctly used commas are written signals that can be as effective as a speaker's pauses and voice inflections. Misplaced or omitted commas can make a sentence difficult or impossible to understand or can totally change the meaning. Knowing when to use or not use a comma and using correct end-of-sentence punctuation are important in every sentence you write.

Place a check in the blank next to each punctuation principle you've learned. Review any you're not sure of.

_____ Use commas between items in a series as well as before the conjunction preceding the last item. If conjunctions are used before each item, do not add commas.

_____ Use a comma between consecutive adjectives if *and* is omitted but understood.

_____ Use a comma before *and, but, nor,* or *for,* and sometimes *yet* when one of those conjunctions joins independent clauses.

_____ Use a comma after an introductory expression—

    _____ that has a verb. _____ that has five or more words.

    _____ when necessary for clearness. _____ that is *yes, no, well,* or *oh.*

    _____ that addresses a person by name or title.

_____ Use commas to enclose a state, province, or country following a city.

_____ Use commas before and after dates that explain preceding days.

_____ Use a comma before and after an abbreviation following a name.

_____ Use commas to enclose direct quotations that don't end the sentence.

_____ Use commas to separate thousands, millions, and billions when the number refers to a quantity. Periods are used instead in some countries.

_____ Use commas between parts of addresses within sentences.

_____ Enclose in commas words that interrupt the main idea. Don't separate a subject from its verb with a comma. Most prepositional phrases should not be enclosed with commas.

_____ Use a comma to replace omitted verbs that are understood.

_____ Use a comma before a sharply contrasting or opposing expression. These often begin with a word such as *but, seldom, never,* or *not.*

_____ Do not use a comma before a quotation that blends in with the rest of the sentence.

_____ When choosing between *which* and *that* to begin a clause, use *which* with commas for "nonessential" words and *that* with no commas for "essential words."

_____ Use periods, question marks, and exclamation marks to end sentences appropriately.

Punctuation is defined in *Microsoft Encarta College Dictionary* as follows: . . . the standardized, nonalphabetical symbols, or marks used to organize writing into clauses, phrases, and sentences.

Word Power

Which sentence below was written by a misogynist and which by a philogynist?

    ___misogynist___ Women are pretty generally speaking.

    ___philogynist___ Women are pretty, generally speaking.

SPECIAL ASSIGNMENT

Insert commas, apostrophes, and end-of-sentence punctuation. Think of the rule before inserting a mark. Write **C** beside four sentences that are already correct.

_____ **1.** Do not tell your supervisor that you've been really busy when you're reminded about something you promised to do but haven't done.

___C___ **2.** Instead apologize and say when you will have it ready.

_____ **3.** I spoke with Ann Rosenblatt today, and you should hear from her by Thursday.

_____ **4.** Shortly before eight, thirty men appeared.

_____ **5.** As soon as we had eaten, the kitten jumped on my lap.

_____ **6.** If the only tool you have is a hammer, you might see every problem as a nail.

_____ **7.** "A man's best friend is his ape," said Tarzan.

_____ **8.** The psychologist said, "The most important asset of a student is the will to study."

_____ **9.** "The law," said the speaker, "is broken every minute of the day."

_____ **10.** Mr. Kuwahara, do you think the cost of medical insurance will increase or decrease?

_____ **11.** We believe that the owner, The Robinson Realty Company, will accept $75,000.

_____ **12.** Each alteration, omission, or addition costs us a great deal of money.

_____ **13.** Henry F. Albert, Jr., of Alberta, Canada, wrote the new book.

_____ **14.** The publisher's price on the book, which is entitled _America's Opportunity_, is $28.95.

_____ **15.** White leather shoes are selling fast, but how do your customers keep them clean?

_____ **16.** Get lost!

_____ **17.** Automobiles without brakes are dangerous and shouldn't be driven in Chaosville after Tuesday, November 31, 2003.

_____ **18.** Did you know that Rolls-Royce Motor Cars Ltd. takes just 30 days to produce each Rolls and Bentley?

_____ **19.** Robert M. Green, who sold drinks at a concession stand, usually prepared a popular drink containing syrup, cream, and carbonated water.

___C___ **20.** He inadvertently invented ice cream sodas one day when he used up his supply of cream and substituted vanilla ice cream in the drink.

___C___ **21.** The first ice cream soda was sold in Philadelphia in 1874.

_____ **22.** Hooray, I made a sale!

_____ **23.** Some years later in Evanston, Illinois, ice cream sodas were banned on Sundays because ministers complained that too many young people were drinking sodas instead of going to church.

_____ **24.** Soda-fountain owners came up with the idea of circumventing* the law by calling them "ice cream Sundays," which were ice cream sodas without the soda.

_____ **25.** The next development was that "Sunday," the spelling of the ice cream dish, was changed to "sundae" because using the Sabbath to name a food was declared irreverent.

_____ **26.** In New York City during the early 1900s the "egg cream" was invented—a drink made of carbonated water (called seltzer), chocolate syrup, and a small amount of milk.

_____ **27.** This drink, which became very popular, sold in two sizes—originally at 3 cents for the small glass and 5 cents for the large size.

---

*circumventing—a handy word to have in your vocabulary

_____ **28.** Of course, the prices increased during later years of the 20th century.

_____ **29.** In a way, I cannot understand how Samantha can fall in love with every Tom, Dick, and Harry.

_____ **30.** The property taxes for March 31, 2005, were $580; for June 30, $489; and for October, $643.

_____ **31.** Mr. Wong, not Ms. Wong, was appointed to the City Council.

___C___ **32.** In Seattle the large amount of rain results in lush foliage.

_____ **33.** Help!

_____ **34.** Should you space once after a period at the end of a sentence or twice?

_____ **35.** Will you please return my lawn mower this week.

 ## PROOFREADING FOR CAREERS

Punctuate with commas and end-of-sentence punctuation according to Chapter 11 principles. Delete unnecessary commas. If you keyboard the corrected letter, decide whether to use one or two spaces between sentences.

Dear Harbor Cove Cafe Owners:

My husband and I welcomed the Harbor Cove Cafe when it opened in our neighborhood, and we have been frequent diners ever since.

Friends often join us at Harbor Cove, and also enjoy the food and service. Some of these friends are recent widows and divorcees. As part of the healing process they need to establish their new identities, and self-esteem. They have learned that one helpful technique, is to pay their own way. They should not be forced to go through the unpleasant routine of dividing a bill to determine their share. Although my husband and I would be happy to include their charges in our bill, we respect their desires and their rights.

The same situation occurs when we need separate checks for expense-account reimbursement from our employers or for tax purposes. When we have asked for separate checks, however, the servers tell us of the policy of one check to a table.

Your care in the operation of Harbor Cove shows you know how essential good customer service is in a restaurant. Please be sensitive enough to understand and adjust your policy, which many other fine restaurants are now doing.

Sincerely,

After verifying your punctuation, evaluate your expertise. Good _____ Not good enough _____ Wrong places? _____ I'm suffering from comma trauma. Yes _____ No _____

PRACTICE QUIZ

www.prenhall.com/smith

If the commas, apostrophes, and end-of-sentence punctuation in the following sentences are correct, write **C** in the blank. Otherwise, write **N** and make the corrections. Capitalize the first word of new sentences you create.

N   1. On Monday, March 6, I expect to meet Mr. Lombard, president of Lake County Industries, Inc., in the town of Kiamesha Lake.

N   2. When you attend a meeting or business lunch, do you turn off the ringing device on your cell phone?

N   3. In a business communication class you can learn oral, written, and nonverbal strategies for success in a career. Can you also learn social skills?

C   4. At business meetings wear your name tag high on your right upper side for maximum visibility.

C   5. An advertising agency acts as an intermediary between a company that wants to advertise and the various media that sell space and time.

N   6. When traveling to other countries for business, know how to say hello, goodbye, please, and thank you in the country's language.

N   7. Of course you should also know the country's money and recognize its flag.

N   8. Do not, however, speak entire sentences in another language if you have not mastered that language. or !

N   9. In India, Senegal, and many other Asian countries, it is considered offensive to eat with your left hand.

N   10. When speaking with Chinese businesspeople, remember that the surname usually comes first and the given name last—unless, however, they were educated in a Western-type school.

N   11. Therefore, when speaking with Mr. Lo Win Chee, call him Mr. Lo, not Mr. Chee!

N   12. George, please avoid commas that are not necessary. also avoid overly long sentences when you write to us from Antarctica.

C   13. Charles Davis Jr. and Charles Davis III worked for Avco Inc. before they joined Metromedia Corp.

N   14. A factory closing means the community loses jobs, tax revenues, and retail sales.

C   15. You had better wear work shoes if you want to leave footprints in the sands of time.

N   16. After I lent Twileen one of my yellow tennis rackets, Jesse threw it into the pool.

N   17. If a dependent clause precedes an independent clause, put a comma after the dependent clause.

N   18. Yes, Jack Smith loved the English language more than flowers and wine.

N   19. "In Dayton, Ohio, we have two factories," explained Skyler's father.

C   20. George wondered whether the winters would be colder in Nova Scotia or Manitoba or Antarctica.

N   21. Dr. Waterman, who was a national authority on Ninja Turtles, developed a new video game.

N   22. Working quickly, he carefully organized the questions and the facts.

N   23. Although the new office furniture arrived, telephone service was not immediately available.

N   24. International Business Machines, General Motors, American Telephone and Telegraph, and several other major business organizations have become more autocratic in their internal relationships.

C   25. Thomas Carlyle referred to music as "the speech of angels."

GLOBAL MARKETPLACE—VIETNAM

Vietnamese	Spell	Say	Currency
GREETING	chao	chow	new dong (D)
DEPARTURE	tam biet	tom byet	
PLEASE	vui long	vooee long	
THANK YOU	cam o'n	come on	
I DON'T UNDERSTAND	Toi khong hieu	Toy kong hue	

Courtesy of Jackie Huynh

Punctuation Potpourri

12

Use ! . ? , as well as (; : " - — ')

"In spoken English, to add emphasis, the speaker must shout, glower, or gesticulate. In written English, the writer may add emphasis by underlining, italicizing, or punctuating (!!!) (?)."

—*William Safire*

After Completing Chapter 12, You Will

> Use 11 punctuation marks with precision: comma, period, question mark, exclamation mark, semicolon, colon, hyphen, dash, quotation mark, apostrophe, parentheses.

Chapter 12 Pretest is in your *Instructor's Resource Kit.*

Chapter 12 provides new information as well as overall review for correct punctuation: topics include comma splices, run-ons, fragments, hyphenating compound words, apostrophe review, comma usage, word division at end of line, and end-of-sentence punctuation—all holistically combined with additional punctuation principles.

"Rotten pot" is the literal translation of the French word *potpourri*, which in American English means a combination or mixture of ingredients. The Chapter 12 ingredients are dots, lines, and curves that when correctly placed enable readers to understand what they're reading. Look up *potpourri* in your college dictionary to see whether you pronounce it correctly; and while you're there, notice its definitions.

Leaving the French now, we move to a 15th-century Italian, Aldus Manutius, who published beautiful editions of the classics. He is the person chiefly responsible for systematizing punctuation. His system has been modernized through the years and adapted to various languages. The objective, however, is still to enable a reader to better understand a writer.

One of the best-known examples of double meaning caused by the absence of punctuation is the prophets' written reply to Roman soldiers. The soldiers asked whether they would return from the war. Since death was often the punishment for an inaccurate prediction, the Roman soothsayers took great care with their words: *ibis et redieris non morieris in bello* (you will go and return not die in war). However, as punctuation had not yet been invented, the reader could interpret it either way—that is, with either a pause *before* or a pause *after* "not." The absence of punctuation enabled the soothsayers to keep their heads no matter how the war turned out.

If you want to "keep ahead" in your career, modern punctuation helps you communicate with precision and without risk of criticism. In a few cases, you're offered choices where you must use judgment. However, most rules are specific, providing you with confidence in the correctness of your writing.

Although e-mail is less formal than other kinds of business communication, accurate punctuation is still required for creating clear messages.

Inaccurate e-mail punctuation leads to misunderstandings and gives the reader an impression of carelessness. Punctuate as carefully and skillfully in a business e-mail as you would in preparing a business communication on paper.

Punctuation affects meaning, clearness, ease of reading, and reader's emotions and mood, as well as how important or unimportant an idea seems to the reader. Punctuation expertise adds professional polish to your business writing.

Read

66

Few minds wear out; most rust out.

My manager is out of town; she left yesterday.

A comma before *then* joining independent clauses is one of the most prevalent comma splices. Mail the card today; then you'll get the gift.

According to our July statement, we filled your six orders; and we shipped them to your Jacksonville, Florida, warehouse.

THE HALFWAY MARK ;

The **semicolon** (;) is a halfway mark because it is midway in "pausing value" between a comma and a period. That's why it consists of one of each.

SEMICOLONS JOIN INDEPENDENT CLAUSES

Without Coordinate Conjunctions

> Use a semicolon between closely related independent clauses that are not joined by a coordinate conjunction—*and, but, or, nor, for,* or sometimes *yet.**

Success is getting what you want; happiness is wanting what you get.

---

*When *yet* joins independent clauses, it may be preceded by a comma, a semicolon, or a period. See page 217 for more transitions.

> **Use a semicolon or a period before a transition expression joining independent clauses. Use a comma after a transition expression of two or more syllables. See "The Transition Trap" on page 217 of Chapter 10.**

When you travel in other countries, your name sounds foreign**;** **therefore,** have a supply of your business cards ready to exchange with people you meet.

> **Omit the comma after short transitions—*then, thus, hence, still, yet,* and *also.***

The early 2000s were unforgiving years; thus marginal employees didn't survive.

The preceding sentences would also be correct with a period (and a capital letter) instead of a semicolon. To **separate** instead of **join,** use a period after the first independent clause. *You* decide whether joining or separating will read more smoothly.

With Coordinate Conjunctions

> **Place a semicolon before a coordinate conjunction that joins independent clauses—IF the sentence already has two or more commas.**

Professor Dennison, who works in Huntington, will show the slides; but we expect others to participate also.

Through the ages languages have gathered new words; but popular use, as Betty explained, often becomes proper use.

Recap

Insert semicolons and commas where needed. Be sure the clauses are independent before inserting a semicolon. Write **C** before two sentences that are already correct.

_____ 1. Job security hardly exists any more; now there is only skill security.

_____ 2. Even if unemployment is high, your skills will enable you to be placed quickly.

_____ 3. Here's a surefire way to double your money; fold it in half, and put it in your pocket.

___C___ 4. Dun & Bradstreet was one of the first to eliminate hyphens in its phone number and replace them with dots.

_____ 5. My employer, Mr. Anton, was upset by the criticism; and he refused to discuss the issue of why salary increases were unlikely.

_____ 6. Banks, he explained, pay depositors interest on their savings accounts; but borrowers must pay interest to the banks.

_____ 7. The Dvorak keyboard was a keyboard simplified to increase typing ease and speed; however, the qwerty keyboard is the one in general use.

___C___ 8. Look at the third row of keys on the left side of your computer keyboard if you don't know what a qwerty keyboard is.

_____ 9. Build a better mousetrap; then the world will beat a path to your door.

_____ 10. When speaking with clients whose English is limited, their nods and smiles do not necessarily mean they understand your English.

Check your answers on page 408.

Sentence 8 needs no commas or other punctuation because an "essential" dependent clause follows an independent clause. Think before placing commas—to avoid excess commas.

SEMICOLONS BEFORE CERTAIN TRANSITIONS

The factory is convenient to needed services; namely, a fast-food restaurant, a bus stop, and a post office.

> **Use a semicolon after an independent clause that precedes *for example, for instance, namely,* or *that is*—if one of these transitions introduces a list or an explanation that ends the sentence.**

When invited to dinner in Egypt, bring a gift; for example, flowers or chocolates.

Be prepared to discuss appropriate topics; that is, topics like Egypt's ancient civilization, Egyptian achievements, or the high quality of Egyptian cotton.

SEMICOLONS BETWEEN SERIES ITEMS BECAUSE OF COMMAS WITHIN

We are open weekdays from 8 to 5; Saturdays, 10 to 5; and Sundays, 1 to 5. [Commas used because *from* is omitted.]

> **If a series has commas *within* the items of the series, use semicolons *between* the items.**

NO AMTRAK stops at Schenectady, New York, West Burlington, Iowa, and Pasadena, California. [It seems as though AMTRAK makes five or six stops.]

YES AMTRAK stops at Schenectady, New York; West Burlington, Iowa; and Pasadena, California. [The reader easily sees just three stops will be made.]

A.   Insert semicolons, commas, and apostrophes where needed. Write **C** for correct sentences.

_____ 1. SHAPE is an acronym; that is, it's pronounced as one word meaning *Supreme Headquarters Allied Powers, Europe.*

_____ 2. England's famous naval hero, Lord Nelson, suffered from seasickness throughout his entire life but did not let it interfere with his career.

_____ 3. Our records show, Mr. Ho, that we filled six orders for you; and every one of them was delivered promptly.

_____ 4. Our records show that we filled six orders for you; every one of them was delivered promptly. (Yes, I know 3 and 4 are almost the same, but look again.—LRS)

_____ 5. It is considered good manners to leave a little bit of food on the plate when you've finished eating at someone's home in Egypt.

_____ 6. The new officers are Louise Fuller, president; Sandra Hall, vice president; and Gina Hecht, treasurer.

_____ 7. Use the semicolon properly; always place it where it's appropriate and never where it isn't.

_____ 8. Gene has one overpowering ambition; namely, to fly one of Willy's planes on a peace mission.

_____ 9. Mr. Kim has many assets that people don't know about; for example, he has a law degree.

_____ 10. As he was graduated from Northwestern University's drama department, he may get the part in the new play.

___C___ 11. The president of a big corporation generally earns a higher salary than the president of the United States.

_____ 12. A large business is highly complex and difficult to understand; it is divided into many departments in which people perform specialized functions.

_____ 13. As long as the government has the power to tax and private citizens still have considerable wealth, the government will not go bankrupt.

_____ 14. Like several other applicants, his computer skills are good; however, they selected him because of his excellent written and oral communication ability.

_____ 15. This year Fast Track, Inc., declared a dividend; that is, a sum of money paid to stockholders out of the company's earnings.

_____ 16. While doing research in Irish history, I learned about the Kilarney elections on 19 March 1941.

___C___ 17. Typewriters were first patented in 1714 but didn't become practical to use until the 1860s.

_____ 18. I am a great believer in luck and find that when I work harder, I have more of it.

_____ 19. I believe, therefore, that you're right about that issue.

_____ 20. It isn't hard work that kills; it's worry.

B. If you don't believe this event took place, insert one semicolon and one comma:
Charles the First walked and talked; half an hour after, his head was cut off.

Please check your answers on pages 408–409. Look closely at each punctuation mark to see whether yours match those on the answer page.

Read
67

AN EASY MARK :

Colon (:) principles are easy to learn.

You can be sure of this: We'll grant Mr.Greene no further credit.

> **Use a colon after an independent clause if a clause, a phrase, or even a single word explains or supplements the original clause.**

Just one word describes him: lovable. [*Lovable* is not capitalized.]

In an 1899 issue of the *Literary Digest*, a prediction was made about the horseless carriage: "Automobiles would never come into as common use as bicycles." [Capitalize the first letter of an independent clause that follows a colon.]

> **Use a colon after a complete sentence that introduces a quotation.**

For a three-or-more-line quotation, single-space and indent five spaces both left and right— instead of using quotation marks.

He added this statement to the contract: "The housesitter must provide food and affection for my 18 cats."

BUT

Since the contract states, "The housesitter must provide food and affection for my 18 cats," and she didn't do so, we will not pay her the agreed-upon amount.

> **Use a colon after an independent clause (a complete sentence) when a series follows.**

His goals were clear: health, wealth, and love.

These traits are required of his employees: initiative, loyalty, and honesty.

BUT

His clear goals were health, wealth, and love.

The personality traits required of his employees are initiative, loyalty, honesty, and dependability.

He likes to cook curry, tamales, and egg rolls.

> **If the items are listed vertically, use the colon whether or not the introduction is a sentence.**

The traits he's interested in are: The traits he's interested in are these:

Initiative   Initiative

Loyalty   Loyalty

Honesty   Honesty

Dependability   Dependability

> **Do not use a colon before a list if another sentence follows the introductory sentence.**

Please send the following people to my office. I need to see them immediately.

    Shuzu Itakara
    Frank Chang
    Sherrill Frank

> **"Standard Punctuation style" for business letters requires a colon after the salutation and a comma after the complimentary close.**

Ladies and Gentlemen: Sincerely yours,

> **"Open Punctuation style" means no punctuation after the salutation and complimentary close.**

Ladies and Gentlemen   Sincerely yours

> **Use a colon between hour and minutes and for proportions in technical writing.**

The fajitas were delivered at 12:30 p.m.

The ratio is 3:1. [When read aloud, this technical notation is "3 to 1."]

Replay 67

A. Insert colons where needed. Write **C** beside correctly punctuated sentences.

_____ 1. Please ship the following: two dozen Style No. 308 and three dozen Style No. 402.

___C___ 2. Ship these items to Birmingham, United Kingdom; Mysore, India; and Provence, France.

_____ 3. He has just one goal in life: revenge.

_____ 4. In three words I can sum up everything I've learned about life: It goes on.—Robert Frost

_____ 5. We plan to visit these cities: Winston Salem, Bisbee, Honolulu, and Rancho Cucamonga.

_____ 6. Two important things can be done to prevent shoplifting: Place mirrors in strategic locations and post special warning signs.

_____ 7. Los Angeles Harbor College is in an interesting harbor community: Nearby are fishing boats, oil refineries, and a home port for huge cruise ships.

_____ 8. Here is something worth thinking about: A small idea that produces something is worth more than a big idea that produces nothing.

_____ 9. Judges have a double duty: they must protect the innocent and punish the guilty.

___C___ 10. Evan Davis wrote, "Appearance counts greatly when an employee is to be chosen from among a number of applicants."

B. Punctuate this paragraph by adding one colon, one apostrophe, commas, period, and capitalization.

Yesterday Anthony Agresta decided to hire a new assistant and requested resumés from those interested. Applicants require at least an AA or AS degree, but preferably a bachelor's. The new assistant must also have these qualifications: excellent skills in Microsoft Word, Excel, communication, and interpersonal relations. The salary will be based on the applicant's education and experience.

Please check your answers on page 409.

Read 68

PLAGIARISM'S ENEMY

If you're not sure of what "plagiarism" is, look it up now to see why **quotation marks** are its enemy.

> **To avoid an accusation of plagiarism, use quotation marks before and after repeating someone else's exact words. Do not, however, use quotation marks when paraphrasing. Paraphrasing means expressing someone's stated or written ideas in your own words.***

*Be sure to include the name of the original speaker or writer whether you quote or paraphrase.

QUOTATION	He added, "The applicant should understand Windows 2002."
PARAPHRASE	He added that the applicant should understand Windows 2002.
QUOTATION	"A business," Elwood Chapman writes, "is an organization that brings capital and labor together in the hope of making a profit for its owners."
PARAPHRASE	The principal aim of a business, according to Elwood Chapman, is to make a profit for its owners.

QUOTATION: In 1905 President Grover Cleveland said, "Sensible women do not want to vote."

Paraphrase: In 1905 President Grover Cleveland thought that sensible, responsible women wouldn't want to vote.

Writing for Your Career

Good writers sometimes start with the quoted words and then find a suitable place to insert the source of the statement, as shown in the Elwood Chapman **quotation** above.

A frequently made error is to place commas and periods *after* quotation marks instead of before.

In an 1899 *Literary Digest* article, the author wrote, "The horseless carriage will never, of course, come into as common use as the bicycle."

HINT: Always place commas and periods **before quotation marks. Colons** and **semicolons,** however, are **typed after** quotation marks.— **Barbara Halperin, Nassau Community College**

"That's wonderful!" he exclaimed.

MORE NOTES ON QUOTES

These "notes" review some of "A Comma Medley" Read (62) and also introduce additional information about quotation marks.

> **For American English, insert closing quotation marks after (not before) the period or comma. British English rules differ.**

"Some publishers are born great, some have greatness thrust upon them, and others merely survive television," said John H. Johnson of *Ebony* magazine.

Mr. Johnson received the Magazine Publishers Association award for "Publisher of the Year."

> **If a colon or a semicolon is needed with a closing quotation mark, place the quotation mark before the colon or the semicolon—no exceptions.**

The following scientists were quoted in the article entitled "Rediscovering the Mind": Carl Sagan, Georgi Lozanov, and Jean Houston.

The accountant explained, "The check was accidentally postdated"; however, we have still not received a correctly dated check from the company.

> **If a question mark or exclamation mark is required with a closing quotation mark, "before-or-after" placement depends on the sentence. See explanations in brackets after each example.**

Jesse said, "Do you love me?" [The question mark is inside the quotation mark because it applies only to the quotation.]

Did he really ask, "Do you love me?" [Both parts of the sentence are questions; the question mark is still inside.]

Did you know that he said, "I love you"? [Question mark is outside quotation mark because it applies to entire statement.]

The pamphlet "Ten Ways to Lose Weight" is out of print. Marian Anderson sang "The Star Spangled Banner" at athletic events.

Information on early typewriters is from *A Brief History of the Typewriter*, a pamphlet from Remington Rand (the world's first typewriter manufacturer).

> **Use quotation marks for titles of subdivisions of published works, such as titles of articles in magazines or titles of chapters. Also use quotation marks for names of short poems, songs, lectures, and so on.**

This is a good time to review "Notes on Quotes," Read 62, and to attend the "Punctuation Points" workshop.

> **Use italics (or, if unavailable, underline) for a word or expression intended to draw attention to itself or that might seem out of place.**

Do you know how to distinguish between *effect* and *affect*? [or effect and affect]

Too many *don'ts* in your life can lead to frustration. [Do **not** add another apostrophe between the *t* and the *s* to make the contraction plural.]

> **Use italics for full-length published books, magazines, newspapers, films, and plays. If italics are not available, underline instead.**

Elwood Chapman wrote the book *Your Attitude Is Showing* (or Your Attitude Is Showing).

According to the *New York Times*, the film *Dances with Wolves*, which is about Native Americans, won several Academy Awards.

A. This Replay requires a potpourri of punctuation marks, including underlines (to show italics are needed); on your mark, get set, go! Capitalization is already correct.

1. He shouted, "Your house is on fire!"

2. "Your house is on fire!" he shouted.

3. Twileen whispered, "Are you sure you love me?"

4. "Are you positive you love me?" she whispered again.

5. Do you know whether Jesse said, "I love you?"

6. Germany's Helmut Kohl has been a strong force behind the euro, the universal currency for many European countries.

7. Jim Rogers wrote, however, that many European politicians had been eager to establish the new currency.

8. What does the phrase negotiable instrument mean? (or italics)

9. He faxed us as follows: "We depart from O'Hare on Unix Flight 23 at 8 a.m. and arrive at Kennedy at 10:30 p.m."

10. Was it Mr. Higgins, the character in My Fair Lady, who said, "Results are what count"?

11. "You need more tacos for the company party!" exclaimed Jim angrily.

12. "This shipment of boys' wear," the manager said, "will arrive in time for your January sale."

13. "Are you all right?" asked Ann. "Yes," groaned Dad, as he lifted Merriam Webster's Unabridged Dictionary.

14. "Bach's Suite No. 2 in B Minor" is first on the program at Laredo.

15. University of South Carolina Professor Benjamin Franklin is tired of people asking him, "Why aren't you out flying your kite?"

B. Capitalization is already correct, but punctuation is needed in the following paragraph:

When the New York YWCA offered typing instruction for women in the late 1800s, the schools managers were called "well-meaning but misguided ladies." The female mind and constitution were considered "too frail" to survive a six months' typing course. Upon completion of the course, the girls still faced strong opposition. The business world was a man's world: Men spoke "male" language, smoked strong "male cigars," and ignored the niceties.

Check your answers on pages 409–410; then take the Pop Quiz on page 355.

HALF A DASH -

Half a dash equals one **hyphen** (-). The hyphen and the dash are two completely different punctuation marks. In computer vocabulary a hyphen is called an *en dash*, and a dash is called an *em dash*. Dashes are introduced in Read 71. Correct **hyphen** use or nonuse is shown below and builds on the compound nouns you studied in Read 26.

WORDS SPELLED WITH A HYPHEN

> **Compound numbers from twenty-one to ninety-nine are spelled with a hyphen when they are not written in figures. The numbers *one hundred, five million,* and the like are not hyphenated. Hyphenate after *self* when it's joined to a complete word or *ex* when it means former.**

self-control   self-respect   ex-president   ex-husband

The only exception is *selfsame.*

> **When *non, over, under, semi,* or *sub* is a prefix, write the word without a space or hyphen.**

nonfat   overpayment   underexposed   semisweet   subhuman

> **If a prefix ending with *e*—such as *re, de,* or *pre*—begins a word that might confuse the reader, use a hyphen.**

re-cover   to cover again

recover   to get better from illness or to get something back that had been lost

> **Most *re, pre,* and *de* words do not require hyphens; however, some are optional if the main part of the word begins with *e*.**

reheat, deplane, predict   reelect or re-elect, preexist or pre-exist

The common error of calling the hyphen a dash often results in keyboarding a hyphen where a dash belongs. Remember that the hyphen is half a dash.

Compound Noun Groaner: A newspaperman doing a story in a South American jungle was captured by cannibals and delivered to their chief. The chief asked, "What you do for newspaper?" He replied, "I'm an editor." The chief said, "Ah, I have good news for you: tomorrow you be editor *in* chief."

non-Russian, sub-Arctic, off-the-wall, and off-white—but offshore and offstage

262   Chapter 12 | Punctuation Potpourri

> **Use a hyphen after a prefix preceding a capitalized word.**

mid-July   un-American

> **When in doubt about other words, consult the dictionary.**

The tendency over the years has been to drop hyphens. Many words that used to be hyphenated are now written solid. The only complete guide is an up-to-date dictionary. Words spelled with hyphens show hyphens in the entry word. When a dot or an accent mark appears between the syllables, write the word solid.

Recap

Consult your dictionary to see how the following entry words appear.

1. brotherinlaw   _brother-in-law_
2. vicepresident   _vice president_
3. semisweet   _semisweet_
4. uptodate   _up-to-date_
5. offlimits   _off-limits_
6. deescalate   _de-escalate_
7. exboyfriend   _ex-boyfriend_
8. rediscover   _rediscover_

Check the answers on page 410 or your dictionary.

COMPOUND ADJECTIVES

In addition to words **spelled** with hyphens, hyphens are also used for a purpose that cannot be verified in the dictionary: to join the elements of a **compound adjective.**

The dictionary is of no help in such cases. (Note how the colon is used in the preceding sentence.)

> **An adjective requiring two or more words is called** *compound* **and has a hyphen if it is a permanent compound shown in the dictionary OR if it precedes the noun being modified.**

These statistics are **up-to-date. OR** These are **up-to-date** statistics. [permanent compound expression as shown in dictionary spelling]

           comp adj   noun   comp adj   noun
You gave a **first-class** report about life in a **two-story** house. [Compound adjectives precede nouns *report* and *house.*]

BUT

You gave a report that was **first class** about life in a house with **two stories**. [Compound adjectives do not precede the nouns, and these are not permanent compounds.]

When the compound adjective follows the noun it modifies, the part of speech often changes; it is no longer a compound adjective. Being aware of this grammar change is unimportant. Just remember to hyphenate when the modifier comes **before** the noun or if the expression is spelled with a hyphen in the dictionary.

Extra Practice:
Do you know the difference between a **small business** owner and a **small-business** owner?

I attended a one-day workshop that included six 45-minute sessions. A hardworking park ranger supervising government-owned lands bought a tax-exempt bond.

A **real estate** agent bought a **Fifth Avenue** shop.

Imagine a word omitted from a possible compound adjective. See whether it makes sense and the remaining word keeps its meaning. In the sample sentence about the report and the house, the meaning is changed completely if you refer to a **first report** or a **story house**. Since **first class** and **two story** cannot be separated, they are compound adjectives. Join them with a hyphen if they precede the noun being described.

If the first word of the compound expression is an *ly* adverb, the hyphen is not required. If, however, the *ly* word is an adjective, do use the hyphen.

The **fashionably dressed** executive carried an **Italian leather** bag. [No hyphens are required because *fashionably* is an adverb modifying the adjective *dressed*; *Italian* and *leather* are both adjectives but not compound because *leather* can be used without *Italian*—see preceding Word to the Wise.]

My employer is a **friendly-looking** man. [Friendly is an adjective; therefore, use the hyphen to create the compound adjective modifying *man*.]

Do not hyphenate common compound expressions that represent a single idea.

A **high school** student found a **Social Security** check in a **mobile home** park.

Recap

Insert hyphens where needed. See your dictionary when in doubt about whether an expression is a permanent compound. Write **C** beside the correct sentence.

_____ 1. Buy your back-to-school clothes now and some Windows-based software.

___C___ 2. If you're a charge account customer, you can buy clothes now for going back to school.

_____ 3. He has a part-time job planning never-to-be-forgotten parties.

_____ 4. This seven-foot-tall basketball player understands the problem-solving process.

Check answers on page 410.

Suspended Hyphens

Suspended hyphens are used within a compound adjective that includes an interrupting word (or words). Space before and after the interrupting word/s.

The financial adviser said, "Both long- and short-term gains must be considered."

I ordered $8\frac{1}{4}$- by $11\frac{1}{2}$-inch paper for correspondence with European businesses.

WORD DIVISION

The third major use of the hyphen is dividing words at the end of a line. Here's the most important principle:

> **If you plan to divide words at the ends of lines, activate hyphenation on your word processor. That will assure that you divide only between syllables.**

If a word is divided at a place other than between syllables (fl-ower), the effect of the entire document is destroyed as the reader wonders, "*Where* did that writer go to school?" Or "*Did* that writer go to school?"

Whether to divide at the end of a line is discussed in Appendix D. If you set your word processor for automatic hyphenation, undo the following automatic word divisions: (a) proper nouns, (b) hyphens on more than two consecutive lines, or (c) the last line of a page.

Word processing dictionaries hyphenate at the end of almost any syllable, which is fine for newspapers and certain other written materials. For professional looking documents, however, use the specific word division rules in Appendix D. If you keep that page and a dictionary close by (and refer to them) when writing, you'll soon know the word division principles for correspondence, reports, and similar documents.

Writing for Your Career

Avoid distracting the reader by word division that might amuse or momentarily confuse:

Please send me your cat-
alog. Just over the horizon-
tal line, you'll see a number. ☹   cat?   horizon?   dog?
He had a blind date with a dog-
matic woman.

A. With the goal of conciseness, rewrite these correct sentences so that they have compound adjectives requiring hyphens. Do not change the meaning. Keep your dictionary handy.

EXAMPLE   Lionel Barrymore was an actor who was well known.

Lionel Barrymore was a well-known actor.

1. I work in a building that has 100 stories. I work in a 100-story building.

2. Do you need a ladder that is 10 feet or one that is 20 feet? [Use suspended hyphens.] Do you need a 10- or a 20-foot ladder to do the repairs at Golden West College?

3. My father is a man who works hard. My father is a hard-working man.

4. The case against the company that is based in Dallas was handled in Seattle. The case against the Dallas-based company was handled in Seattle.

B. Some words in the sentences below need a hyphen or space between the parts. Use the dictionary. Write the correctly spelled word in the blank, or write **C** if no correction is needed.

5. The artist feels that recreation of the entire scene is possible. ___re-creation___

6. What is your favorite form of recreation? ___C___

7. Our overall objectives are similar, but our methods differ. ___C___

8. You should report underpayments as well as overpayments. ___C___

9. My fatherinlaw acts like a commanderinchief. ___father-in-law___
   commander in chief

10. Professor Foster made an offhand comment about a selfmade millionaire.
    self-made

11. Do you think our country will ever produce enough oil to be selfsufficient?
    self-sufficient

12. The newly developed procedure is unavailable for general use.
    C

13. Her goal is owning one hundred pairs of shoes when she is twenty one.
    twenty-one

14. This financier is a Johnny come lately whose effect on the market is
    overstated. ___Johnny-come-lately___

15. A person who acts as though intellect and reason are unimportant to
    solving world problems is called an antiintellectual. ___anti-intellectual___

C. Refer to *Read About Word Division* in Appendix D and verify syllables in your dictionary. Use a vertical line (|) to show the preferred place to divide the following words at the end of a line. If a word should not be divided at all, place the vertical line at the end of the word.

EXAMPLE   catallog headed|

function   believe   horizontal   wouldn't| thousands   punctuation
aligned| impossible   interrupt   syllables   stopped| guesswork

Answers are on page 410.

WILD APOSTROPHES '

Our punctuation potpourri could not be complete without a reminder to replace wild apostrophes with tamed ones so that they'll be exactly where they belong. Here's a quick review of possessive nouns, as studied in Chapter 9:

A possessive noun shows ownership, authorship, place of origin, type of use to which something is put, and time relationship. A possessive noun always ends with an 's or s'.

SINGULAR AND PLURAL POSSESSIVES

> **To make a singular noun possessive, add 's.**

The boss's office	Franklin's notebook
Ms. Jones's secretary	a semester's work

> **To make possessive a singular proper noun with two or more syllables that ends in an *s* sound, you could omit the added *s* to avoid a hard-to-pronounce word.**
>
> Socrates' disciples   Ms. Perkins' report   Dr. Martinez' prescription
>
> **To make a plural noun possessive, first look at the last letter of the plural noun. If the last letter is *s*, add only an apostrophe; if the last letter is not *s*, add 's.**

According to Judy's report, Mr. Atkins' sisters had several years' experience making children's clothes for New Haven's boys and girls. Ms. Hendrix' letter is in Dr. Jacobs' office.

Last Letter of Plural Is *S*

Adamses	The Adamses' factory is closed.
weeks	Three weeks' work was wasted.
ladies	He designs ladies' clothes on the 5th floor.

Last Letter of Plural Is Not *S*

alumnae	Who collected the alumnae's contributions?
men	The men's fortunes were lost at these gaming tables.
children	Our children's room is neat and clean.

POSSESSIVES VS. PLURALS

> **Use an apostrophe only if the possessive relationship of a noun is evident. Do not use an apostrophe for a nonpossessive plural.**

The Joneses own factories all over the world. [*Joneses* is the plural subject of the sentence; *own* is the verb; no possessive relationship is shown.]

BUT

The Joneses' factories are all over the world. [*Factories* is the plural subject of the sentence; *Joneses'* tells whose factories and is therefore possessive.]

JOINT OR SEPARATE OWNERSHIP

> **To show something is possessed jointly, add the apostrophe to the final owner. To show separate ownership, add the apostrophe to all owners.**

Rozini and Marino's factories [jointly owned]

Rozini's and Marino's factories [individually owned]

Replay 70

Insert apostrophes where needed and correct any *s* errors based on Chapter 9 and Read 70 principles.

1. This sales representative's approach is to get her assistant's opinion of the new hardware and software. [The sales representative has one assistant.]

2. The comments are taken from Patricia Hill's pamphlets. [Her name is Hill.]

3. Our women's and girls' jackets are on sale for prices that fit your pocketbook.

4. Here's a statement from Donna and Gary's book: "It takes more than money's worth to satisfy buyers—a fact that must be implanted in every businessperson's mind."

5. Ms. Watkins' memos are about scheduling dates for next year's programs. [Her name is Watkins.]

6. Germany's Thyssen Co. was sold to America's Giddings & Lewis.

7. Mercury and Gemini astronauts attended last weekend's National Science Teachers' Association convention.

8. My wife's aunt is my favorite in-law; she lives on New York's upper east side.

9. The Lopezes have bought an Internet provider called Cyberwire.

10. Dr. Lopez's husband is Baruch Meeks; they are co-editors of an online news service.

Answers are on page 410.

Read 71

GOOD MARKSMANSHIP—( )

These contracts—and let no one underestimate their importance—must be fulfilled to the letter.

We are not interested (not now, at least) in your offer.

A business writer with good marksmanship hits with precision even the less frequently used marks—the dash and the parentheses. When handwriting, insecure writers tend to use the dash like a security blanket. When not sure whether to use a period, comma, semicolon, or colon, they use a dash and hope the reader assumes the writer was in a hurry. Although parentheses errors are rare, some suggestions for appropriate use are included.

> **Dashes emphasize nonessential (as defined in Read 63) expressions ordinarily enclosed with commas. Parentheses de-emphasize such expressions.**

ORDINARY
The president of this company, a man who once earned $10 a week as a janitor, is one of the richest men in the world.

EMPHASIS
The president of this company—a man who once earned $10 a week as a janitor—is one of the richest men in the world.

DE-EMPHASIS
The president of this company (a man who once earned $10 a week as a janitor) is one of the richest men in the world.

Of the preceding sentences, a good writer would probably choose dashes to emphasize an interesting—but nonessential—point. However, the commas and parentheses also result in correctly punctuated sentences. In the next example you can emphasize the *main* idea of the sentence by choosing parentheses to de-emphasize the nonessential information.

DE-EMPHASIS   Our supervisor (a new employee) is a holography expert.

However, if you wish to emphasize that the supervisor is new on the job, choose dashes for the nonessential expression.

EMPHASIS   Our supervisor—a new employee—is a holography expert.
Commas would also be correct.

> **Use dashes for emphasis or parentheses for de-emphasis before and after a nonessential expression containing one or more commas. Do not use commas around the nonessential expression; they make such a sentence harder to understand.**

Some of these sentences are correct with either parentheses or dashes. Exclamation marks and dashes "shout" and parentheses "whisper." Shout too much, and people stop paying attention. Whisper too much, and people think you're afraid. See your *Instructor's Resource Kit* for additional dash and parentheses examples.

NO   Mr. Slimton, director of the Sebastopol, Scottsdale, and Albany offices, said abuses have been common in ads for many weight-loss products.

YES   Mr. Slimton (director of the Sebastopol, Scottsdale, and Albany offices) said abuses have been common in ads for many weight-loss products. [de-emphasis of nonessential expression]

NO   My plan saved the company thousands, no, it was nearer tens of thousands, of dollars last year.

YES   My plan saved the company thousands—no, it was nearer tens of thousands—of dollars last year. [Emphasize this "nonessential" expression.]

> **Use a dash after a word or word group that precedes an independent clause.**

YES   Dependability, good attitude, and efficiency—those are qualities required in our employees.

OR   These are the qualities required in our employees: dependability, good attitude, and efficiency.

OR   The qualities required in our employees are dependability, good attitude, and efficiency.

> **Use parentheses to enclose directions.**

The profits (see chart, page 7) were the highest in the history of the company.

Writing for Your Career

Keyboard a dash on your computer by accessing an *em dash*, not the *en dash*: For the *em* dash, type two hyphens with no space before or between or after.

Em Dash   A diamond is the hardest stone—to get. (two hyphens)
En Dash   Use for a range, such as *from 100–200*. (one hyphen)

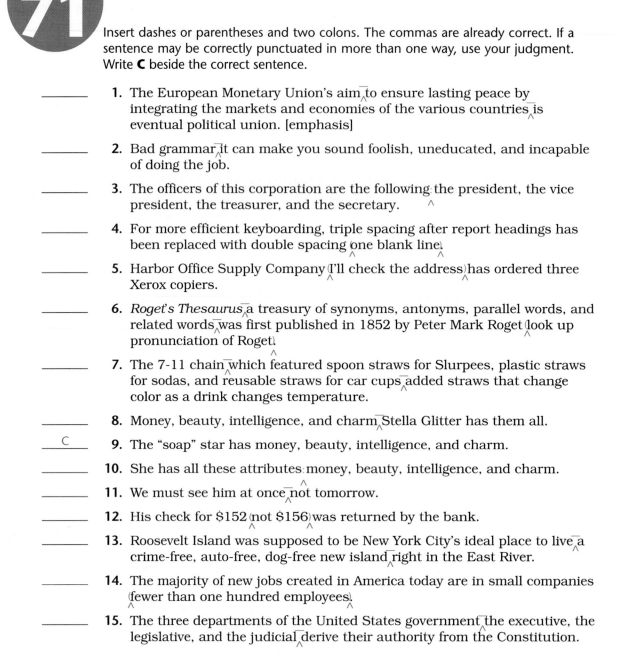

Replay 71

Insert dashes or parentheses and two colons. The commas are already correct. If a sentence may be correctly punctuated in more than one way, use your judgment. Write **C** beside the correct sentence.

_____ 1. The European Monetary Union's aim to ensure lasting peace by integrating the markets and economies of the various countries is eventual political union. [emphasis]

_____ 2. Bad grammar it can make you sound foolish, uneducated, and incapable of doing the job.

_____ 3. The officers of this corporation are the following the president, the vice president, the treasurer, and the secretary.

_____ 4. For more efficient keyboarding, triple spacing after report headings has been replaced with double spacing one blank line.

_____ 5. Harbor Office Supply Company I'll check the address has ordered three Xerox copiers.

_____ 6. *Roget's Thesaurus* a treasury of synonyms, antonyms, parallel words, and related words was first published in 1852 by Peter Mark Roget look up pronunciation of Roget.

_____ 7. The 7-11 chain which featured spoon straws for Slurpees, plastic straws for sodas, and reusable straws for car cups added straws that change color as a drink changes temperature.

_____ 8. Money, beauty, intelligence, and charm Stella Glitter has them all.

__C__ 9. The "soap" star has money, beauty, intelligence, and charm.

_____ 10. She has all these attributes money, beauty, intelligence, and charm.

_____ 11. We must see him at once not tomorrow.

_____ 12. His check for $152 not $156 was returned by the bank.

_____ 13. Roosevelt Island was supposed to be New York City's ideal place to live a crime-free, auto-free, dog-free new island right in the East River.

_____ 14. The majority of new jobs created in America today are in small companies fewer than one hundred employees.

_____ 15. The three departments of the United States government the executive, the legislative, and the judicial derive their authority from the Constitution.

The answers are on pages 410–411. A Pop Quiz for Reads 69–71 is on page 356.

CHECKPOINT

Whenpeoplestartedwritingtheyputonewordafteranother. Later, writers separated words with spaces. Eventually they began to mark their writing with dots, dashes, and curves. Today we know punctuation is the most important single device leading to easy reading. With punctuation, writers imitate spoken language on paper.

All punctuation commonly used in workplace writing except for the comma is explained in this chapter. Write checks in the blanks below for the punctuation (and italics) principles you are confident about, and review the appropriate Reads for those you're not sure of.

_____ Use a period, question mark, or exclamation mark at the ends of sentences or expressions that stand for sentences.

_____ Use a period after a courteous request, even if it is worded like a question.

_____ Four reasons are given for correct semicolon use. What are they?

_____ Use a colon after an independent clause when (a) words that explain the clause follow it, (b) a series or list follows it, or (c) a quotation follows it.

_____ Use a colon after the salutation of a letter with standard punctuation, between the hour and minutes, and in ratios (or proportions).

_____ Use quotation marks around direct quotations, subdivisions of published works, and other published materials.

_____ Use italics for a quotation within a quotation; for emphasis; to indicate a foreign word; for titles of published books, articles, newspapers, magazines, etc.; and for an expression used to draw attention to itself.

_____ Three rules determine whether the quotation mark is before or after another mark of punctuation. What are they?

_____ Use a hyphen (a) as part of the spelling of a word hyphenated in the dictionary entry, (b) in a compound adjective preceding the noun being described, or (c) for word division at the end of a line.

_____ Apostrophes show ownership, authorship, kind, and time relationships. How do you know whether the apostrophe goes before or after an _s_? (see Chapter 9)

_____ Read 50 Chapter 9 also shows apostrophe uses for other than possessive nouns. Review these now.

_____ Use dashes to set off (a) a nonessential expression to be emphasized or one that has commas in it, (b) a series that has commas within the items, or (c) a word or word group preceding a complete thought.

_____ Use parentheses to enclose (a) directions, (b) a supplementary expression, or (c) a word or words to be de-emphasized.

SPECIAL ASSIGNMENT

Select ten punctuation principles from Chapters 9, 11, and 12. Keyboard the principles concisely, but completely. Then compose a sentence to apply each rule. By applying more than one principle within the same sentence, you may write fewer sentences. Submit your Special Assignment on _____. (date)

EXAMPLE   **Principle 1:**   Use a semicolon before a transition expression that joins two independent clauses.

Principle 2:   Use a comma after a transition expression except for the very short ones.

Sentence for 1 and 2:   An executive needs to make decisions quickly; in fact, decisions are often necessary before all the facts are available.

PROOFREADING FOR CAREERS

Proofread and correct this letter, which has various errors including the need for punctuation. Start by using the spelling checker. Then use a closed pen below each word as you read it on the screen. Read for content as well as for punctuation, grammar, and spelling accuracy so that you'll spot the various errors. Next print the letter and proofread it from the printed copy. If you tend to overlook errors, the final step is to read the document from the bottom up. Often one finds additional "typos" and spelling errors that way.

Dear Ms. Shay:

We appreciate your inquiry about socialy responsable investing; unfortunately, we are not licensed in your State. However, if you plan to be in Washington State in the future, we'll be happy to be of service.

If you call our office at 1.206.448.7737, we can probably refer you to a Social Investment Forum member in your area. Otherwise, you might want to contact the Social Investment Forum for the names of advisers and financial planers in your locale:

Social Investment Forum, PO Box 57216, Washington, DC 20037, 202.872.5319

We would be pleased to hear from your friends and family in the states we are currently licensed in: Washington, Orlgon, New York, California, Colorado, Kansas, Massachusetts, and Florida. We specialize in socialy, environmentaly, and financialy responsible investments specificaly taylored to fit each investor's needs.

Thank you for your interest.

Sincerely,

Eric A. Smith, CFP

After checking your corrected letter with a key or with your instructor, answer these questions: What kinds of errors if any did you make in completing this proofreading practice? e.g. punctuation? spelling? carelessness? _____
How do you evaluate your English and proofreading skills?
Excellent _____ Good _____ Fair _____ Needs Improvement _____ Other _____

PRACTICE QUIZ
www.prenhall.com/smith

This quiz is based on Chapters 9, 10, 11, and 12. If a sentence is correctly punctuated, write **C** on the left; otherwise, correct it. Take the Practice Quiz as though it were a real test. Don't look back through the book until you've finished the test.

_____   **1.** Obviously upset by the criticism, he refused to make necessary changes.

___C___   **2.** Have you read the article entitled "Increase Your Vocabulary"?

_____   **3.** Jennifer Crystal Imports, Inc., is conveniently located, that is, just two blocks south of the Long Beach exit of the Long Island Expressway.

___C___   **4.** Although I am a great believer in luck, I find that the harder I work, the more I have of it.

_____   **5.** In the Hartford office, for example, this plan saved the company thousands of dollars every year for the past five years.

_____   **6.** His itinerary includes sales calls in Springfield, Massachusetts, Urbana, Illinois, and Galveston, Texas.

_____   **7.** Regarding unneeded commas, Professor Lorraine Ray of Ohio University wrote, "Develop the courage to just say no!"

_____   **8.** Elizabeth, who is a good listener, never once interrupted while I read what must have seemed to be a neverending story.

_____   **9.** When Good Queen Bess, who could swear like a sailor, enjoyed one of her rages, the whole Court trembled.

_____   **10.** This booklet contains hard-to-find facts about Aldus Manutius, a 15th-century Italian.

_____   **11.** His boss shouted, "You're fired!"

___C___   **12.** He had allowed the usual discount; namely, 8 percent off for cash payment.

___C___   **13.** "The sale of men's and boys' coats will be held next week," said the manager.

_____   **14.** Keep a space for study on a table or desk; also set aside a time for study each day.

_____   **15.** One thing about the Information Age is certain: The technology changes affect us all.

___C___   **16.** "The man who lies down on the job," so the lecturer said, "deserves to get run over."

_____   **17.** Our enclosed brochure is self-explanatory and may interest you and your employees.

_____   **18.** PLEASE CLEAN TUB AFTER BATHING—LANDLADY

_____   **19.** Father's Day neckties often get stains because new silk loves spaghetti sauce.

___C___   **20.** John Wanamaker believed that the American system of storekeeping was the most powerful factor discovered to compel minimum prices.

Insert parentheses or dashes where needed, and write **C** beside the one correct sentence.

_____   **21.** In the American system only one person—the president—can be commander in chief of the armed forces. (emphasis)

_____   **22.** The prices of greeting cards and desk sets (see page 46 of the catalog) are reduced.

_____   **23.** Forget the advertisers' promises, it will take a lot more than three minute's a day on the "ab" machine—that you'll drop from a size 14 to an 8. (several advertisers)

_____   **24.** We were cold—freezing is more accurate—when we walked on the glacier. (emphasis)

___C___   **25.** Wade, a gentleman with esoteric interests, collects 1962 Seattle World's Fair memorabilia.

THE GLOBAL MARKETPLACE—FRANCE

French	Spell	Say	Currency
GREETING	bonjour	bone.joor	franc (F)
DEPARTING	au revoir	oh.rev.wahr	
PLEASE	s'il vous plait	seel voo play	
THANK YOU	merci	mare.SEE	
I DO NOT UNDERSTAND.	Je ne comprend pas.	Juh nuh, com-pran' pa.	

Random House Unabridged Dictionary

Getting Your Act Together

Write and Speak Clearly, Correctly, Logically, and Concisely

After Completing Chapter 13, You Will

> "Get Your Act Together" by reviewing the most common writing faults and replacing them with clear, correct, and logical language.

> Reinforce the grammar, word choices, and punctuation you practiced in Chapters 1–12.

"If language be not in accordance with the truth of things, affairs cannot be carried on to success."

—Confucious

See *Instructor's Resource Kit* for Chapter 13 Pretest and for Supplementary Practices.

Sentence faults result in problems for readers and prevent writers from achieving their objectives.

- Readers are amused by an error that sounds funny and are distracted from the document's important message.

- Readers must write or phone for clarification, thus wasting time of several people.

- Readers misunderstand a message, and the result is a failed or an incorrect transaction.

- Readers feel rushed and pressured figuring out a poorly written message when other communications are competing for their attention.

- Readers get the impression that the company attempting to communicate is not efficient, knowledgeable, or competent.

Written workplace communications must be clear, correct, logical, concise, and courteous—for quick, easy comprehension without the sentence faults pointed out in previous chapters and this one.

Chapter 13 reviews basics of writing correct sentences as in Chapter 10 and introduces advanced instruction in effective writing. Students are alerted to various writing faults and how to correct them: choppy writing style, gobbledygook, vague pronoun reference, misplaced modifiers, parallel structure, active/passive voice, and dangling verbals. I believe it's the most challenging, yet the most important, chapter in the course. Additional features of good writing style for business are in Chapter 14.

Read
72

How you handle Chapter 13 depends on course length and students' language level. You could treat it like other chapters, omit it if you lack time, or allow more time by building upon it with various writing experiences, such as Special Assignments I and II at the end of this chapter and Supplementary Practices in your *Instructor's Resource Kit.*

SECRET REVIEW

The "Secret Life of a Sentence" was revealed in Chapter 10, which stresses correctly written sentences vs. fragments, run-ons, and comma splices. Here's a quick review of Chapter 10 sentence faults before beginning Chapter 13 advanced principles.

THE DIFFERENCE BETWEEN A SENTENCE AND A FRAGMENT

> **A sentence is a word group with a subject, a verb, and independence. A sentence requires at least one independent clause and may have one or more dependent clauses and/or phrases.**

┌─── independent clause ────┐ ┌─ dependent clause ─┐

SENTENCE We can substitute nonbiased words for words that may offend.

> **A fragment does not have an independent clause and is misused when posing as a sentence.**

┌─ dependent clause ─┐ ┌─ dependent clause ─┐ ┌prepositional phrase┐

FRAGMENT When we use words that identify people by their work. [This word group has subjects and verbs but lacks independence; it consists of two dependent clauses and a prepositional phrase—preposition *by* + object of the preposition—*work.*]

SENTENCE We often use words that identify people by their work.

See FRAGMENT: The first dependent clause begins with *when* used as a dependent conjunction. The second dependent clause begins with *that,* also used as a dependent conjunction. The third word group begins with a preposition. By deleting the dependent conjunction *when,* we create a SENTENCE.

Write **S** in the blank for the sentences and **F** for the fragments.

Poor: Jose's father is a successful doctor and I'm sure he will be rich someday. [Who will be rich, Jose or his father?]
Good: Jose's father is a successful doctor, who I'm sure will be rich someday. OR Jose's father is a doctor, and Jose will surely be rich someday.

1. When identifying people by what they do, avoid sexist language; for example, say *police officer* instead of *policeman*.   S
2. Hope to hear from you soon and that you're having a great vacation in Hawaii.   F
3. When you accept the idea of limitation.   F
4. Do you accept the idea of your ability being limited?   S
5. By accepting the idea of limitation, you limit yourself unnecessarily.   S

See page 411 for answers.

RUN-ONS AND COMMA SPLICES

> **Run-ons and comma splices consist of two or more independent clauses incorrectly joined.**

RUN-ON   Heather suggested this item we approved it. [Nothing joins the independent clauses but space.]

COMMA SPLICE   Heather suggested this item, we approved it. [Only a comma joins the independent clauses.]

AVOIDING RUN-ONS AND COMMA SPLICES

> **Write two or more independent clauses as separate sentences instead of joining them.**

CORRECT   Chad suggested this item. We approved it.

> **Insert a semicolon, or a semicolon followed by a transition expression, to join independent clauses.**

CORRECT   Chad suggested this item; we approved it.

Chad suggested this item; **therefore,** we approved it.

> **Insert a comma plus *and, but, or, nor, for,* or *yet* to join independent clauses. Very short sentences (up to about ten words) don't require a comma if the conjunction is *and* or *or*.**

CORRECT   Chad suggested this item, but the auditor rejected it in his October 21 e-mail.

Chad suggested this item and we approved it. [no comma]

> **Change one of the independent clauses to a dependent clause or a phrase; then the word group is not a run-on or comma splice.**

CORRECT   Since Chad suggested this item, we approved it. [dependent clause followed by independent clause]

Because of Mr. Valdez' suggestion, we're making the change. OR We're making the change because of Mr. Valdez' suggestion. [phrase followed by independent clause or independent clause followed by phrase]

Write **C** for correct, **R** for run-on, or **CS** for comma splice; correct the **R** and **CS** sentences with a period and a capital letter or with a semicolon.

_____R_____ 1. The word *marketing* needs explanation you may have first heard it used about grocery shopping.

_____CS_____ 2. In business, marketing is everything connected with distributing a product or service therefore, it includes transporting and advertising.

_____R_____ 3. The professional term *marketing* means everything connected with distributing a product or service therefore, it includes transporting and advertising.

_____C_____ 4. *Marketing* is everything connected with distributing a product or service. Therefore, it includes transporting and advertising.

_____C_____ 5. Since the professional term *marketing* means everything connected with distributing a product or service, it includes transporting and advertising.

_____R_____ 6. Many college students major in marketing it is a field of study encompassing varied activities.

_____CS_____ 7. Many college students major in marketing it is a field of study encompassing varied activities.

_____C_____ 8. Many college students major in marketing, a field of study encompassing varied activities.

Check your answers on page 411.

LET'S SUM IT UP

- A complete sentence has a subject, a verb, and independence. A word group lacking these sentence requirements is a fragment.

- Both a run-on and a comma splice consist of independent clauses joined **without** a coordinate conjunction OR a semicolon. A comma joins the independent clauses of the comma splice—but not the run-on. Other correct commas, however, may be elsewhere in the sentence.

COORDINATE CONJUNCTIONS   *and, but, or, nor, for, so,* and sometimes *yet*

TRANSITION EXPRESSIONS   See box in Read 54, page 217.

- To avoid a run-on or comma splice, do one of the following:

    (a) Separate independent clauses with a period followed by a capital letter.

    (b) Join independent clauses with a comma and coordinate conjunction.

    (c) Join independent clauses with a semicolon, either with or without a transition expression following the semicolon.

    (d) Make one of the independent clauses dependent.

Writing for Your Career

To de-emphasize an idea in a sentence with two or more clauses, choose option **d;** that is, create a dependent clause or a phrase for the idea you want to seem less important. To give equal emphasis to both clauses, choose **a, b,** or **c** depending on length and ease of readability.

For further help in avoiding fragments, run-ons, or comma splices, review Read 59 (Independents' Day) and Read 66 (The Halfway Mark).

Replay 72

A. Use commas, periods, and capital letters to correct this business letter (names changed) from an insurance company to its public relations company. Notice the conversational tone and easy-to-understand language after you punctuate correctly. Notice also that Gerber's address is printed in a "letterfoot" instead of in the letterhead. Both stationery formats are correct.

Gerber Life Insurance Company

*A separate subsidiary of Gerber Products,
the baby food people you have known
since you were an infant*

April 13, 0000

Mr. Bruce Finlay, President
The Piers Agency
7 North Maryland Street
Philadelphia, PA 19106

Dear Mr. Finlay:

As you requested, I looked over your suggestions to let you know our preferences. Since your list indicates an understanding of our newsletter, I'm sure your selection of topics is valid.

I'm holding the article "Five Steps to a Healthy Heart" for an issue close to Valentine's Day. Use the other articles, however, as you see fit. Just provide a good mix of subjects.

Will you please also look through the enclosed manual about our logo. Then you can see the various sizes available. Call either Heather Scanlon or Harold Teesdale, our graphic artists, to request negatives of the sizes you need.

Sincerely,

Janet B. Sugg

Janet B. Sugg
Creative Director

enclosure

c Heather Scanlon
  Harold Teesdale

Home Office: 66 Church Street • White Plains, New York 10601
914.761.4404

B. Correct the punctuation in the following sentences. Write **C** beside the sentences that are already correct.

_____ 1. I no longer maintain an office downtown; therefore, please call me at home.

_____ 2. Regarding the map, I do have people who are supposedly working on it; however, so far nothing has happened.

_____ 3. I would be happy to have your friend try his luck with the map; ask him to call me when he's in Colombia next month.

__C__ 4. Twileen is working on this project, but nothing has been completed yet.

_____ 5. The movement of money is the fuel that keeps the machinery of our economic system working; economists call this phenomenon "money flow."

__C__ 6. The U.S. business scene is a very complicated process, and few people understand all aspects of it.

_____ 7. Most people agree, however, that the system *produces.*

_____ 8. When you go out for a business lunch, avoid ordering food that is messy to eat; for example, whole lobster, pasta, or thick sandwiches.

_____ 9. At a business meal, you need to put business first; if you're still hungry, you can always eat again later.

__C__ 10. If you order too much to eat or foods that are clumsy to eat, you're likely to have trouble answering questions and keeping up your end of the conversation.

C. Write **F** for fragment, **CS** for comma splice, or **C** for correct. Correct the comma splices by inserting a coordinate conjunction; correct the fragment by crossing out one word.

__CS__ 1. Answer your workplace phone within one to three rings, and include a greeting such as hello, good morning, etc.

__C__ 2. Also identify yourself by your organization's name and your full name, not just your first or last name.

__CS__ 3. Using just your first name is too informal and unprofessional sounding, but using just your last name sounds too abrupt.

__C__ 4. Corporations must meet their responsibilities to their employees, stockholders, and society and still show a profit.

__F__ 5. The prices of our kitchen utensils, which as you know, have not increased.

Answers are on page 411.

Write sentences more concisely by omitting an **unnecessary,** but understood, *that*—if you're sure the resulting sentence is clear and easy to understand. Another technique is to use contractions from time to time.

Good	We are glad that the system works.
Better	We are glad the system works. **OR** We're glad the system works.

Read
73

TLC FOR SENTENCES

When young children write a letter or a composition, they are taught to compose short, simple sentences like those in their beginning reading textbooks. This writing style is important in learning written expression. However, it may result in "choppy" writing habits for adults who may not have been taught to write smoothly. To improve writing style, vary the length of sentences. For workplace writing, however, rarely use sentences with more than 25 to 30 words. Here are *tender loving care* techniques to convert "choppy" writing into polished writing.

CHOPPY*-STYLE SENTENCES VERSUS TLC

> Consider joining choppy sentences of equal importance with a coordinate conjunction. The result might be a single smoothly written sentence.

The "choppy" examples that follow are, of course, exaggerations to make a point—and not how you, as an adult, would really write.

CHOPPY	Jane likes her work. Dick likes to play.
TLC	Jane likes her work, but Dick likes to play.

> Consider combining two choppy sentences by making one of the clauses dependent.

CHOPPY	Jane likes to work. Dick likes to play.
TLC	Although Jane likes to work, Dick likes to play.

> Consider writing more concisely and gracefully by using a phrase, an adjective, or an adverb instead of only short clauses.

CHOPPY	Jane is wearing a new dress. She looks pretty. She approached Dick. Jane is shy.
TLC	Jane, looking pretty in her new dress, shyly approached Dick.

---

*Shifting constantly and abruptly—*Webster's New World College Dictionary*

> **Consider using a transition expression to join choppy sentences.**

A transition is like a bridge enabling the reader to cross over from one thought to the next.

CHOPPY Dick and Jane like to read. Dick and Jane like to dance.

TLC Dick and Jane are very much alike; for example, they both enjoy reading as well as dancing.

> **The following TLC revisions show several techniques for writing more smoothly as well as more concisely. You might think of other ways.**

CHOPPY He wears glasses for reading. They have a tortoise-shell frame.

TLC For reading, he wears glasses with a tortoise-shell frame.

CHOPPY You write in code. That is how you convey the information secretly.

TLC By writing in code, you convey the information secretly.

CHOPPY The dean's office is next to the reception room. It is on the first floor.

TLC The dean's office is next to the reception room on the first floor.

CHOPPY Measuring industrial output is comparatively easy. Measuring the education system's "output" is difficult.

TLC While measuring industrial output is comparatively easy, measuring our education system's "output" is difficult.

CHOPPY Some manufacturers engage in wholesale trade. They are not regarded as wholesalers. Their primary function is that of manufacturing.

TLC Some manufacturers engage in wholesale trade. They are not, however, regarded as wholesalers because their primary function is manufacturing.

Recap

Apply TLC to these sentences to make them smoother, more concise, or both.

1. He finished dinner. He returned to work. *He finished dinner and returned to work. OR After finishing dinner, he returned to work.*

2. You might go on a trip to China for your work. Do not give your hosts presents that cost a lot. They might be embarrassed. They might not accept them. *If you go to China on a business trip, avoid giving your hosts expensive gifts; they might be embarrassed or even refuse them.*

3. Someone in China might invite you to dinner. You will get rice. Hold the rice bowl next to your mouth. They will give you other food too. Eat from every plate on the table. *When you dine in China, hold the rice bowl close to your mouth and sample every dish offered.*

See page 411 for suggested solutions.

TLC AND A KISS VERSUS GOBBLEDYGOOK

The preceding examples suggest ways to **join** ideas for more effective writing. Too much joining, however, can result in sentences too long and complicated for efficient reading. Such writing, especially when combined with needlessly technical or long words, is *gobbledygook*.

> Gobbledygook is speech or writing with needlessly long words, superfluous words, and complicated sentence structure. Gobbledygook is pompous and hard to understand.

Research shows that big words and long sentences do not impress readers favorably. A simple language style that expresses ideas clearly, correctly, and concisely is more likely to get results in the workplace. Unfortunately, too much technical writing is gobbledygook. Technical organizations desperately need individuals who not only understand the technology but can also explain it clearly in writing.

Gobbledygook

Distributors of merchandise for profit in the Middle Ages kept numerical records of the merchandise they sold on "tally sticks," on which they produced a notch on a stick which was then broken in half, with the merchant retaining half and the other half being presented to the customer in order to have a record of the data for the merchant and the individual who made the purchase.

KISS—Keep It Simple, Sweetheart

Replace gobbledygook with a KISS (Keep It Simple, Sweetheart) and TLC. Separate the one long sentence into three clear sentences and use fewer and simpler words: Merchants in the Middle Ages recorded sales on "tally sticks." The data was notched on a stick, which was then split. The merchant kept half and gave the matching half to the customer.

TLC—for Pronouns

Apply tender loving care by providing the reader with clear, immediate reference to the pronouns you use.

> Each pronoun should mean to the reader precisely what you want it to mean. If necessary, rephrase a sentence or replace a vague pronoun with an appropriate noun.

POOR Ms. May is a good writer, **which** she acquired from a Watterson College communication expert. [Did she acquire **good writing** from the expert?]

GOOD Ms. May is a good writer, a **skill** she acquired from a Watterson College communication expert.

POOR If washing machines have been tearing your fine linens and laces, let us do **it** for you by hand.

GOOD If washing machines have been tearing your fine linens and laces, let us do **your laundry** for you by hand.

POOR They keep the streets clean in Auburn Hills. [Avoid vague use of *they*.]

GOOD The streets are clean in Auburn Hills. OR The sanitation department keeps Auburn Hills' streets clean.

Replay 73

Rewrite the following sentences to improve the sentence construction.

1. The beautiful movie star smashed a bottle of champagne over her stern as she slid gracefully into the sea. The beautiful movie star smashed a bottle of champagne over the stern as the ship slid gracefully into the sea.

2. We got up at 6 o'clock. We had a quick breakfast. _We got up at 6 o'clock_
and had a quick breakfast.

3. A college pennant is in the student's room. There is also a picture by Monet. Both of these hang on the wall. _A college pennant and a Monet_
painting are on the wall of the student's room.

4. A house sits far back among the trees, and it is in need of painting.
A house in need of painting sits far back among the trees.

5. Dennis and Mike walked to the telephone company office, from there they walked to the gas company, then they went to the electric company. They were paying their bills, it took them all afternoon. _Dennis and Mike spent all_
afternoon walking to the telephone, gas, and electric company offices to pay their bills.

6. An animal paced restlessly back and forth in the cage, it appeared to be a hyena. _An animal that appeared to be a hyena paced restlessly back and_
forth in the cage.

7. We realize that people who are traveling by airplane are going to make a choice of an airline based on the quality of its service to the people who fly on that airline's planes and we are extremely regretful that we have not been able to meet your expectations. _We realize travelers choose_
an airline based on the quality of its service, and we're sorry we let you down.

8. We received your May statement. We are enclosing a check in payment. The check is for $635.23. _Enclosed is our check for $635.23 in payment of_
your May statement.

9. I missed the final exam because I had been out late the night before, nevertheless, they wouldn't give me an "Incomplete" or "Withdrew" grade, they failed me instead. _I missed the final examination because I had_
been out late the night before. Nevertheless, the instructor and the dean refused
to change my F grade to Incomplete or Withdrew.

10. Newspaper ad: Respectable lady seeks comfortable room where she can cook herself on an electric stove. _Respectable lady seeks_
comfortable room where she can cook on an electric stove.

11. Newspaper ad: FOR SALE: The ladies of the First Presbyterian Church have discarded clothing of all kinds. They may be seen in the church basement any day after six o'clock. _FOR SALE: Discarded clothing_
collected by women of the First Presbyterian Church. The clothing may be seen in
the church basement any day after 6 p.m.

12. For those of you who have small children and didn't know it, we have a nursery downstairs. _For those of you who didn't know it, we have a_
nursery downstairs for small children.

13. This is the friend of Professor Loup, who lives in Binghamton. _This is_
Professor Loup's friend, who lives in Binghamton. [or rephrase if it's Professor Loup
who lives in Binghamton.]

14. We can't recommend that he go to Mr. Anderson's office uninvited because he is so unkind. __Because Mr. Anderson is so unkind, we don't recommend that Mr. Sedirko go to his office. [or rephrase if it's Mr. Sedirko who is unkind.]__

15. Despite our repeated reminders regarding nonpayment of your account, which is long overdue. __Despite our repeated reminders about nonpayment of your long overdue account, we have still not received your check.__

Possible solutions are on pages 411–412.

Read 74

LADIES WITH CONCRETE HEADS AND PARALLEL PARTS

The late James McSheehy, a member of the San Francisco Board of Supervisors, addressed a group of women about his work on a finance committee. "Ladies," he said, "I have here some figures I want you to take home in your heads, which I know are concrete." Of course, Mr. McSheehy really meant that the *figures* were concrete. To avoid this kind of error in your writing, look for **misplaced words** when proofreading. If the words in a sentence are not in correct order, the reader or listener may be confused or amused. Either way, concentration on your message is lost.

MISPLACED WORDS

> Proofread to be sure all words are in the best place for getting the message across. Avoid the sentence fault of *misplaced words.*

MISPLACED   I have some figures that I want you to take home in your heads, which I know are concrete. [*which I know are concrete* are the misplaced words]

CORRECT   I have some figures, which I know are concrete, that I want you to take home in your heads.

MISPLACED   Irene hung a picture on the wall painted by Rembrandt. [Rembrandt painted the wall?]

CORRECT   The picture Irene hung on the wall was painted by Rembrandt.

OR   Irene hung a Rembrandt in the president's office. [Unnecessary words are omitted, and additional information is included in one concise sentence.]

MISPLACED   *He only* had $5 when he arrived in Kenansville. [No one else, only *he*?]

CORRECT   He had only $5 when he arrived in Kenansville.

MISPLACED   On the bulletin board of a factory building—WANTED: Worker to sew buttons on 4th floor.

CORRECT   WANTED ON 4TH FLOOR: Worker to sew buttons.

Find the misplaced words in each sentence and move them to where they belong. Write the correct sentence in the blank.

1. Sign in a Santa Fe gas station: We sell gas to anyone in a glass container. __We sell gas in a glass container to anyone.__

2. We sat there listening to his singing in awed silence. ___We sat there in___
___awed silence listening to his singing.___

3. Ms. Grigg worked for IBM during her vacation in the Information Systems Department. ___Ms. Grigg worked for IBM's Information Systems___
___Department during her vacation.___

4. Genevive Astor died in the home in which she had been born at the age of 96. ___Genevive Astor died at the age of 96 in the home in which she had been born.___

5. The fire was brought under control before much damage was done by the fire department. ___The fire department brought the fire under control before___
___much damage was done.___

See page 412 for possible solutions.

See page 412 for possible solutions.

PARALLEL PARTS

> **Express parallel ideas—that is, similar sentence elements—in the same grammatical form.**

Parallel construction enables readers to understand immediately how two or more parts of a sentence are related. In the following *parallel* example, "The cat chased the mouse" in three places, all expressed in prepositional phrases.

PARALLEL — The cat chased the mouse **through the barn, over the fence, and into the yard.**

NOT PARALLEL — He was tall, dark, and had a handsome face. [Make all three describing expressions parallel by using three adjectives and omitting the unneeded words.]

PARALLEL — He was **tall, dark,** and **handsome.**

NOT PARALLEL — His ambitions were to join a fraternity and becoming a football player. [Use *to with a verb* (the infinitive) before each ambition. **OR** use the *ing* form for each ambition.]

PARALLEL — His ambitions were **to join** a fraternity and **to become** a football player.

OR — His ambitions were **joining** a fraternity and **becoming** a football player.

Sidebar (left margin):

PARALLEL PARTS—
I believe in the right to be able to live, being free, and pursuit of happiness.
CORRECTION: I believe in the right to life, liberty, and the pursuit of happiness.

You'll get better results by allowing it rather than if you deny it. [. . . . rather than by denying it.]

Recap

Rewrite the following sentences so that the parallel ideas are parallel in construction; that is, have the same grammatical form.

1. With the new software we hope to improve response time, reducing input errors, and see that systems problems are identified more readily.
___With the new software we hope to improve response time, reduce input errors, and___
___identify systems problems more readily.___

2. Keyboarding accurately can be more important than to keyboard fast.
___Keyboarding accurately can be more important than keyboarding fast.___

3. Linda is a full-time securities analyst, and her husband is working part time as an insurance agent. ___Linda is a full-time securities analyst, and___
___her husband is a part-time insurance agent.___

4. We would appreciate learning your views on how to introduce change, controlling quality, and the motivation of employees. _We would appreciate learning your views on how to introduce change, control quality, and motivate employees._

5. Please bring to the meeting our new production schedule, list the names of all new employees, and when the administrative assistants are taking their vacations. _Please bring to the meeting our new production schedule, the names of all new employees, and the vacation dates for the administrative assistants._

6. Ophthalmologists and optometrists may examine eyes, and prescriptions for glasses may be issued by them also. _Ophthalmologists and optometrists may examine eyes and may also issue prescriptions for glasses or contact lenses._

See page 412 for possible solutions.

MEMO FROM THE WORDSMITH

Little Girl   I know a man with a broken leg named Jones.
The Nanny   Oh really? What's his other leg named?

What is the name of the little girl's sentence fault? _misplaced words_

Replay
74

A.   Write **P** for lack of parallel parts, **M** for misplaced words, and **C** for correct.

_M_ 1. Oranges are a valuable source of vitamins, which are not mentioned in your report.

_M_ 2. The English teacher was sitting by the fireplace with his dog reading Shakespeare.

_P_ 3. He is interested in science, math, and he likes to read good books.

_M_ 4. Mr. Gorjus was frantically searching for the telephone number in his office that was missing.

_P_ 5. The woman suggested that we fill out the form and to leave it with her.

_P_ 6. Ad for a famous cosmetic: Lady Ester Dream Cream is recognized by leading dermatologists as highly effective in improving skin's texture, smoothness, and for counteracting aging.

_P_ 7. I believe that playing a good game of chess is a better accomplishment than to play a good game of bridge.

_P_ 8. Keyboarding business documents requires excellent grammar, and they should be punctuated very well also.

_P_ 9. Her hobbies are painting, to go to concerts, and communication on the Internet.

<u> C </u>   **10.** A wholesaler's function is to buy in large quantities and sell in small quantities.

<u> M </u>   **11.** Father sat down in an easy chair to tell his children about his childhood after dinner.

<u> C </u>   **12.** According to a "Human Development Index" that includes incomes, education, and life expectancy, Norway ranks third in the world.

<u> P </u>   **13.** A major Norwegian industry is production and transporting oil and gas from offshore petroleum deposits.

<u> M </u>   **14.** Norway is heavily dependent on world trade conditions with a population under five million.

<u> P </u>   **15.** We would like to hear your ideas on motivating employees and how to introduce change.

<u> C </u>   **16.** Some employees react quickly, get things done promptly, and beat deadlines.

<u> P </u>   **17.** That program has too much sex, violence, and the language is bad.

<u> C </u>   **18.** Friendship means forgetting what one gives and remembering what one receives.

<u> P </u>   **19.** She believed him as well as having faith in him.

<u> M </u>   **20.** At the age of 7, my father gave me a kitten.

      B. Keyboard the number of each item above that you marked **P** or **M**, and write the sentence correctly.

Solution to A is on page 412. Solution to B is in *Instructor's Resource Kit.*
Take the Pop Quiz on Reads 72–76 on page 357.

Read
75

THE BEELINE

Only transitive verbs have *voice*. Explaining this and other aspects of active and passive voice structure is unnecessary to achieve improved writing. The "Instant Test" on page 290 is fail-safe. If further information will help your students, however, see "Grammar for the Expert" in your *Instructor's Resource Kit.*

Do you know someone who makes a beeline for the snack table upon arriving at a party? This food fancier moves in a straight, direct path just as a bee flies to its chosen flower. A sentence written in beeline style is direct and to the point and is the type of sentence most often written by good business writers. In the workplace, people usually react more favorably to a direct writing style.

In addition to your friend who makes a beeline for the refreshments, you may have other friends who eventually get there but take their time about it. Skillful business writers sometimes use an indirect style, but they get there eventually too.

VOICE OF VERBS

To choose whether to write a sentence in the direct or indirect style, you need to know about the **voice** of verbs. Certain verbs have what is called *voice*—which may be *active* or *passive*.

Active Voice

> A direct writing style is called *active voice*, which means the subject *does* the verb's action.

George kissed Twileen. [The subject *George* did the action—*kissed.*]

Norway won 26 medals at the 1994 Winter Olympics. [The subject *Norway* did the action—*won.*]

Passive Voice

> An indirect writing style is called *passive voice*, which means the subject *receives* the verb's action.

Twileen was kissed by George. [The subject *Twileen* received the action—*was kissed.*]

Twenty-six medals were won by Norway at the 1994 Olympics. [The subject *medals* received the action—*were won.*]

Since the required ingredients of a sentence are subject, verb, and independence, we can write complete sentences in the passive voice without mentioning who did the action.

PASSIVE Twileen was kissed. [This sentence has a subject and a verb and is independent; it is passive because the subject **received** the action.]

Twenty-six medals were won at the 1994 Winter Olympics. [This sentence is in passive voice because the subject is the receiver even though the doer of the action, Norway, is not named.]

ACTIVE Jesse accurately prepared a spreadsheet of the accounts receivable.

PASSIVE The accounts receivable spreadsheet was prepared accurately by Jesse.

The accounts receivable spreadsheet was prepared accurately.

ACTIVE George presented the report to the stockholders.

PASSIVE The report was presented to the stockholders by George.

The report was presented to the stockholders.

ACTIVE OR PASSIVE VOICE IN WORKPLACE WRITING

> Use active, rather than passive, voice for most workplace writing. It is often more concise.

Fewer words are usually required for the active voice, it is more efficient, and it takes the reader from Point A to Point B in a straight line.

A—————————→ B

> Use passive voice for tact or emphasis.

Tact

You may want the reader to know about an error but prefer to omit the culprit's name.

PASSIVE An error was made on the report.

Emphasis

Use the passive voice if you don't know who the doer is, or if you want to emphasize the receiver of the action rather than the doer. The passive enables you to begin the sentence with the receiver; you can omit the doer or end with the doer. For example, the report is more important than the auditor in the following sentence:

PASSIVE   A report was presented to the stockholders by the auditor. OR

               A report was presented to the stockholders.

ACTIVE   The auditor presented a report to the stockholders.

INSTANT TEST TO DISTINGUISH ACTIVE FROM PASSIVE VOICE

> **If "by someone" makes sense after the verb, the voice is passive. If "by someone" is already there, then you *know* it's passive.**

PASSIVE   The book was purchased last week. [**Instant Test:** The book was purchased (by someone) last week. "By someone" makes sense after the verb.]

               The book was purchased by my mother last week. [**Instant Test:** "By my mother" makes sense after the verb.]

ACTIVE   Dorothy Larson bought the book today. [**Instant Test:** Dorothy Larson bought (by someone) the book today. "By someone" after the verb doesn't make sense.]

Write **A** next to the sentences with active voice verbs and **P** next to those with passive voice verbs.

_P_   **1.** Antarctica is not owned by any country.

_A_   **2.** Interested nations from around the globe have signed an agreement to preserve Antarctica as a zone of world peace.

_P_   **3.** Military activity, nuclear testing, and disposal of radioactive waste are prohibited in Antarctica by the Antarctic Treaty.

_A_   **4.** Constantly chattering penguins bellyride down icy slopes and dive into the placid waters of Antarctica.

_A_   **5.** The Atlantic, Pacific, and Indian Oceans surround Antarctica.

_P_   **6.** Antarctica is surrounded by parts of the Atlantic, Pacific, and Indian Oceans.

_A_   **7.** When doing business in England, use the term *British* rather than *English—except* for referring to the language.

_A_   **8.** A great many jobs now involve interactions with the people of various countries.

_A_   **9.** Importers and exporters constantly move merchandise from one country to another.

_P_   **10.** Some knowledge of geography, history, and foreign languages is needed to obtain or advance in many of the better jobs.

_A_   **11.** Companies such as Avon Products and the Fuller Brush Company used to employ sales representatives who sold products door-to-door.

P	**12.** In today's economy and culture, few products can be sold that way.
A	**13.** Many people who can afford to buy are at work, and some others won't open the door to a salesperson because of fear of crime.
P	**14.** The plants were watered every day during your vacation.
P	**15.** All freight charges must be verified before they are paid.
A	**16.** He misspelled "Mississippi" in every paragraph.
P	**17.** "Mississippi" was misspelled in every paragraph.
P	**18.** These tools are manufactured in Toronto.
A	**19.** Networking engineers take care of problems day or night.
A	**20.** She sold all her stock during a bear market.

Answers are on page 413.

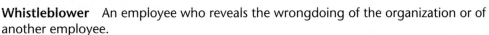

Word Power

Workplace Jargon

Short list   The "finalists" called back for a second interview; the one to be offered the job is selected from the "short list."

Whistleblower   An employee who reveals the wrongdoing of the organization or of another employee.

Read 76

DON'T LET YOUR VERBALS DANGLE IN PUBLIC

More danglers and possible corrections: Being bad grammar, good writers don't use dangling verbals. [Correction: Because dangling verbals are bad grammar, good writers don't use them.]

A dangling verbal is a phrase containing a **verbal** that hangs loosely, or dangles, in a sentence. A verbal may be one of the following:

- **An Infinitive:** *to* plus a verb, such as *to work, to go, to eat*
- **A Past Participle:** *worked, gone, eaten*
- **A Present Participle:** *working, going, eating*
- **A Combination:** *having worked, to have gone, to be eating*

> A verbal is not a verb; it looks like a verb but isn't functioning as a verb. If the word group in question has a real verb, the word group is not dangling.

A dangling verbal phrase usually—but not always—opens a sentence, and a clause follows it.

> The subject of the clause following a verbal phrase must tell who does the action referred to in the verbal phrase; if it doesn't, the verbal phrase is dangling.

If the subject doesn't tell who does the action of the verbal, the verbal phrase is dangling. The resulting sentence may amuse, confuse, or distract the reader.

SENTENCES WITH DANGLERS

DANGLER	┌───── verbal phrase ─────┐ ┌─ subj.─┐ To get the most out of our time, the session will include discussion of our day-to-day problems.

The verbal is *To get*. It looks like a verb, but it isn't because it can't have a subject: You wouldn't say, "I to get the most out of our time." It would have to be "I **want** to get . . . " with *want* as the verb. After noting that the sentence begins with a phrase containing a verbal, find the subject of the clause that follows the phrase; the subject is *session*.

Then ask, "Is the **session** to get the most out of our time?" No, it's **we** who want to get the most out of our time. Therefore, the verbal phrase is dangling. To stop the dangling, the subject must identify **who** is to get the most out of our time:

The house was 57 miles beyond Portland, heading East. Change *was* to *is* if the house hasn't moved; lowercase *e* for *east?*— *heading East* is "dangling." Is Portland heading east? [Correction: The house is 57 miles beyond Portland if you're heading east.]

CORRECT	To get the most out of our time, we will include in the session a discussion of our day-to-day problems.
DANGLER	┌────── verbal phrase ──────┐ ┌──── subj. ────┐ Having made too many errors on the test, the personnel director did not hire Joe. Since it was not the personnel director who made the errors, the opening phrase is dangling. Save the verbal (having made) from dangling by changing the subject to the one who did make the errors.
CORRECT	Having made too many errors on the test, **Joe** did not get the job. *Having made* is a verbal—not a verb—because you can't give it a subject.
ALSO CORRECT	Because Joe made too many errors on the test, he did not get the job.
DANGLER	While flying over the jungle at an altitude of 2,000 feet, the villagers could be seen hunting and fishing.

The opening phrase includes the **verbal** "flying." Notice that *flying* can't be a verb unless a helping verb precedes it. You can't say, "They flying over the jungle." The **subject** following the phrases is **villagers.** To decide whether the opener is dangling, ask, "Are the villagers flying over the jungle?" Since the answer is "no," correct the sentence so that the subject tells who is flying:

CORRECT	While we were flying over the jungle at an altitude of 2,000 feet, we could see the villagers hunting and fishing.
DANGLER	After looking at the cars for a while, the salesperson approached me.

After looking at the cars for a while is dangling. Was the salesperson looking at the cars? To make this sentence clear and avoid a dangler, the **subject** must be the one doing the looking:

CORRECT	After looking at the cars for a while, I was approached by the salesperson.
DANGLER	Topped with whipped cream, the minister devoured a piece of Dad's delicious pecan pie. [The minister is topped with whipped cream?]
CORRECT	Topped with whipped cream, Dad's delicious pecan pie was devoured by the minister. [subject is *pie*]

Recap Rewrite this sentence so that the verbal doesn't dangle. [Make the **subject** of the in–dependent clause do the verbal's action; for example, the subject could be *you* or *I*.]

Dangler While strolling along the beach, unusual shells and pebbles can be found.

Correct ___Possible Answer: While strolling along the beach, you can find unusual shells___ and pebbles.

Check your answer on page 413.

> **Another way to correct a dangling verbal is to change it to a subject and verb—that is, a clause; clauses don't dangle.**

DANGLER While swimming in a river near his farm, his clothes were stolen. [Since his clothes were not swimming in the river, change the dangling verbal phrase to a dependent clause; that is, a dependent conjunction (While) followed by a subject and a verb. Then the second clause does not have to be changed.]

CORRECT While **he** was swimming in a river near his farm, his clothes were stolen.

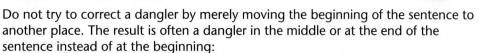

WORD TO THE WISE

Do not try to correct a dangler by merely moving the beginning of the sentence to another place. The result is often a dangler in the middle or at the end of the sentence instead of at the beginning:

Dangler. His clothes were stolen while swimming in a river near his farm. [This still reads as though his clothes were swimming in the river, but the verbal is now midsentence.]

DANGLER Driving too fast to the busy intersection, the brake was applied quickly.

CORRECT After driving too fast to the busy intersection, I applied the brake quickly.

After I had driven too fast to the busy intersection, a police officer stopped me.

DANGLER Having been sick for two weeks, Enrique's father took him to the doctor. [*Father* is the subject; *Enrique's* is a possessive noun. FATHER was not sick for two weeks. Therefore, *having been sick for two weeks* is dangling.] The correction below changes the opener to a dependent clause.

CORRECT After Enrique had been sick for two weeks, his father took him to the doctor.

Recap Rewrite these sentences so that the verbals are no longer dangling. Replace the verbal phrase with a clause.

1. **DANGLER** Being one of our most discriminating customers, we invite you to attend this private showing. [**we** are not our own customer]

   CORRECT ___Since you are one of our most discriminating customers, we invite___ you to attend this private showing.

2. **DANGLER**   Turning the corner, the new building was right in front of him.

   CORRECT   When he turned the corner, the new building was right in front of him.

3. **DANGLER**   While using the computer, the cursor became stuck in the middle of the screen.

   CORRECT   While I was using the computer, the cursor became stuck in the middle of the screen.

4. **DANGLER**   Unlike many others who became millionaires in the 1990s, her talents in fine art and classical music resulted in her financial success.

   CORRECT   Unlike many others who became millionaires in the 1990s, she became financially successful through her talents in fine art and classical music.

5. **DANGLER**   Before having dinner with the woman he met through Ultimate Encounters Dating Agency, his table manners need improvement.

   CORRECT   Before he has dinner with the woman he met through Ultimate Encounters Dating Agency, he needs to improve his table manners.

See page 413 for possible solutions.

Writing for Your Career

The preceding "dangler sentences" can be corrected in several ways besides those shown above. Just don't let your verbals dangle.

Replay 76

If the sentence has a dangler, underline it and write **D** in the blank. Otherwise, write **C** for correct.

D 1. <u>Having keyboarded just half the report,</u> the phone began ringing insistently.

D 2. <u>On examining the goods,</u> we found them to be defective.

D 3. <u>Like many people living in Alaska,</u> the summer months are our favorites.

D 4. <u>Having recovered from his illness,</u> his mother took him to Israel.

C 5. To keep the machine running in perfect condition, we oiled it once a month.

D 6. <u>Before going to lunch,</u> this report must be typed.

D 7. This report must be typed <u>before going to lunch.</u>

D 8. <u>Walking quickly down the aisle,</u> her skirt caught on a nail.

_____D_____ 9. While doing the daily chores, a fire started in the farmer's barn.

_____D_____ 10. Handing me the $50,000 order, his face broke into a broad smile.

_____D_____ 11. Having produced a printout, the text was stored on the floppy.

_____D_____ 12. After looking the cars over for a while, a saleswoman approached me.

_____D_____ 13. Looking marvelously glamorous in a midnight green gown adorned with baguette rhinestones, Mr. Martinez accompanied Stella Glitter to the performance.

_____D_____ 14. Upon landing in Dallas, his assistant picked him up at the airport.

_____C_____ 15. Good communication skills and interpersonal effectiveness are qualities employers look for when interviewing job candidates.

_____C_____ 16. When looking up a word in the dictionary, notice its pronunciation as well as definitions.

_____D_____ 17. Being in dilapidated condition, she bought the building cheaply.

_____D_____ 18. If invited to dinner at a British home, chocolates or flowers are a suitable gift for the hosts.

_____C_____ 19. If you decide to give flowers, avoid white lillies as they suggest death.

_____C_____ 20. To be a serious student of business, you should understand advertising, promotion, and marketing.

_____C_____ 21. Confused by crowds of students rushing around the campus, a new student welcomes a familiar face with a sigh of relief.

_____D_____ 22. After standing and repeating the pledge, the meeting began.

_____D_____ 23. While walking home, a hundred dollar bill suddenly appeared before me.

_____C_____ 24. While I was walking home, a hundred dollar bill suddenly appeared before me.

_____D_____ 25. It began to rain after being on vacation for two hours, and it didn't stop for two weeks.

Answers are on page 413. Please turn to page 358 for Pop Quiz.

CHECKPOINT

Overcoming common sentence faults is a major step to improve your workplace writing ability and increase your opportunities for a successful career. Place a check in the blank when you can recognize and correct these sentence faults.

_____ run-on sentence

_____ fragment

_____ unclear pronouns

_____ nonparallel construction

_____ dangling verbals

_____ comma splice

_____ "choppy" writing style

_____ misplaced words

_____ gobbledygook

_____ active and passive voice

SPECIAL ASSIGNMENTS

I. Writing That Works

In 100 to 150 words write an explanation of how to do something you understand very well. The process should be simple enough so that most of your colleagues at work or in class could perform the task by following your written instructions. Examples of such tasks are sewing on a button, changing a tire, checking the oil in a car, or answering a business telephone and taking a message. Get right into your subject. Use complete sentences with smooth transitions. The sentences need to be clear, concise, logical, and correct.

SAMPLE OPENING SENTENCE

Sewing a button on a shirt is easy if you have the right tools and follow these steps:

_____

Submit this assignment by _____. (date)

II. Write What You Mean   Solution is in *Instructor's Resource Kit.*

Rewrite these ads so that they say what they mean and mean what they say:

- Wanted: Student to deliver fish and oysters with good references.
- Now you can buy six different products to protect your car from your Mobil dealer.
- Wanted: Someone to take care of horses who can speak German.
- Now on the market: a Norco Shaver for women with three heads.

_____

_____

_____

_____

Submit this assignment on _____. (date)

PROOFREADING FOR CAREERS

I have altered the author's well-written article to provide practice in correcting word and sentence errors. Please make the corrections according to what you have practiced in Chapters 1–13. Either correct the errors while keyboarding the article, or use a red pen to show the corrections below.

A BRIEF HISTORY OF INFORMATION TECHNOLOGY

Working under a IBM grant, the Mark I computer was produced by Howard Aikens, a Harvard University mathematic's instructor, in '44. This machine, which combined mechanical, electrical and electronic principals was the 1st to perform all the functions of a true computer.

In 1945, University of Pennsylvanias Professor's John W. Mauchly and J. Presper Eckert completed ENIAC—**E**lectronic **N**umerical **I**ntegrator **and C**alculator—the first all-electronic computer. 2 years later they formed there own Corporation, the corporation constructed UNIVAC. UNIVAC means **Univ**ersal **A**utomatic **C**omputer. UNIVAC was the world's 1st commercial computer. UNIVAC was delivered to the United States Bureau of the Census in the Spring of 1951.

After developing the first UNIVAC, information technology did not progress smooth and steady. Instead, technological break throughs propelled computers foreword in a series of grate leaps. Each of the break-throughs have created computers far advanced over previous ones. These developments have brought about radicle changes in are concepts of information processing the new family's of computers became known as computer generations.

Now in the 21st century, computers continue to grow in number, and memory capacity and decrease in size, and price. Voice and other new methods of input and output make it easier for people all over the world to use computer technology for a variety of job and leisure activities. Vast new areas of information technology have put us on the information superhighway to make our daily lives more productive, convenient and enjoyable.

After verifying corrections with your instructor or with a key, evaluate your English for careers and proofreading skills.
Excellent _____ Good _____ Fair _____ Need improvement in spotting errors _____ Need to review the following chapters _____

PROOFREADING FOR SENSE From a caption below a newspaper photo: Salvation Army Cooking Students. The grammar is fine, but the reader may picture a scene quite different from what the writer intended.

PRACTICE QUIZ
www.prenhall.com/smith

Take the Practice Quiz as though it were a real test. Don't look back through the chapter while you take this Practice Quiz.

I. Write the appropriate letter and number in the blank according to these instructions:

A Properly constructed sentence

B (1) Fragment or (2) Unclear pronoun

C (1) Run-on or (2) Comma splice

D (1) Lacks parallel parts or (2) Is choppy or (3) Has a dangling verbal

E (1) Misplaced words or (2) Gobbledygook

<u>D-2</u>   **1.** Marianne is smart. She likes to study. She also likes her job.

<u>C-1</u>   **2.** The new drug has proved to be highly effective it has no side effects.

<u>B-2</u>   **3.** Mara heard Beth talking to her boyfriend on the phone.

<u>C-2</u>   **4.** He is an extremely capable worker, however, he lacks seniority.

<u>D-1</u>   **5.** The angry manager began stamping his foot and to pound the desk.

<u>A</u>   **6.** Managers tend to promote people who can make decisions even at the risk of being wrong.

<u>B-2</u>   **7.** He is very much interested in science, which he acquired from his cousin who is a chemist.

<u>D-1</u>   **8.** Working accurately is more important than to work fast.

<u>B-2</u>   **9.** The main problem for people who grow African violets is that they stop blooming.

<u>D-3</u>   **10.** When completing the invoice, one item was omitted.

<u>D-3</u>   **11.** After voting on several issues, the meeting was adjourned.

<u>E-1</u>   **12.** The bank approves loans to reliable individuals of any size.

<u>B-1</u>   **13.** Even though George's love life is in shambles.

<u>D-1</u>   **14.** The steps for making peanut butter are picking the nuts, roasting the nuts, and to squeeze the oil from the nuts.

<u>D-1</u>   **15.** Learning to fly is challenging and a thrill.

<u>A</u>   **16.** Think of all the beauty still left in and around you, and be happy.—*The Diary of Anne Frank*

<u>E-2</u>   **17.** Market research and analysis is assuredly and definitely a specialized field of endeavor, and especially excellent career opportunities for being productive exist in these fields of productivity for human beings who seek to be gainfully employed.

<u>D-3</u>   **18.** To get the most from your employees, it is important to be sympathetic.

<u>D-3</u>   **19.** While going over the work more carefully, more errors were discovered.

<u>A</u>   **20.** A symbol of life, wealth, and fertility since ancient times, rice sustains half the world.

II. Write **A** if the verb is in the active voice and **P** if it is in the passive voice.

_____P_____ **21.** The people of Scotland are called *Scots,* not *Scotch.*

_____A_____ **22.** They named the alcoholic beverage *Scotch.*

_____A_____ **23.** Scots speak *Scottish,* not *Scotch.*

_____P_____ **24.** "Cheese" should be said by everyone who wants to be in the picture.

_____P_____ **25.** Allegiance to the flag is pledged by me.

THE GLOBAL MARKETPLACE—IRAN

FARSI	Spell	Pronounce	Currency
GREETING	Salaam	sa LAM	rial
DEPARTURE	Khodahafez	ko da ha FEZ	
PLEASE	Khahesh meekonam	caw his me ko nam	
THANK YOU	Tashakor meekonam or Merci	tau shaw kor me ko nam or mer SEE	
I DON'T UNDERSTAND	Man ne'me'fah'mam		

by Sholeh Khorooshi

14

Take Your Show on the Road

Technology, English, and communication styles are in a constant state of change. Nevertheless, strength in all three remain vital requirements to meet personal and career goals.
—Leila R. Smith

Use Today's Vocabulary, "Grammar for Grownups," Formatting, and Presentation Techniques

After Completing Chapter 14, You Will

> Clearly, concisely, and correctly compose and format business letters, envelopes, memoranda, e-mail, multimedia presentations, and oral presentations for your career.

See *Instructor's Resource Kit* for Chapter 14 Pretest. Chapter 14 deals with (a) composing and formatting letters, memos, faxes, and e-mails; (b) writing for multimedia; and (c) producing oral presentations on the job.

Some Chapter 14 information might differ from that in a particular word processing text or office reference manual. More than one way is OK, if an alternate method is recommended by an expert in the field. However, some employers issue correspondence manuals with formatting preferences.

Good communication—via speech, writing, and multimedia—may be the most important attribute needed for a successful career. However, employers report that communication skills are lacking among many of today's job applicants.

Multimedia communication enables businesspeople to provide information effectively to viewers. **Written communication** provides a permanent record for both sender and receiver, and arrival of written communications doesn't interrupt the receiver. Despite fax and Internet technology, postal services around the world continue to deliver enormous amounts of mail, including first class. Successful business careers also require personal contacts—including **oral communications** with clients, customers, and co-workers.

Today's workplace thrives on information, and much of it is still communicated on paper—despite earlier predictions for a "paperless society." Well-prepared written, multimedia, and oral communications often result in considerable profit for the sender. However, poor communication (regardless of technology) may result in lost income. Employees and entrepreneurs who are effective communicators are vital to an organization's success and are often rewarded by career advancement.

Chapter 14 includes formatting and composing letters, memos, faxes, and e-mail; making oral presentations; and preparing multimedia presentations.

LOOKING GOOD

A first consideration in effective communications is to be sure they look good. While Read and Replay 77 is primarily about business letters, much of the information is also applicable to other documents.

STATIONERY

> Use appropriate stationery for written communications.

It's not good business policy if associates, customers, or vendors are distracted by the appearance of communications that aren't in a generally accepted style. Students expecting assistants to take care of business communication needs presented in Chapter 14 should learn it's risky to assume all office support personnel have been trained.

Most business letters are on attractive, good quality $8\frac{1}{2}$ - by 11-inch letterhead stationery (in Europe $8\frac{1}{4}$ by $11\frac{1}{2}$—and called A4) and mailed in matching envelopes. Smaller stationery is used in some professional, top executive, and diplomatic offices. Although some letterhead stationery is in a color, white or off-white paper is typical. Information printed at the bottom of letterhead stationery is called the *letterfoot*.

Some business letters you write deal with your personal transactions, not those of your employer. You might write to your insurance company, the motor vehicle department about your license, an employer when you're looking for a job, a department store about billing errors still uncorrected after your phone calls, and so on. Keyboard personal business letters on plain $8\frac{1}{2}$ - by 11-inch paper—not **company letterheads.**

ENVELOPES

> Most business letters are mailed in business envelopes known as No. 9 or No. 10. For personal business letters, use either small envelopes or the No. 9 or 10.

The sender's name and address is printed in the upper left corner of the envelope. For personal business letters, keyboard your name and address in that position.

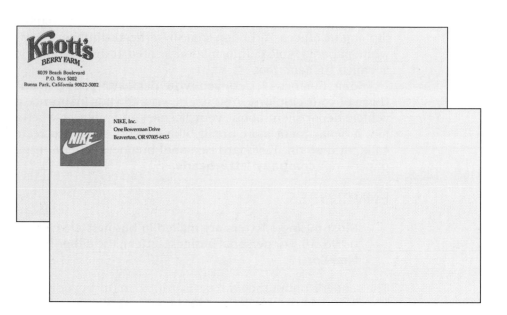

Take a moment now to look at the stationery and envelope samples throughout the chapter.

FOLDING LETTERS

Fold letters correctly before inserting them into envelopes.

To fold an $8\frac{1}{2}$- by 11-inch page for a No. 9 or 10 envelope, fold the sheet in thirds as follows: Bring the bottom third of the letter up, and make a crease. Then fold the top of the letter down, but stop a fraction of an inch before the crease you already made. Now make the second crease. Holding the second crease, insert the letter into the envelope.

For a No. $6\frac{3}{4}$ envelope, fold the page almost in half from the bottom up so that the bottom edge is a fraction of an inch below the top edge. Then from the right, fold over about one third. Finally, from the left, fold the remaining third so that the left edge does not quite meet the crease.

Take a moment now to fold a sheet of $8\frac{1}{2}$- by 11-inch paper for both sizes of envelopes.

Management and professional staff need Chapter 14 information in order to give intelligent approval before signing or OKing documents prepared by assistants.

FAXING LETTERS

If you plan to fax a letter, type *FAX* between the date and the inside address.

After typing "FAX" above the inside address, complete the letter in the usual way. Then fill out a cover sheet—a form with spaces for name, company, and fax number of addressee; telephone number of sender; number of pages to be faxed, including cover page with a brief message if needed; and instructions to call or fax you if all pages are not received and legible. Faxing is often faster and cheaper than phoning or special mail services for rapid delivery, and it enables you to communicate instantly. Confidentiality, however, is lacking as fax machines are often in a common area. In addition, the reader doesn't get the psychological effect of your quality stationery; and type size and clearness may be reduced by the fax process. If you use a pen for a document to be faxed, be sure the pen has black ink.

PROOFREAD, PROOFREAD, PROOFREAD! THEN SIGN

Before signing, proofread letters and other documents not only to detect and correct errors but also to improve writing style.

See Proofreaders' Marks on inside back cover.

After initial proofreading on your monitor, print the document on draft paper and proofread from the "hard copy." You will often find an error you missed or a way to improve the communication. If so, make the change, proofread again, and then print the final copy. If possible, proofread correspondence a final time before inserting it in the envelope.

Sign letters with a pen, never a pencil.

Writing for Your Career

Skillful business writers often do several drafts before producing a well-written, correctly formatted document. On the job, most people compose at the keyboard but change the wording a few times before being satisfied. If you don't keyboard by "touch," take a course in it.

1. What is the standard size stationery for American letters? _____$8\frac{1}{2}$_____ by _____11_____ inches

2. A small envelope is called a No. _____$6\frac{3}{4}$_____.

3. A large envelope is called a No. _____9_____ or a _____10_____.

4. What should be handwritten in pen on a letter before mailing or faxing it? _____signature_____

5. When a properly folded letter is removed from a large business envelope, it has _____2_____ creases; from a small envelope, it has _____3_____.

6. After keyboarding a letter, p _____roofread_____ it on the screen and then on the h _____ard copy_____.

7. If you are faxing a three-page document, how many pages should be sent? _____4_____

8. Because of e-mail and fax, first-class mail is rarely used now in the United States. (true or false) _____False_____

9. Information printed at the bottom of a letterhead is called the _____letterfoot_____.

10. Where is the return address on a business envelope usually printed? _____upper left corner_____

Keep an office reference book, a college dictionary, and this text on your desk.

Read
78

Check answers on page 414.

THE RIGHT PARTS FOR THE RIGHT PLACES

The basic parts of typical business letters from top to bottom are: printed letterhead, date, inside address, salutation, body, complimentary close, writer's signature, writer's typed name and title or department, and typist's initials.

Letterhead and Date

> **Keyboard the date two or three lines below the printed letterhead. Neither abbreviate the date nor use *nd, th,* or *st.***

AMERICAN   June 2, 2006 [Note the space between the comma and the year.]

INTERNATIONAL AND MILITARY   2 June 2006 [no comma]

Inside Address and Salutation

Students should read the sample letters to learn format specifics and conversational writing style.

> **The inside address is the name and address of the person and/or company to receive the letter. Begin 4 to 10 spaces below the date, depending on the letter's length. The salutation (a greeting) follows. The first line of the inside address and the salutation should agree. Double-space between the inside address and the salutation.**

Precede the name with a professional title such as *Dr., Captain, Father, Rabbi, Senator, Professor,* etc., or with courtesy title *Mr.* or *Ms.* if a professional title is

To help ESL and international students, ask for examples of first names that may be male or female; e.g., Leslie, Robin, Chris, Pat.

not known. If no professional title is known and you don't know whether you're addressing a man or a woman, use the complete name in the address and the salutation. For example, Can Nguyen is a Vietnamese name that many Americans cannot identify as masculine or feminine.

Professor Mary Rowe
Miami Dade Community College
11380 NW 27 Avenue
Miami, FL 33179

Can Nguyen, Service Manager
Computerland of La Mirada
12345 Imperial Boulevard
La Mirada, CA 92702

Dear Professor Rowe:

Dear Can Nguyen:

Avoid out-of-date salutations such as *Dear Sirs, To Whom It May Concern, Dear Sir or Madam*. If you know the person's name, the salutation is the same as in a personal letter: *Dear Jonathan* or *Dear Ms. Lovely*. If you don't know the individual's name, use an appropriate official title, such as *Dear Sales Manager, Dear Lacy's Linens, Dear Credit Department, Ladies and Gentlemen*—but NOT *Dear Ladies and Gentlemen, Dear Sirs, To Whom It May Concern*, or *Dear Sir or Madam*. If the organization consists of all men or all women, use either *Ladies* or *Gentlemen*. Avoid abbreviations in the inside address, except for Mr., Dr., and official state abbreviations, as listed in Appendix D. One way to avoid salutation questions is to use the "simplified" style, which requires no salutation (and no complimentary close). Instead of a salutation, use a subject line in all capital letters.

Use of first names is prevalent among American businesspeople; a comma (instead of a colon) after the first-name salutation reflects informality.

International correspondence is more formal. When working with people from other countries, you often need to learn which part of the name gets the courtesy title, since naming systems differ around the world.

Body

> **The body of the letter is the message. Begin it a double space after the salutation (or subject line if there is one). Single-space the body, but double-space between paragraphs.**

Leaving a blank line between paragraphs eliminates the need to indent the first line of a paragraph. While it isn't wrong to indent paragraphs, begin paragraphs at the left margin unless your supervisor requests otherwise. (Paragraph indention was useful when correspondence was handwritten.)

Complimentary Close

> **The complimentary close, either traditional or personalized, signals the end of the message. Double-space the complimentary close after the last line of the body; capitalize only the first word.**

Traditional complimentary closes are *Sincerely, Yours sincerely, Sincerely yours,* or *truly* closings such as *Very truly yours* or *Yours very truly. Cordially* closings are also correct. *Respectfully* is inappropriate for ordinary business letters but may close a letter to a judge or high-ranking government, school, or church official. Informal or personalized closings—such as *Regards, Best wishes, Thanks again, Happy holidays*—are appropriate, especially if you know the reader well enough to use a first-name salutation.

Handwritten Signature

> **With a pen that writes blue or black, sign in the space after the complimentary close. Do not use a title (Mr., Miss, Mrs., Ms., Dr.) with your signature in business letters.**

In workplace correspondence, indicating a woman's marital status (Mrs. or Miss) is unnecessary and out-of-date.

Writer's Name and Title or Department Name

Only the most frequently used styles are shown here. See a reference manual for other styles; e.g., simplified letter and simplified memo.

> **Type the writer's name and official job title or department about four spaces below the complimentary close, thus leaving space for the signature.**

Avoid including a courtesy title in both the signature and the typed name. To indicate whether the writer is male or female, spell out names such as Chris to Christine or Christopher or Pat to Patrick or Patricia. Another possibility is to include the writer's middle name—Whitney Dawn Blake or W. Dawn Blake. Those devices let the reader know whether to use Mr. or Ms. when responding. Using only initials before your last name (E. J. Anderson) is inconsiderate and pompous-sounding.

Type the official job title or department name below the writer's typed name. If the writer's name and job title are short, they may be on the same line with a comma between them.

Sincerely,   Very truly yours,

Larry Lutsky   *JoAnn R. Michels*

Larry Lutsky, President   JoAnn R. Michels
                                                   Accounts Payable Department

Identification Initials

> **Key typist's initials (if other than the author of the letter) in lowercase letters at the left margin a double space below the name and title or department name.**

Betty Van Meter
Business Skills Center

lrs

OPTIONAL LETTER PARTS SOMETIMES NEEDED

Writer's Address

> **If you don't have letterhead stationery, key your street address about an inch from the top of the page. The city, state, and ZIP are on the next line. Type your name in the usual signature position at the end of the letter.**

Here is an example of the writer's address on a personal business letter. Begin about one to two inches from the top of the page. The date follows immediately, and other letter parts are the same as a letter on letterhead stationery.

2166 Clinton Avenue
Bronx, NY 10406
May 31, 0000

Confidential/Personal or Mailing Notations

> **Double-space below the date special notations like PERSONAL, CONFIDENTIAL, FAX, BY MESSENGER, REGISTERED, SPECIAL DELIVERY, FEDERAL EXPRESS, etc.**

January 2, 0000

FAX TO 405-311-2222

Attention Line

> **Attention lines are not used much any more.**

Knowledgeable business writers omit the "attention line" and address letters directly to the name of the organization, an individual, a department, or a title (Human Resources Manager). Type the envelope address just like the inside address, both without the word *Attention*. See the inside addresses on the sample letters that follow.

Subject/Reference Lines

> **A subject or reference line may be used to state the letter's main topic and serve as a heading for the message. Type it between the salutation and the body as in the letter to Billy in Read 79.**

A subject or reference line showing the topic of the letter is a considerate addition to some business letters. Before reading the letter, the receiver might review related documents.

Dear Mr. Wallace: (salutation)

JOB PLACEMENT OF GRADUATES (subject line below salutation)

The five United graduates listed below had high scores on our employment test and. . . . (body)

Some stationery has the word *Reference* printed below the date. A name or transaction number may be keyed beside *Reference*.

Reference: Policy No. 26382 (reference line below date)

Company Name (at End of Letter)

> **The name of the organization sending the letter is sometimes keyed in all capital letters a double space below the complimentary close.**

Following the complimentary close, double-space, and type in all capital letters the name of the organization sending the letter. Then leave three or four blank lines before the name and title (if any) of the person who will sign the letter in the blank space.

Sincerely,

SNYDER COMMUNICATIONS, INC.

Handwritten signature here

Joshua Snyder, Personnel Director

Enclosure Notation

> **If you enclose something with a letter, add an enclosure notation at the left margin a double space below the typed name and title of the sender of the letter.**

Enclosure: Check No. 268

Copy Notation

> **If someone other than the addressee is to receive a copy, add a notation at the left margin a double space below the typed signature line.**

If you don't want the addressee to know you've sent copies, key *bc* (for blind copy) on the copies only—not the original—followed by the name/s of those to receive a copy. A colon after the *c* is optional.

c: Professor Nina Hixon

bc: Professor Foster

cc stands for courtesy copy to . . . ; one *c* simply means *copy to*; *bc* means "blind copy," which means the copy notation appears only on the file copy, not on the original.

Postscript (PS)

> **If a postscript is used, it is the final item on the page. The abbreviation P.S. or PS is correct, but not required, before the message.**

Avoid using a postscript for something you forgot to include in your letter. Instead, insert the information where it belongs within the letter. Use a postscript for emphasis (particularly in sales letters), occasionally as a personal message, or for information on a topic different from the rest of the letter.

PS Please fill out the enclosed reply card and return it today. (for emphasis in a sales letter)

PS Have a great time on your Trinidad trip. (personal message added to a letter to a client)

Replay 78

A. Write **T** (true) or **F** (false) in each blank.

F	1. Whenever a copy of a letter is made, key a copy notation below the initials.
F	2. The typist's initials should be in capital letters.
F	3. A courtesy title should be used before the letter writer's name.
T	4. A *bc* notation should not appear on the original of a letter.
T	5. One correct way to type the date in a business letter is 6 June 2004.
T	6. The letters PS are not required before a sentence that is a postscript.
T	7. If a subject line is used, key it a double space after the salutation.
T	8. The company name is not required as part of the closing information.
F	9. Most well-written business letters include an attention line.
F	10. If the inside address begins *Mr. Peter Settle*, use *Dear Sir* as the salutation.

B. Fill in the blanks with the names of the letter parts numbered on the letter to Professor Shay.

1.	Date	6.	Writer's name
2.	Inside address	7.	Writer's title
3.	Salutation	8.	Typist's initials
4.	Body	9.	Enclosure notation
5.	Complimentary close	10.	Blind copy notation

Check answers on page 414.

Eric Smith FINANCIAL SERVICES
4464 Fremont Avenue No. • Seattle, WA 98103 • (206)632-3337

(1)   September 18, 2006

(2)   Professor Adell Shay
      Los Angeles Harbor College
      1111 Figueroa Place
      Wilmington, CA 90744

(3)   Dear Professor Shay:

(4)   Joe Santona gave us your name as someone who likes to see profit, good
      business practices, protection of the environment, and social responsibility
      all working hand in hand.

      Joe thinks you'll be interested in learning more about how you can have a
      healthy profit, growth, and safety while investing wisely in activities you believe
      in. Few had even heard of socially responsible investing in 1986 when we began
      to develop SRI programs. Individuals like you, small businesses, and large
      pension funds are now benefiting from our experience in the social investment
      field.

      We invite you to a complimentary initial consultation during which we'll discuss
      your present needs and financial goals. We'll determine whether we can help
      you meet your objectives for short-term profit, long-term growth, and safety
      through socially responsible solutions. A brochure is enclosed that shows how
      we help our clients achieve peace of mind while their money works for them.

      To arrange a no-obligation appointment, please fill out the enclosed card and
      return it to us.

(5)   Sincerely,

                                                       Eric A. Smith

(6)   Eric A. Smith, CFP
(7)   Certified Financial Planner

(8)   Irs

(9)   Enclosure

(10) bc Joseph Santona

Avoid out-of-date jargon. Examples: We trust, aforementioned, aforesaid, herein, hereto, herewith, in due course of time, for
your perusal, this will acknowledge your request for . . ., I am writing to inform you that, this is to advise you that, etc.

PRODUCING WRITTEN COMMUNICATIONS

Successfully and *personally* **are** often redundant and can often be left out; for example, *successfully* avoided, *successfully* withstand, *successfully* captured, I *personally* believe, I *personally* require. . . . The italicized words can be omitted for better writing style. Sign on a California dry cleaner shop—"Open 7 Days a Week, *Sundays, Too.*"

Produce professionally prepared written business communications: letters, interoffice memorandums, business envelopes, and e-mails. You can save time by using templates and wizards to standardize your letters, memos, envelopes, and faxes, as well as e-mails. With the instructions that follow, however, you can produce correspondence with or without word processing templates.

BLOCK LETTER, STANDARD PUNCTUATION

The letter to Billy shows the block layout and standard punctuation. In addition to basic letter parts, notice the subject line and the PS. Please read the message since it is really for your information, not Billy's.

TELCOM SERVICE

TELEPHONE SYSTEMS - SALES AND SERVICE - SINCE 1979

February 1, 200-

Mr. Billy Crystal
Entertainment Productions, Inc.
711 Wilshire Boulevard
Hollywood, CA 90210

Dear Billy,

THE BLOCK LETTER STYLE

Two types of letter layouts are used most often: block and modified block. This letter is formatted in block style; that is, all lines begin at the left margin. Here are some reasons for choosing this style:

A contemporary appearance results from beginning all lines at the left margin. In addition, it's faster to keyboard a letter in block style than in other styles. The clean, trim arrangement tends to carry over to the psychological image of the company and its products or services. Finally, it looks good with this letterhead design.

Choose either open or standard punctuation. The punctuation in this letter is standard—which means a colon or comma after the salutation and a comma after the complimentary close. Use a comma after a first-name salutation; otherwise, use the colon. A letter with "open punctuation," has no punctuation after **both** the salutation **and** the complimentary close.

Perhaps you'll decide on the block letter style when replying to your fans.

Best wishes,

Leila

Leila R. Smith
Professor of Office Administration

mht

PS Regards to Janice, Jennie, Mike, Lindsay, and Ella.

10221 Slater • Fountain Valley, CA 92708 • (714) 964-1600

MODIFIED-BLOCK LETTER—OPEN PUNCTUATION

In the following letter, modified-block layout and open punctuation are used. This format is explained in the letter.

GOLDFINGER & SIMON
ACCOUNTS PAYABLE AUDITS
99 WEST 42 STREET
NEW YORK, NY 10030
TELEPHONE 212-384-1248
FAX 212-384-7890

JACK I. GOLDFINGER
MARTIN L. SIMON

March 19, 2008

President Keshia Mary Washington
The White House
Washington, DC 20500

Dear Madam President

In the modified block letter, the date and closing begin at about the horizontal center of the line. The first word of each paragraph may begin at the left margin as in this letter, or it may be indented five to ten spaces for a more conservative look. We recommend you block the paragraphs for the more current look shown here.

Some people prefer modified block format because it is more conservative and traditional looking than the full block style. Therefore, modified block may suggest a more appropriate image for your fiscal policies or political philosophy.

This letter, Madam President, has open punctuation, which means no punctuation after either the salutation or the complimentary close. Standard punctuation would be equally correct but more conservative and traditional looking.

I hope this information helps your office professionals respond to the many letters you receive from the people of America. Congratulations on being the first female president of the United States of America.

Respectfully

Margaret H. Taylor

Margaret H. Taylor
Office Technology Supervisor

lrs

c: Secretary of State Sean Chandra

Avoid needlessly
negative expressions,
such as: You neglected
to, You failed to (never
tell clients they "failed"),
You are in error, We are
surprised at you for, It
surely must be obvious
to you that, If you
actually did, You
claim, etc.

PERSONAL BUSINESS LETTER

This modified block personal business letter on non-letterhead paper has standard punctuation. Your address above the date is the preferred format unless you use personal letterhead stationery, in which case, simply begin with the date. Phone and fax numbers and e-mail address are optional.

30 Christanna Street
Alberta, VA 23921
462.531.8700
January 4, 200-

Ms. Kerry Gambrill
Designing Women Ltd.
11901 Beach Boulevard
Winston Salem, NC 27103

Dear Ms. Gambrill:

It's a pleasure, Ms. Gambrill, to provide you with information about the personal business letter.

Use this format with confidence when corresponding about personal business, such as your insurance, credit, job applications, charitable functions, and comments to government officials.

Since you use plain paper with no letterhead, start with your street address, followed by city, state, and ZIP. Next type the date. The rest of the letter is the same as any other business letter, except do not use identification initials. **NEVER use your employer's letterhead stationery for your personal business communications.**

I expect to be in Winston Salem next month and hope to visit you in your new Forsyth Avenue showroom.

Sincerely,

Al Stewart

Al Stewart

SUMMARY OF LETTER PLACEMENT

> Show your professional expertise in setting up business letters. Using block or modified block style, arrange letters on the page with approximately equal left and right margins.

One suitable font for business letters is 12 point Times New Roman. Avoid right-margin justification (all lines ending at exactly the same point). Use margins of about 1 inch to $1\frac{1}{4}$ inches left and right. Spell out the month when you key the date—two or three spaces below the letterhead. Leave a bottom margin of at least 1 inch. For stationery with a "letterfoot," leave at least a half inch before the first line of the letterfoot.

If you hyphenate for a trim right margin, apply word division rules (pages 00 and 000) in addition to your word processor's automatic hyphenation. You can improve the appearance of especially short letters by adding an extra blank line or two before and/or after the date and by enlarging the margins. To fit long letters on one page, decrease the left and right margins to about $\frac{3}{4}$ of an inch. To center short or long letters, reduce or enlarge the font size and/or change the font style to one providing either more or fewer characters to the inch.

For letters of two or more pages, key a heading with recipient's name, page number, and date beginning about 1 inch from the top of the second and succeeding pages. Print second pages either on plain paper matching the quality and color of the letterhead stationery or specially printed second-page stationery.

Ms. Betty G. Dillard

page 2

January 4, 2006

PLACEMENT OF ENVELOPE PARTS

> Single-space addresses on envelopes; use either the envelope default placement or begin at about the center of the envelope.

Most organizations use envelopes with a printed return address. For personal business letters, keyboard your return address or use your name and address sticker. Capitalize the first letter of each proper noun and use conventional punctuation as shown in the illustrations. Type the "outside address" the same as the inside address that's on the letter. For addresses not in your country, add the country's name in all capital letters on a line alone. Key or rubber stamp mailing instructions such as SPECIAL DELIVERY just below the postage corner. A PERSONAL notation, if desired, may be typed a little below the return address.

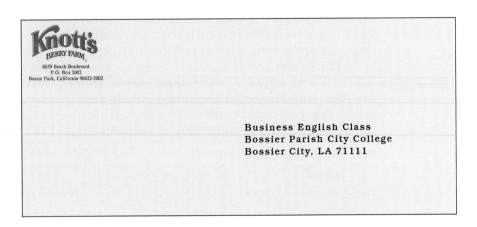

INTEROFFICE MEMORANDUMS

Memorandums are principally for communicating within an organization, although they may occasionally be sent outside because of speed of preparation. Some interoffice memos are formal communications to employees, and others are informal. Memorandums can be important to career progress. Your colleagues or superiors who receive yours judge you, perhaps subconsciously, on the content and on how you express yourself in writing.

You might use preprinted interoffice memorandum forms or e-mails. Regardless of how the form is produced, a memorandum (also called *memo*) is designed to be keyboarded quickly and requires no vertical centering. Just begin the message a blank space after the heading if it's a printed memo form or a blank space after the salutation if it's an e-mail. Following are two samples. Perhaps the first one, written with tongue-in-cheek, will bring a smile.

VALLEY COLLEGE
INTEROFFICE MEMORANDUM

TO: Coach I. M. Musselman, PE Department

CC: Professor U. R. Jockette, PE Department

FROM: Ben Wisenheimer, English Department *BW* [sender's handwritten initials]

DATE: September 1, 200-

SUBJECT: Every Coin Has Two Sides

Remembering our discussions of your football players who were having troubles in English, I have decided to ask you, in turn, for help.

We feel that Paul Barebones, one of our most promising scholars, has a chance for a Rhodes scholarship, which would be a great thing for him and for our college. Paul has the academic record for this award. We find, however, that the candidate must also have other excellences and, ideally, should have a good record in athletics. He does try hard at sports, but it has always been difficult for him.

We propose you give some special consideration to Paul as a varsity player, putting him, if possible, in the backfield of the football team. In this way, we can show a better record to the committee deciding on the Rhodes scholarships. We realize that Paul will be a problem on the field, but as you have often said, cooperation between our department and yours is highly desirable. Of course, we do expect Paul to try hard. During breaks from study, the English teachers will coach him as much as we can.

His work in the English Club and on the debate team will force him to miss many practices, but we intend to see that he carries an old football around to handle (or whatever one does with a football) during intervals in his work.

We expect Paul to show entire goodwill in his work for you, and though he will not be able to begin football practice until late in the season, he will finish the season with good attendance.

EXPRESSIONS TO AVOID: To be perfectly honest, Enclosed herewith, Please find, As you know, As per our conversation, Please be advised, At your earliest convenience, I am writing this letter to inform you, Please rest assured.

PACIFIC RIM IMPORTS, INC.

DATE: June 30, 200-

TO: New Employees

FROM: Ronald Tawa, Director of Human Resources *Ron* [Sender handwrites
                                                        initials or name.]

SUBJECT: Employment Policies

COPY TO: Yasmin Salazar, President

Welcome to Pacific Rim Imports, Inc. We hope you'll find your employment here enjoyable and personally rewarding. These policies will help you understand our operation and help you realize the importance of your job:

- **Security** Each new employee should promptly obtain a permanent identification card from the Personnel Department. A guard is on duty around the clock. Everyone who enters the building is required to show this card.

- **Absenteeism** Employees are expected to be in their assigned departments and ready to begin work at 8:30 a.m. If you are unable to come to work, please call my administrative assistant, Jim Harrison, at Extension 711. He will inform your department manager.

- **Loyalty** Loyalty is expected from all employees. Information about new products or financial matters should not be disclosed to the public. Anyone who passes on such information is subject to immediate dismissal.

- **Smoking** Because of the dangers of secondhand smoke, smoking is permitted only on the terrace outside the cafeteria—and nowhere else in the building.

If you have any questions, please see Jim Harrison or me. Stop by my office at any time. I look forward to meeting you.

E-MAIL COMMUNICATION

- **DO** provide a meaningful subject line for your reader.
- **DO** double-check your message for grammar and spelling accuracy.
- **DO** prepare e-mail messages that make you appear efficient and courteous.
- **DO NOT** write e-mails with all capital letters.
- **DO NOT** send confidential messages via e-mail.
- **DO NOT** write anything you wouldn't want made public or forwarded to others.

Because most businesspeople read and reply to a great many e-mails each day at work, most business e-mails should be short and preferably about one topic. Special abbreviations may be used to shorten messages; for example, BTW (by the way), GMTA (great minds think alike), or FAQ (frequently asked questions). Use them sparingly and only if you're sure your reader knows what they mean. Begin with a salutation, but if you know the person well, express it informally; for example, "Hi Doreen," "Doreen," "Thanks, Ms. Lieberman," or "Dear Joseph." A complimentary close may be used, omitted, or informal such as "Thanks," or "Cheers" or may even be an emoticon (**emot**ion + **icon**), although some people think they're "too cute" or otherwise objectionable: **@-->** (rosa) **:-O** (shocked) **:-~)** (user has a cold). Avoid using them if in doubt.

Word Power

Workplace Jargon

blogs Internet sites enabling people to get in touch with others who share interests.

dot-com company A company that does business on the Internet.

factoid Small bit of information frequently repeated; likely to be unproven.

Replay 79

Write **T** (true) or **F** (false) in each blank.

T	**1.** In a block style letter, all lines begin at the left margin.
F	**2.** In a modified block letter, the paragraphs must be indented.
F	**3.** For standard punctuation, use a semicolon after the salutation.
F	**4.** Put the ZIP code on the line below the state on the envelope.
T	**5.** If you use a subject line in a letter, place it below the salutation.
F	**6.** The body of a business letter should usually be double spaced.
F	**7.** Most business letters require attention lines.
F	**8.** Typical left and right margins for business letters are 2 inches.
F	**9.** Type envelope addresses in all capital letters with no punctuation.

T	10. The most popular letter styles are block and modified block.
F	11. The only difference between open punctuation and standard is that a comma follows the complimentary close in the standard style.
T	12. In a modified block letter, begin the date and the closing information at about the horizontal center.
F	13. Type the salutation on an interoffice memo a double space above the body.
F	14. *Very truly yours* is an appropriate close for most e-mail business messages.
F	15. Before making a copy of a letter for your file, type a lowercase *c* below the initials.
T	16. If you keyboard a letter for someone else's signature, type your initials at the left margin a line or two below the author's name and title.
F	17. To communicate confidential information within your company, use e-mail rather than interoffice mail.
F	18. To format a business letter correctly, justify the right margin.
T	19. Begin the address at about the center of an envelope.
T	20. Repeat the date and addressee's name on the second page of a two-page letter.

See page 414 for answers.

Read 80

SOUNDING GOOD

USE A FRIENDLY TONE

A good letter or e-mail could consist of just one sentence; e.g., Would you please send your mail-order catalog to me at the address on this letterhead. (Notice the period at the end, not a question mark.)

Compose letters, memos, and e-mails in a friendly, conversational tone—but use complete, clear sentences and Standard English. Business communications shouldn't look like you wrote them for an antique collection (see the letter below!) nor as though "you ain't seen no schools."

Dear Sir and Madam:

Enclosed herewith please find your life insurance policy; we trust it is in order. If our company can be of any further service along the insurance line, kindly feel free to advise us. Please do not hesitate to contact us in the event that you have any questions.

Respectfully yours,

Below is today's version of the note shown above.

Dear Mr. and Mrs. Marvin:

Your new life insurance policy is enclosed. Please call us at 800.345.6789 if you have any questions about this policy or if we may help you with any other insurance matters. I enjoyed meeting you both and wish you a happy holiday season.

Sincerely,

Make the reader feel a warm human being wrote the message, not an impersonal corporation or a computer. Don't select words because they seem "businesslike" or because they have many syllables and sound impressive. Untrained businesspeople may think a formal tone sounds successful and educated, but they end up sounding pompous and insecure. Select words that deliver your message clearly and completely and that help build a good relationship between you (or your company) and the reader. Plain (but correct) English can improve productivity, improve customer relations, and reduce costs caused by misunderstandings.

POMPOUS GOBBLEDYGOOK   This company sincerely regrets any inconvenience caused you by our inadvertently miscalculating the extensions on our invoice to you.

FRIENDLY AND CLEAR   We're sorry about the error on our Invoice No. 2482.

POMPOUS GOBBLEDYGOOK   This letter is to advise you that the merchandise you ordered is out of stock. We regret to inform you, therefore, that we cannot ship same to your establishment until some future date.

FRIENDLY AND CLEAR   Thank you for your order for one dozen gold birthday charms. Although we are temporarily out of the charms, we expect them from our factory early in March. We'll rush them to you as soon as they arrive.

Here are a few examples of expressions to avoid—along with suggestions for writing concisely and clearly without sacrificing courtesy.

AVOID	USE
at an early date, in the near future	soon or a specific date
at your earliest convenience	omit, be specific, or give approximate time
please do not hesitate to call	please call
we regret to inform you that	we're sorry that
allow me to state that	omit; just go ahead and say it
may I take the liberty of	omit or just do it
we are in receipt of	we received or thank you for
attached please find	here is/are, enclosed is/are, or attached is/are
reached the conclusion	concluded
for the purpose of	to, for
in a satisfactory manner	satisfactorily
utilize	use
in the event that	if

MEMO FROM THE WORDSMITH

Blaise Pascal, an eminent 17th-century French philosopher and mathematician (for whom Pascal computer language was named), ended a letter to a colleague with these words: "I have made this letter so long only because I have not had the time to make it shorter."

It takes time and careful proofreading to cut excess verbiage from business communications as well as academic writing, such as term papers.

Recap

Rewrite the following sentences to sound friendly, conversational, and clear.
Possible answers

1. I am writing to inform you that I have received the book that you sent me, and I sincerely appreciate your kindness. ___Thank you for sending___
   Writing Better Letters.

2. In accordance with your request, we are herewith enclosing the price list.
   Enclosed is the price list you requested.

See page 414 for possible answers.

EMPHASIZE POSITIVE IDEAS

> **Try to write about what you or your company _can_ do, not what you _can't_ do.**

People respond better to positive ideas. Even if you have something negative to write about, you can often find a positive way to express it.

WORD TO THE WISE

Success comes in cans.

NEGATIVE We cannot conduct this seminar with fewer than ten students.
POSITIVE We can conduct this seminar with ten or more students.

NEGATIVE We're sorry we can't extend more than $3,000 credit to you.
POSITIVE You can charge up to $3,000 worth of widgets and whatnots.

NEGATIVE Our catalogs are not free.
POSITIVE Please send your check for $3 in the enclosed envelope; your catalog will be mailed immediately.

Recap

Use a positive tone to rewrite these negatively expressed sentences. Possible answers

1. We hope you will not be disappointed. ___We hope you will be pleased.___

2. We are not open after 8 p.m. on weekdays or on weekends after five.
   We are open from 10 a.m. until 8 p.m. on weekdays and until 5 p.m. on weekends.

3. You can't have a refund. We only give credit slips. ___We'll be glad to give___
   you a credit slip that you can use for any other item in the store.

See possible answers on page 414.

CONCISENESS VERSUS REDUNDANCY

> **Unless repetition is for special effect, express an idea just once. Purpose-less repetition is called *redundancy*.**

Concise writing means each word contributes to the purpose of the message. In addition to conciseness, however, courtesy, clearness, correctness, and completeness are essential. Redundancy means a document has more words than **necessary** to convey a desired meaning and tone. A concise document may be three lines or three pages, depending on the nature of the message.

REDUNDANT	CONCISE
repeat again	repeat
return back	return
cooperate together	cooperate
round in shape	round
yellow in color	yellow
modern and up-to-date	modern
visible to the eye	visible
final outcome	outcome
advance planning	planning
past history	history
true facts	facts
my personal opinion is	my opinion is or I believe
new innovation	innovation
free gift	gift (A gift is a gift is a gift; if you pay for it, it's not a gift.)

NO It is absolutely essential that each and every widget be round in shape.

YES Each widget must be round.

NO The consensus of opinion is that although the object is 4 feet long in size, we can see it visually from a distance of 50 feet.

YES We believe that although the object is 4 feet long, we can see it from 50 feet.

NO We requested final completion of the project by the year of 2004.

YES We requested completion of the project by 2004.

Recap

Rewrite these sentences to eliminate the redundancies.   Possible answers

1. We want to suggest to you that first and foremost you pack each and every basic essential. _We suggest you pack the essentials first._

2. Final completion of the research investigation study revealed and showed that the UFO is small in size, triangular in shape, and purple in color. _The investigation revealed a small, triangular, purple UFO._

3. I am of the opinion that in the majority of instances the event is held during the month of December. <u>I believe the event is usually in December.</u>

4. Please repeat the instructions again. <u>Please repeat the instructions.</u>

For possible answers, see page 414.

PARAGRAPHING BUSINESS LETTERS

Paragraphing is based on judgment. Since many well-written documents are about one subject only, begin a new paragraph when you want a longer pause than occurs after a period. Opening and closing paragraphs are often short, sometimes just one sentence each. A brief and clear opening tells the reader what the letter is about and is followed by the main message. The closing sentence or paragraph might summarize what you want the reader to do, such as pay the bill or place an order.

Dear Mr. Goldman:

The dozen gold charms you ordered January 3, Order No. 268, are on their way to your shop today. We apologize for the delay but believe you'll be pleased with the rapid turnover of these delicately engraved charms.

Our sales representative, Kim Silverstein, will be in to see you soon to show you our newest sterling bracelets—designed by Pagliano, the famous Italian silversmith.

Sincerely,

There is no "right" or "wrong" way to paragraph business letters, only good or poor judgment. Avoid a "chopped-up" look of too many short paragraphs one after the other. Long paragraphs, however, look hard to read and discourage busy people from concentrating on the message. Use the preceding hints plus your common sense to vary paragraph lengths sensibly.

FIRST DRAFT

Even the most experienced business writers often make changes on their first draft of a message. Improve your first draft on screen; then print and reread it. You will usually think of additional changes to further improve your communication. Make the improvements and reprint. Conserve paper by using the marked-up first printout as your file copy. Follow this procedure even for e-mails important to your work.

Replay 80

A.   Write **T** (true) or **F** (false) in each blank.

_____ F   **1.** Long words that sound businesslike give readers the impression you are well educated.

_____ F   **2.** Conversational-style writing should be avoided in written business communications.

_____ F   **3.** Good letter writers follow specific paragraphing rules.

_____ T   **4.** The first and last paragraphs in good letters are often the shortest.

_____ F   **5.** To be sure a letter states your policy clearly, stress what your company cannot do.

B.   Improve the following sentences according to Read 80 suggestions. Possible solutions.

1. About half the flowers in my garden have died. ___About half the flowers in___
   ___my garden are blooming.___

2. This letter is to advise you that the parts you ordered were shipped
   today. ___The widget parts you ordered were shipped today via Airborne Express.___

3. We believe you won't have any trouble with our newly designed
   widgets. ___Our newly designed widgets have been carefully tested for accuracy.___

4. We have forwarded the document to you by fax. If you have any
   questions about any of the items, please do not hesitate to contact us
   at your earliest convenience. ___We faxed the price list to you yesterday. Let us___
   ___know if you have any questions.___

5. Upon investigation, it has come to our attention that we didn't receive
   your check in the amount of $6,453.00 to pay your bill. ___To maintain___
   ___your good credit rating, please send your check for $6,453 in the enclosed___
   ___envelope today.___

Possible answers are on pages 414–415. Take the Pop Quiz on page 360.

Read 81

WRITING FOR MULTIMEDIA
BY BARBARA MORAN

A new form of business communication has developed in recent years. With the
evolution of the World Wide Web and presentation software such as MS Power-
Point, writing for multimedia has become a necessary skill for your career.

> **Ultimately—words are the reason a Web site exists, words are the reason
> for the slide show, and words *equal* information in the Information Age.**

Content is **key;** that is, primary! Graphics, images, photos, and design must
complement—not overwhelm—words, but the reverse is also true: The words
must not overwhelm the graphic parts of the presentation. More than ever, however, words have the power to inform, to sell, to entertain, to enlighten, to up-
lift, to change, or to inspire. Don't make words an afterthought when you put
up a Web site or prepare a slide presentation. Consider written content during
the first stages of planning.

TWO WRITING STYLES

Traditional

Writing is *read* by a *reader.* Letters, memos, and reports are written on paper
and held in the reader's hands. The recipient reads the documents at his or her
own pace from beginning to end. Long and detailed sentences, paragraphs, and
columns of text are the norm.

Multimedia

Writing is *viewed* by a *viewer.* Text is presented on a screen. It isn't held, but
rather *viewed.* The viewer has less time to digest the text and may not read it

from beginning to end. For that reason, briefer sentences, paragraphs, and columns of text are preferred.

PLAN YOUR PRESENTATION

As with any quality business communication, planning is important. Decide what you want to say, why you want to say it, and with whom you wish to communicate. With a slide presentation, your audience may be limited to one specialized group. But with a Web site, your audience is the world. Be aware of language that is too colloquial for nonnative speakers of English or that may inadvertently prove offensive to an international visitor. Follow these steps to plan your presentation:

- **Summarize the goal** of your Web site or slide presentation in one or two paragraphs, often dubbed "nut graphs." If you can't do this, think about it more.

- **Describe your audience** in one or two sentences. Are you trying to reach a certain age or demographic* group or a specific business clientele or customer base? Since this will influence your writing style, know your audience well.

- **Gather your information and ideas** Write them down in order of importance, grouping related topics together. Particularly for a Web site, you may want to draw a chart, mapping out what information will be on your "homepage" (the main page of your site) and where links from your homepage will go.

PREPARE YOUR ROUGH DRAFT

After planning and deciding on content, prepare a rough draft. Then leave it for awhile, and come back later to edit and revise. If possible, have someone review it with you. Ask yourself these questions:

- Does it flow in a clear, organized way?

- Is it accurate? (Your credibility is at stake.)

- Are spelling, grammar, and punctuation correct? Any inaccuracy reflects poorly upon the professionalism of your entire effort.

- Is it concise yet complete?

- Is the tone appropriate for the subject matter and audience—that is, formal vs. informal, humorous vs. serious, introductory vs. advanced, objective vs. persuasive, etc.?

- If information is from another source, did you credit the source? If not, you may have committed plagiarism, which is unlawful. Always identify your sources.

DESIGN MULTIMEDIA EFFECTIVELY

The Click of a Mouse

Although you control the speed and order of your slide presentations, visitors can hop from any part of your Web site to another with the click of a mouse. Make sure all your Web pages tie in with one another. Visitors may first land in the middle of your site (which happens courtesy of search engines). Prepare a

---

*If in doubt, please look it up **now.**

Web site so that visitors quickly perceive what your site is about and how to use it. Otherwise, they may simply jump to another site.

Involve the Viewer

Use active words and tone as well as short sentences and phrases.

BORING Learn How To Cook Chinese Food
BETTER Create Chinese Cuisine

BORING Find Out About Mexico
BETTER Explore Mexico

BORING A message is best conveyed with carefully chosen and effectively applied words.
BETTER Choose the right words to convey your message.

BORING When writing, always try to tailor your business presentation for your specific audience.
BETTER Write for your audience.

CREATE STRONG VISUAL IMPACT

> In multimedia presentations, words are graphic elements. The amount of text, arrangement of text, font choices (style, size, and color), movement (animation), and sound all affect your audience—and therefore your message.

Amount of Text

Avoid crowding words. On a screen a bundle of words can be time-consuming and hard to view and comprehend. Long words and sentences take more time to digest than short words and sentences. Using short phrases in a slide presentation is totally acceptable. A large, blocky chunk of text is not inviting to a viewer. Keep sentences and paragraphs short.

Arrangement of Text

Leave space between lines. Make use of bullets and text boxes. The placement of words affects their importance and tells viewers where to look first. Blank space provides impact too. Open space—or "negative space"—is used for balance or a dramatic effect.

Font Choices

The style, size, and color of font you use **affects the viewers and their mood.** The font **size** affects the importance of the **words**. We generally notice **bigger** text before smaller text. The color of text also impacts the viewer. Choose aesthEtically pleasing colors. Other techniques for emphasis include **boldface,** *italics,* underlining, changing portions **of text** to **different** fonts, putting text in boxes, and text that moves—animation. Be careful not to **overuse *emphasis* techniques.** Choose what you use carefully, or your multimedia effort will look meSSy and unprofessiOnal. A little goes a l o n g way.

It is your job as business writer to capture and keep the viewer's attention with brightly composed, effectively displayed text with emphasis where needed.

SPECIAL CONSIDERATIONS

Slide Shows

Don't go too fast. Speak slowly, clearly, and loudly enough to be heard if you don't have a microphone. Pause for questions. Don't rely on an automatic timing device to determine the pace. Review slide presentations periodically to make sure the content remains up-to-date. Unlike traditional business communications, multimedia content can be readily changed, updated, or deleted.

Web Sites

When you publish a book, it's finished. But when you publish a Web site, you've just begun to write! Plan to update your Web site regularly—monthly, weekly, or even daily—to keep visitors coming back.

When you read the front page of a newspaper, the most important headlines are "above (or over) the fold"—to use newspaper jargon. Most Web pages rely on a scroll bar. When you plan a Web page, therefore, be sure your most important and interesting information is "above the fold." This means the viewer doesn't have to scroll down the page to see it. The less important the information, the farther down the page you can put it. Also remember that sentences may configure differently from screen to screen depending on the size of the browser window. Your Web content won't always look just the way you formatted it.

Don't crowd too much text in one place or your page will look "too gray" (industry jargon). Viewers mustn't feel lost, confused, irritated, or bored trying to use your Web site. Control the value of your visitors' experience through the quality, organization, and appearance of your content.

A. Write **T** (true) or **F** (false) in each blank.

T	**1.** In multimedia writing, briefer is better.
T	**2.** With a Web site, your audience is the world.
F	**3.** Visitors to a Web site always start at the first page (or homepage).
T	**4.** Using phrases rather than whole sentences is fine in a slide show.
F	**5.** "Negative space" should be avoided in multimedia writing.
F	**6.** Always use as many emphasis techniques as possible.
T	**7.** The experience of visitors to your Web site depends on the quality and organization of your written content.

B. List six ways to emphasize text. Possible answers

1. underline, negative space
2. italics, animation
3. color, bullets
4. bold, text boxes
5. large font size
6. font style, sound

Please see answers on page 415.

LET'S TALK BUSINESS—AGAIN

The four P's of oral presentations: Know your PURPOSE; Adequately PREPARE; Practice, Practice, PRACTICE; PRESENT with confidence.

Read 12 in Chapter 2 "Let's Talk Business" deals with vocabulary and pronunciation. The Read 82 title here refers to oral presentations. As you progress in your career, you'll occasionally need to address a group of colleagues, customers, or competitors. Did I just hear someone say, "Oh, I couldn't speak in front of a group of people"? It seems frightening and intimidating the first time. However, the tips in Read 82 enable you to know how to go about it.

We all want to be liked and respected and fear appearing incompetent or foolish. As with many other fears, knowledge can eliminate this fear.

Write **T** (true) or **F** (false) in the blank.

_____T_____ **1.** Giving a work-related oral presentation can contribute to success in your career.

_____F_____ **2.** You can't learn to make effective presentations on the job because this ability comes naturally to certain people.

_____T_____ **3.** We all fear appearing incompetent or foolish.

Answers are on page 415.

FORMULA FOR A SUCCESSFUL PRESENTATION

A successful presentation requires you to write a first draft in which you do the following: Determine your precise objective/s; that is, exactly what you expect to achieve as a result of this presentation. Then plan your presentation—introduction, ideas, supporting facts, and specific examples. Conclude your presentation with a summary of how your company will benefit and specifics of when, where, and how to start. Prepare handouts if they will help. Finally, practice your delivery, timing, and techniques.

Always be aware that members of the audience are thinking, "What's in it for me?" **Tell them how they will benefit.**

Objective

Write out your objective; that is, what you want to accomplish as a result of this presentation. For example: My objective is to convince supervisors and managers that the company should purchase QXR Software because it will save the company money and time and increase productivity.

Plan Your Presentation

Everything you plan to do or say during the presentation must help you meet your objective. Plan what the presentation will include, where and when it will take place, where you will stand or sit, and what equipment and handouts you

In your own mind, be able to "think" the central idea of your presentation in one sentence; e.g., The use of XYZ software in our offices will increase productivity, save the company money, and improve sales.

Use visual aids with your presentation, but choose them carefully so that they're easy to handle. Practice how to use them.

Pronunciation hints for presenters: **There is no:** X in espresso, I in Iraq, wine in genuine, foe in forward, berry in library, cue in coupon, Tia in Tijuana, chief in handkerchief, noise in Illinois, pair in incomparable, purr in prerogative. When in doubt about other words you plan to use in an oral presentation, avoid embarrassment by checking your dictionary for pronunciation.

Your language is you, just like the clothes you wear, the job you have, and the degrees you've earned. We usually form opinions of a person as soon as we hear him or her speak. You are judged by how smart you sound when you speak.

will need. When you complete the plan, prepare an outline of your presentation (include space for notes) for distribution to the audience along with any other useful documents. Then arrange for equipment, seating, and handouts. If you will need equipment, such as a microphone, practice using it as you recite your speech aloud.

Plan to make direct eye contact with as many listeners as possible, dress appropriately, have good posture, and vary your tone of voice and volume appropriately for the content. Above all, speak clearly, distinctly, and enthusiastically.

Content

Tell them what you're going to tell them (introduction), tell them (body), and tell them what you told them (conclusion).

Introduction

If possible, begin with a story or joke relating to the subject.

> **Then tell them what you're going to tell them.**

Body

> **Get down to business and tell them.**

Provide supporting facts that explain what the software is, how it works, and how it will save the company money and time and make their jobs easier. Use visuals if possible, and give specific examples of what this software will do. Each fact you state must support your objective.

Conclusion

> **Tell them what you've told them.**

Summarize what you explained in the "body"—how the company will save money and time and increase productivity. Make specific recommendations as to where, when, how much, and how to purchase the items. Invite questions from the audience.

Cautions

Avoid jargon, details, and vocabulary some listeners may not understand. Explain and simplify as needed, without making listeners feel foolish. Avoid introducing new ideas in the conclusion!

Replay 82

CONGRATULATIONS! HERE'S THE FINAL "REPLAY!"

I'll make it easy—true or false (but mostly false) on the subject of oral presentations. Write **T** (true) or **F** (false) in each blank.

_____ F   **1.** Most people say, "Oh, I love to speak to groups at work—especially before a group of my co-workers and supervisors."

_____ F   **2.** Never begin a presentation with a joke; if your boss doesn't think it's funny, you'll probably get fired.

F	**3.** If you haven't prepared a good speech, speak softly so they won't hear you.
F	**4.** Use as many big words and technical vocabulary as you can when you make a presentation at work.
F	**5.** Don't waste paper and time preparing handouts before giving a presentation.
F	**6.** So that you'll sound natural, don't practice your timing and techniques before making a presentation to your colleagues.
F	**7.** Avoid direct eye contact with your audience as you might strain your eyes.
F	**8.** If you have a new idea for improving sales of your company's products, tell the audience about it in your conclusion.
T	**9.** Making a good presentation to supervisors and co-workers can improve your chances for promotion.
F	**10.** This is the hardest Replay in the whole course!

CONGRATULATIONS! You have finished the last Replay. Answers can be found on page 415. Now take the Pop Quiz on page 361.

CHECKPOINT

Chapter 14 provides you with technical skills to compose and format business communications, whether transmitted electronically, manually, or orally. Check each item when you understand it and know where and how to apply it to communications on the job.

_____ Printed letterhead or letterfoot; return address; date: United States and international/military styles.

_____ Inside address, attention line [optional], salutation, subject line [optional], body.

_____ Company name [optional], signature, writer's typed name, title, and/or department name.

_____ Typist's initials [Don't include if you are both the author and the typist.]

_____ Copy notation or blind copy notation [if needed]; enclosure notation [if needed].

_____ Postscript [when appropriate; not for something you forgot to write in the letter].

_____ Block and modified block letters; punctuation styles are _open_ and _standard._

_____ For business writing, use conversational style, express ideas positively, avoid redundancies, and use judgment for paragraphing.

_____ If faxing a letter, type FAX in the heading; prepare a fax cover sheet to accompany the document being faxed. Avoid using fax or e-mail to transmit confidential materials.

_____ Compose e-mail messages in less formal style than those on paper.

_____ Use interoffice memorandums to send messages within an organization.

_____ When composing correspondence, multimedia presentations, and oral presentations, prepare a first draft and evaluate it before completing the final copy.

_____ If someone else's words are used, quotation marks or credit must be given to the author.

SPECIAL ASSIGNMENT

A. Draw a line through the redundant words in the following expressions:

1. ~~free~~ gift
2. at ~~the hour of~~ 3 p.m.
3. square ~~in shape~~
4. at ~~the present time~~
5. each ~~and every~~
6. enclosed ~~herewith~~
7. few ~~in number~~
8. in ~~the state of~~ Missouri
9. ~~we want to~~ thank you for

B. Replace these expressions with a single word: Possible answers

10. a large number of _____many_____
11. are of the opinion that ___believe, think___
12. in the event that _____if_____
13. at the present time _____now_____
14. made the announcement that ___announced___
15. are in a position to _____can_____

PROOFREADING FOR CAREERS

I don't know whether sending a letter like the following is a good idea, but it's fun to read. Proofread and correct the format, punctuation, and other errors. Then redo it in block style with open punctuation.

Mr.
Steve Gates, Vice President, Intn'l Widget Co., 103 Null St. NY NY 71111   (sp. out)   (sp. out)

Dear Mr. Gates, Thank you for your letter of March 7th. after careful consideration, I regret to inform you that I am unable to accept your refusal to offer me employment. This year I have been particularly fortunate in receiving an unusually large number of rejection letters. With such a varied and promising field of candidates, it is impossible for me to accept all refusals. Despite International Widget Co.'s outstanding qualifications and previous experience in rejecting applicants, I find your rejection, does not meet my needs at the present time, therefore, I will initiate employment with your firm immediately following graduation. I look forward to seeing you then. Sincerely, Bill Jobs.

Dear Student, I hope the proofreading practice in each chapter has been a challenge as well as fun. Please remember: Before submitting papers to an instructor, a colleague, or a client—Proofread, **Proofread, PROOFREAD.** Put an important document aside for an hour or more. Then reread it with "fresh eyes." Best wishes, Leila.

Practice Quiz
www.prenhall.com/smith

Most dictionaries have charts showing currencies of the world. Such a chart might be found near the listing for currency, money, or monetary. For example, *Microsoft Encarta College Dictionary* has an alphabetic list showing currencies throughout the world, alphabetically arranged by the currency's name.

Write **T** (true) or **F** (false) next to each sentence.

_F_ **1.** In writing for multimedia presentations, content (words) is less important than graphics.

_T_ **2.** If you write to a Mexico City address, type MEXICO in all capital letters on the last line to be typed on the envelope.

_F_ **3.** The principal advantage of writing instead of telephoning is that it is less expensive than long-distance calls.

_T_ **4.** Most business letters in the United States are typed on $8\frac{1}{2}$ - by 11-inch paper.

_F_ **5.** It's a good idea to use your employer's letterhead stationery to write to your insurance company about an automobile accident.

_T_ **6.** If the writer's typed name ends a letter, a handwritten signature is still necessary.

_T_ **7.** Before inserting a letter in a No. 9 envelope, fold it in thirds starting from the bottom up.

_F_ **8.** When writing a personal business letter, put your address below the date.

_F_ **9.** Words are the least important component of a Web site or slide show.

_F_ **10.** Include as many words as possible on each slide in a slide show.

_F_ **11.** When making an oral presentation to your colleagues at work, you'll impress them by using as many technical terms as possible.

_T_ **12.** The recommended place for a subject line is below the salutation.

_T_ **13.** *25 September 2005* is an acceptable style for the date in a letter.

_F_ **14.** The modified-block arrangement means you indent the first line of each paragraph.

_T_ **15.** If the first line of a letter's inside address is Turbotax Video, Inc., a correct salutation could be *Ladies and Gentlemen.*

_T_ **16.** Place your most important information "above the fold" on a Web page.

_F_ **17.** *Very truly yours* is currently the preferred complimentary close.

_T_ **18.** It's correct to use the default margins when keyboarding letters.

_T_ **19.** "Let's meet at 4 p.m. in the afternoon." is an example of redundancy.

_F_ **20.** The first paragraph of a business letter should usually be the longest.

_F_ **21.** When possible use a courtesy title in the signature of a letter.

_F_ **22.** Good business writers use more formal language than was used in the past.

_T_ **23.** "To keep your good credit rating, send your check for $843 today." is an example of emphasizing the positive in workplace writing.

_F_ **24.** When making an oral presentation to colleagues at work, avoid repeating your ideas; once is enough.

_T_ **25.** It is not advisable to have a Web page that looks "too gray."

The Global Marketplace—South Korea

Korean	Spell	Say	Currency
GREETING	an hyoung haseyo	an young ha.SAY.yo	won (Wn or W)
DEPARTURE	jal ga	chal kah	
PLEASE	butakkhamnida	boo.tuck.ham.knee.dah	
THANK YOU	kamsahamnida	kum.sah.ham.knee.dah	
I DON'T UNDERSTAND	Juh-neun eehae-haji mot		

Courtesy of Kumsoon Kim

Pop Quizzes A

Score Sheet for _____
Student's Name

Chapter	Read No.	Page No.	Date Completed	Total Questions	No. Right
1	1–6	334			
2	7–9	335			
2	10–12	336			
3	13–16	337			
4	17–19	338			
4	20–22	339			
5	23–25	340–341			
5	26–28	342			
6	29–31	343			
6	32–34	344			
7	35–41	345–346			
8	42–44	347			
8	45–47	348			
9	48–50	349			
10	51–53	350–351			
10	54–56	352			
11	57–60	353			
11	61–65	354			
12	66–68	355			
12	69–71	356			
13	72–76	357			
13	72–76	358–359			
14	77–80	360			
14	81–82	361			

POP QUIZ FOR READS AND REPLAYS 1–6 (CHAPTER 1)

A. Please insert the appropriate Chapter 1 word in the blank. Yes, spelling counts!

1. A person who creates or runs a Web log is often called a ___blogger___ .

2. When a third party is appointed to settle a dispute between two groups, the process is called ___arbitration___ .

3. The field of law relating to computers, information systems, and networks is called ___cyberlaw___ .

4. Someone who works as an assistant to residents of a hotel or an apartment building is called a ___concierge___ .

5. The acronym for the cost of living adjustment for wages or Social Security payments is ___COLA___ .

6. The rate at which currency of one country can be exchanged with currency (money) of another country is the ___exchange rate___ .

7. The initials EU stand for ___European Union___ .

8. The currency unit used in many European countries is called the ___euro___ .

9. What is the opposite of the business term *appreciate*? ___depreciate___

B. In the blanks, show in capital letters the syllable that should be stressed (accented) when pronouncing the word.

10. accessories ___CES___   11. debris ___BREE___   12. Des Moines ___MOYN___

13. affluent ___AF___   14. preface ___PREH___   15. statistics ___TIS___

C. Write T (true) or F (false).

___F___ 16. The business term AMEX is an abbreviation for a person visiting the United States from Mexico.

___T___ 17. The *t* in *mortgage* is a silent letter.

___F___ 18. The American Stock Exchange is the largest stock exchange in the United States.

___T___ 19. A corporation with branches in several countries is referred to as a multinational corporation.

___T___ 20. Antitrust legislation prohibits most monopolies.

___F___ 21. An offer to buy or sell services at a certain price is a bill of lading.

___F___ 22. Time during which equipment cannot be used until it has been repaired is called *uptime*.

___T___ 23. A special software preventing employees from visiting certain Web sites while at work is called e-blocker.

___T___ 24. The amount remaining after all deductions have been made is called the *net* amount.

___T___ 25. The following words are all spelled incorrectly: balence sheet, morgage, dificet, lible
Please respell them here: ___balance, mortgage, deficit, libel___

Name: _____ Date: _____ Number right: _____ out of 25

POP QUIZ FOR READS AND REPLAYS 7–9 (CHAPTER 2)

Circle the incorrect words in the following sentences. Write the correct words in the blank at the end of each sentence.

1. No-one accept you would think its alright to ware such bazaar close. _No one, except, it's, all right,_
   wear, bizarre, clothes

2. They defered about witch devise wood be illegible for the trip to the dessert. _differed, which, device,_
   would, eligible, desert

3. In this physical year, its necessary for parents of miners to be conscience of everything there
   children due. _fiscal, it's, minors, conscious, their, do_

4. I wish to sit besides you and have desert with you weather or not the proceeds go to hour
   organization. _beside, dessert, whether, our_

5. Moral is low among the personal, but they listened to my lecture with wrapped attention.
   Morale, personnel, rapt

6. The realty of the situation is that were not sure about weather Yule be illegible for taking the
   tour of the capital. _reality, we're, whether, you'll, eligible, capitol_

7. We respectively request that you make a through investigation of how the wry bread is baked.
   respectfully, thorough, rye

8. In the throws of passion, he wrote an owed to a beautiful marquise. _throes, ode_

9. Were quiet certain that the reason is because his prospective was off balance. _We're, quite, reason is_
   that, perspective

10. She was dyeing to meat the principle because she herd he is the air too a fortune. _dying, meet,_
    principal, heard, heir, to

Name: _____ Date: _____ Number right: _____ out of 10

Pop Quiz for Reads and Replays 10–12 (Chapter 2)

A. Fill in the blanks with words from Reads and Replays 10–12.

1. To estimate the value of an item is to _____appraise_____ it.
2. A _____bloc_____ of angry voters will prevent the law from passing.
3. A book's _____foreword_____ includes information about several of the authors.
4. One should be _____wary_____ of picking someone up under that theater's _____marquee_____ .
5. The new _____stationery_____ is only half as expensive as the old.
6. The _____indigenous_____ people of Alaska are called Inuits.
7. Babies' toys should not be _____flammable_____ .
8. We need _____ingenious_____ employees to make a success of this company.
9. Someone who seems not to care but really does might be _____ingenuous_____ .
10. Circle the word that is NOT Standard English: ⟨irregardless⟩ regardless

B. Many Chapter 2 words are frequently mispronounced. Join with one or more partners, listen critically to one another as you pronounce the words, and request help from your instructor or other experts as needed. It won't take long to be confident you're pronouncing each one correctly.

Name: _____ Date: _____ Number right: _____ out of 10

POP QUIZ FOR READS AND REPLAYS 13–15 (CHAPTER 3)

1. What is the name of your dictionary? _Answers will vary._

2. What is a lexicographer? _A lexicographer is one who writes, edits, and compiles dictionaries._

3. Dictionaries *describe* how English is used. They don't *prescribe* how to use English. True or False? _T_

4. To interpret the abbreviations used in your dictionary, look in the _preface—or front matter_ .

5. If you're having difficulty thinking of just the right word for something you're writing, look up a related word in a _thesaurus_ .

6. Show the pronunciation for *faux pas* and say it aloud. _foh pah—symbols vary with dictionary used_

7. *Abstract* is used as what parts of speech? _adjective, noun, verb_

8. The treasurer dis__rsed (pe or bu?) all the funds. _disbursed_

9. What does the abbreviation ILGWU mean? _International Ladies Garment Workers Union_

10. Like *Webster* in dictionaries, the famous name in thesauruses is _Roget pronounced RO.JAY_ .

11. What do you find in a thesaurus? _synonyms, antonyms, and other related words_

12. Spell the other word that sounds like *its* but has a different meaning. _it's_

13. How old was Beethoven when he died? _57_

14. What was the nationality of the poet and playwright Dante? _Italian_

15. Is a college dictionary abridged or unabridged? _abridged_

16. What is a *boom box*? _a large radio, cassette, or CD player with built-in speakers_

Referring to your dictionary, divide these words into syllables by rewriting the word and drawing a diagonal (/) at the end of each syllable. For a one-syllable word, draw the diagonal at the end.

17. worked ___worked/___
18. getting ___get/ting___
19. contagion ___con/ta/gion___
20. beetle ___bee/tle___
21. fullness ___full/ness___

22. Write a brief definition obtained from your dictionary for:
    (a) protein engineering _____
    (b) pisciculture _____
    (c) voxel _____
    (d) primavera _____ or _____
    Answers vary somewhat depending on the dictionary used and how the definition is condensed.

23. Correctly pronounce No. 22 b, c, and d.

24. What does it mean if a word **doesn't** have a usage label? _It means the word is Standard English, and there are no particular restrictions regarding its use._

25. What is or are the usage label/s in your dictionary for *bombed*, and what is the slang or colloquial meaning? _slang or colloquial; meanings vary somewhat with dictionary used_

26. In which decade of the 1900s was the word *cybernation* first used? _1960s_

27. What do we call words that mean almost the same? _synonyms_

28. Why is it important to recognize the etymology in your dictionary? _so that you don't confuse the etymology with the definition_

Name: _____ Date: _____ Number right: _____ out of 28

POP QUIZ FOR READS AND REPLAYS 17–19 (CHAPTER 4)

A. Please insert nouns that make sense in the blanks.

1. The _____manager_____ interviews all the _____applicants_____.
2. The _____water_____ in Lake _____Gilroy_____ is extremely deep.
3. _____Jonathan_____ wants a new red _____car_____.
4. _____Allen_____ was a conscientious _____student_____.
5. Our _____office_____ is closed on _____Sundays_____.

B. Please insert pronouns that make sense in the blanks.

6. Paralegals _____who_____ are word processing experts get the best jobs.
7. Thank _____you_____ for faxing _____me_____ _____your_____ résumé.
8. _____They_____ wrote to us about _____her_____.
9. When _____anything_____ goes wrong with _____it_____, we'll blame _____ourselves_____.
10. Give _____this_____ to _____whoever_____ _____will_____ take _____it_____.

C. Please insert verbs that make sense in the blanks.

11. Many Americans _____speak_____ another language in addition to English.
12. Vince _____proofread_____ the letter and then _____signed_____ it.
13. In Sweden don't _____drink_____ your beverage until the host _____says_____ "Skoal."
14. Diablo Valley students _____built_____ the float that _____won_____ the prize.
15. The IRS _____audited_____ our tax return and _____asked_____ for a refund.

D. Please insert suitable adjectives in the blanks.

16. Our _____new_____ copier is _____easy_____ to use.
17. Please make _____ten_____ reservations for a _____great_____ show.
18. Professor Taylor had an _____innovative_____ plan for teaching the Vietnamese language.
19. The _____chrome_____ chairs and the _____black_____ desk look _____great_____ in your _____new_____ office.
20. Write a _____short_____ report outlining your _____special_____ plan.

E. Please convert these adjectives to adverbs.

21. happy _____happily_____
22. joyful _____joyfully_____
23. real _____really_____
24. busy _____busily_____
25. generous _____generously_____

Name: _____ Date: _____ Number right: _____ out of 25

Pop Quiz for Reads 20–22 (Chapter 4)

A. Please insert appropriate conjunctions in the blanks.

1. More _____and_____ more firms offer software training to employees.
2. He will qualify for a good job _____if_____ he passes the keyboarding test.
3. You qualify for a job as an administrative assistant _____or_____ an executive secretary.
4. Career paths may be limited _____because_____ management positions may be filled at this time.
5. _____When_____ you apply for a job, it's best to dress professionally.
6. Krista, Ashley, _____and_____ Shawn like snowboarding _____but_____ not skiing.
7. Larry neither uses traditional skis _____nor_____ likes to snowboard.
8. _____Since_____ the check didn't arrive, please call the customer.
9. One in six Americans has a disability _____that_____ interferes with daily tasks.
10. _____Even though_____ the clock is ticking, the time shown is incorrect.

B. Insert prepositions in the blanks, and draw parentheses around the prepositional phrases.

11. The book shelf is _____(above_____ the desk) that's _____(near_____ the copier)
12. New carpeting is needed _____(in_____ our office) and _____(on_____ the first floor)
13. _____At_____ 5 p.m. the contract will be delivered _____(to_____ him)
14. Everyone _____(except_____ him) walked (_____by_____ the water cooler)
15. He looks _____(like_____ his brother) _____(from_____ head _____to_____ toe)

C. In the blank write the part of speech of the underlined word as it is used in the sentence.

16. Shopping is his favorite pastime. _____noun_____
17. Suzanne, however, prefers swimming. _____noun_____
18. They provided her with swimming lessons. _____adjective_____
19. Professor Teufel is shopping for a modem and a printer. _____verb_____
20. They must raise the money before they hire an attorney. _____conjunction_____

D. Fill in each blank with at least three words. (possible answers)

21. Write three coordinate conjunctions. _____and, but, or, nor, for_____
22. Write three words often used as dependent conjuctions. _____since, when, as, because_____
23. Write three words often used as prepositions. _____under, over, in, to, with, by_____
24. Write three masculine pronouns. _____him, he, himself, his_____
25. Write three second-person pronouns. _____you, your, yourself, yourselves_____

Name: _____ Date: _____ Number right: _____ out of 25

Spell the plurals of these nouns. If a word has more than one correct plural, show both. If the printed singular is incorrectly written, write it correctly in the plural form. When in doubt, use your dictionary.

1.	inventory	inventories	**13.**	proxy	proxies
2.	brick layer	bricklayers	**14.**	wolf	wolves
3.	memento	mementos	**15.**	attorney	attorneys
4.	genius	geniuses	**16.**	notary public	notaries public
5.	editor in chief	editors in chief	**17.**	alto	altos
6.	premises	premises	**18.**	chassis	chassis
7.	zero	zeros, zeroes	**19.**	Vietnamese	Vietnamese
8.	Flores	Floreses	**20.**	formula	formulae or formulas
9.	diagnosis	diagnoses	**21.**	medium	mediums or media
10.	phenomenon	phenomena, phenomenons	**22.**	curriculum	curricula, curriculums
11.	hero	heroes	**23.**	scarf	scarves
12.	entry	entries	**24.**	index	indices or indexes

Name: _____ Date: _____ Number right: _____ out of 24

Pop Quiz for Reads and Replays 23–25 (Chapter 5)

A. Draw a line through the incorrect words in the parentheses.

1. Many (companies/~~companys~~/~~companyies~~) don't allow smoking in the workplace.
2. Statistics (is/~~are~~) a difficult subject for many people.
3. These statistics (~~is~~/are) hard to understand.
4. Why (are/~~is~~) the clothes always all over the floor?
5. The recipe calls for two (cupfuls/~~cupsful~~/~~cups ful~~) of oat bran.
6. Two (notaries public/~~notarys publics~~/~~notary publics~~) are available. [Choose preferred form.]
7. Did you order the new (~~check book~~/~~check-book~~/checkbook)?

B. Write **S** or **P** in the blank to show whether these nouns are singular or plural or used for both.

8. __S__ diagnosis
9. __P__ data
10. __P__ hypotheses
11. __P__ criteria
12. __S__ analysis

C. Correct the noun errors in the following sentences. Write **C** beside the two correct sentences.

_____ 13. Did you know that my U̸ncle was the P̸resident of his C̸ompany?
_____ 14. He became a G̸eneral at the age of 27.
_____ 15. Please send the merchandise to our r̲e̲c̲e̲i̲v̲i̲n̲g̲ d̲e̲p̲a̲r̲t̲m̲e̲n̲t̲.
_____ 16. I was born in the e̲a̲s̲t̲, but now I am a w̲e̲s̲t̲e̲r̲n̲e̲r̲.
__C__ 17. They were runners-up in the contest.
_____ 18. We need several good N̸urses to work at this H̸ospital.
_____ 19. We drove N̸orth last S̸ummer until we reached Kansas City.
__C__ 20. The plaintiff's attorney was disbarred.
_____ 21. The company's head‿hunter is looking for an assistant who likes text‿books.

D. Spell the plural of each of the following words.

22. sister-in-law	sisters-in-law		28. addendum	addenda or addendums
23. memento	mementos		29. proxy	proxies
24. itinerary	itineraries		30. solo	solos
25. corps	corps		31. formula	formulae or formulas
26. chassis	chassis		32. diagnosis	diagnoses
27. proceeds (noun)	proceeds		33. deer	deer

Name: _____ Date: _____ Number right: _____ out of 33

Pop Quiz for Reads and Replays 26–28 (Chapter 5)

A. Test your spelling and vocabulary skills. The first few letters of each word are given; you complete the word.

1. The choir is made up of al ___tos___ and 2. sop ___ranos___ who also play 3. ban ___jos___.

4. My co-___workers___ 5. on the prem ___ises___ 6. asked both 7. attor___neys___ for two

8. le___tters___ of 9. cr___edit___ for trade with Spanish and Russian companies.

10. Did you know that Swa ___hili___ is Tanzania's official language, and 11. Hi___ndi___ is Northern India's official language? It is appropriate to refer to American Indians as 12. Na___tive___ Americans. 13. Are you bi___as___ ed against hiring 14. draf___ting technicians___ instead of architects for the new project?

B. Underline the preferred choice to eliminate workplace biases.

15. Please invite the (girls/gals/women/ladies) in our office to our committee meeting.

16. The preferred term for restaurant personnel is (waiter/waitron/server).

17. (Man-made/Synthetic) fabrics are in high demand in some communities.

18. He wants to be a (male nurse/nurse).

19. We don't have enough (firefighters/firepeople/firemen) if an earthquake should occur.

20. This hotel has special rates for (businessmen/businesspeople).

21. The (author/authoress) of this book is Leila R. Smith.

22. The store next to ours is owned by (Orientals/Asians), either from Thailand or Vietnam.

23. Paparazzi are (cameramen/photographers/camerapersons) who take pictures of celebrities and sell them to the media.

C. Although workplace biases dealing with age, race, religion, nationality, etc., have diminished greatly, various prejudices still exist to some extent. Have you, a friend, or relative experienced prejudice in the workplace because of language, nationality, color, age, physical or mental disability, sexual orientation, or other biases? Have you behaved in a prejudicial manner toward someone? Write a short report on this subject—not more than one page—double-spaced.

Name: _____ Date: _____ Number right: _____ out of 23

Grade for Part C: _____

POP QUIZ FOR READS AND REPLAYS 29–31 (CHAPTER 6)

Please make the necessary pronoun corrections. Write **C** in the blank if the sentence is correct.

_____ 1. Let me know when Charles West and ~~myself~~ (I) can meet you for lunch.

___C___ 2. Do you really think he's smarter than she?

___C___ 3. Ms. Nguyen and I will use the modem to send our documents.

_____ 4. ~~You're~~ (Your) application for credit has been rejected.

___C___ 5. Lou Knorr regrets this decision even more than I (do).

___C___ 6. Nancy Moody is better qualified than she for the job.

__they__ 7. No accountants are more respected than ~~them~~/they. (are)

_____ 8. If you prefer hers, you could request an exchange.

__me__ 9. Carla Boone should be seated between Chris Lundin and ~~I~~/me.

___C___ 10. The doctor asked the nurse and me to help.

_____ 11. Those reports are our's.

___C___ 12. He and Mr. Ko will discuss it at the forum.

___C___ 13. Phyllis Alderdice and I discussed lighting for the computer lab.

___C___ 14. We new employees appreciate advice from the senior staff.

__he__ 15. Yoshiko is a better listener than he/~~him~~.

___C___ 16. After talking with Henry and them, he stormed out of the office.

___C___ 17. We students study in the library every night.

___C___ 18. Please give more homework to us students.

___C___ 19. You will hear from either the treasurer or me within two weeks.

___C___ 20. Ruth Ellen and he are the committee's chair and secretary respectively.

_____ 21. Our competitors want to win the bid just as much as ~~us~~ (we).

___C___ 22. Betty married someone who is as organized and friendly as she.

___C___ 23. Mr. Earle had a long talk with each of us committee members.

___C___ 24. He asked Peter and me to change the toner for him.

___C___ 25. Ms. Bogdanowicz and I are rushing to get the documents notarized.

___C___ 26. When you finish the letter, give copies to Shawn and me.

Name: _____ Date: _____ Number right: _____ out of 26

POP QUIZ FOR READS AND REPLAYS 32–34 (CHAPTER 6)

Underline the preferred wording for written English.

1. Someone left their books here./Someone left his or her books here./<u>Someone's books were left here.</u>
2. We may hire a credit manager (<u>whom</u>/who) Pizza Hut just laid off.
3. (<u>Who</u>/Whom) do you believe will lead the training session?
4. From (who/<u>whom</u>) did you buy the telephone system?
5. (<u>Who</u>/Whom) do you think we should hire as the Web site manager?
6. Did he say (who/<u>whom</u>) he wants to work with?
7. The program was designed by Barbara Moran (who/<u>whom</u>) we all know to be an excellent writer.
8. We don't know (<u>who</u>/whom) is to blame.
9. Marg said (everyone/<u>every one</u>) of her students did well on the test.
10. Each person took (their/<u>his or her</u>) place in line.
11. All (<u>who</u>/whom) were eligible took (<u>their</u>/his or her) places in the queue.
12. We all respect the man to (who/<u>whom</u>) we dedicated this building.
13. Today we met President Gina Hector (whom/<u>who</u>) had not been here before.
14. (Every body/<u>Everybody</u>) in this district is being asked to donate blood.
15. (Each employee is to indicate when he or she wishes to take their vacation./<u>All employees are to indicate when they wish to take their vacation.</u>)
16. Every student should improve their grammar so that he or she can communicate correctly at work. <u>Students should improve their grammar so that they can communicate correctly at work.</u>
17. Each department has (its'/it's/<u>its</u>) own methods of dealing with absenteeism.
18. Please send the proceeds to (<u>whoever</u>/whomever) made most of the payments.
19. Each student was eager to see (<u>his or her</u>/their) test results.
20. (Whom/<u>Who</u>) do you think will be the next president of Disneyland?
21. The committee members finally made (its/<u>their</u>/his or her) decisions.
22. Nordstrom moved (<u>its</u>/it's/their) headquarters to Lubbuck, Texas.
23. Rosenberg, Sica, & McDaniel, Inc., filed (<u>its</u>/it's/their) case last week.
24. The one who deserves the prize is (<u>she</u>/her).
25. The guilty one appeared to be (him/<u>he</u>/himself).

Name: _____ Date: _____ Number right: _____ out of 25

POP QUIZ FOR READS AND REPLAYS 35–41 (CHAPTER 7)

A. Circle the verb errors in the following sentences. Write the correct replacement verbs in the blanks.

1. One of the messages (were) on my voice mail. _____was_____
2. Neither of these two telephone systems (are) acceptable. _____is_____
3. The lecture series (were) a great success. _____was_____
4. If he (was) a better interviewer, he would have hired her. _____were_____
5. A person with many skills and talents (are) fortunate. _____is_____
6. All the accountants except Evan (works) on receivables. _____work_____
7. They (seen) Broadway and won't go back to the farm. __have seen—or saw__
8. They (have ran) a business in the past. __have run—or delete have before ran__
9. I (use) to drive race cars in Indianapolis. _____used_____
10. A carton of books (have) been there since last week. _____has_____
11. Xerox (manufacture) and (sell) electronic equipment. __manufactures and sells__
12. Good grammar and spelling (is) important in workplace e-mail. __change *is* to *are*__
13. Neither Charles nor Dolores (leave) at 5 p.m. on Fridays. _____leaves_____
14. Each man and woman (need) an application form. _____needs_____
15. Many a boy (hope) to be a professional baseball player. _____hopes_____
16. Ann, along with her sisters, (have) a court reporting agency. __change *have* to *has*__
17. (Where's) your roller blades? __change *Where's* to *Where are*__

B. Each of the following sentences has a collective noun; underline the collective noun and the correct verb choice.

18. The <u>staff</u> (attend/<u>attends</u>) training sessions on Mondays at 8 a.m.
19. The law <u>firm</u> of Graham & James (have/<u>has</u>) been chosen to represent us.
20. Each summer my <u>family</u> (<u>takes</u>/take) its vacation in Kona, Hawaii.

C. In the blanks write the present participle of the following verbs:

21. forget _____forgetting_____   22. cost _____costing_____   23. pay _____paying_____

D. In the blanks write the simple past participle of the following verbs:

24. fall _____fell_____   25. pay _____paid_____   26. freeze _____froze_____

E. In the blanks write the past participle of the following verbs:

27. shake _____shaken_____   28. cost _____cost_____   29. forget _____forgotten_____

F. Correct the verb form in the following sentence:

30. The child (had fell) down. Change _____*fell* to *fallen* or delete *had*_____

G. Please insert appropriate verb forms in the blanks. Possible Answers

EXAMPLE   Those students _should study_ more assiduously. (Where's the dictionary?)

31. Many Americans _speak_ another language in addition to English.

32. Vince _had written_ the letter before he _received_ the call.

33. In Sweden you _should drink_ your beverage after the host _says_ "Skoal."

34. Diablo Valley students _built_ the float that _won_ the prize.

35. The IRS _received_ our tax return and then _asked_ for more money.

36. U.S. Financial, Inc., _was_ headquartered in Dallas. (*was?* or *were?*)

37. The jury _was_ sequestered by the judge. (*was?* or *were?*)

Name: _____ Date: _____ Number right: _____ out of 37

POP QUIZ FOR READS AND REPLAYS 42–44 (CHAPTER 8)

A. For 1–5 write the four pointing adjectives and one three-letter article; follow each by the noun *kind* or *kinds*.

1. _____this kind_____
2. _____that kind_____
3. _____these kinds_____
4. _____those kinds_____
5. _____the kinds_____

B. Write *a* or *an* in the blank.

6. __a__ union member

7. __a__ one-inch line

8. __a__ hairline fracture

9. __an__ honest job

10. __an__ heir

C. Correct verb, adjective, or adverb errors. If the sentence is correct, write **C** in the blank.

_____ 11. He is ~~angryer~~ [angrier] than I have ever ~~saw~~ [seen] him.

_____ 12. The ~~most~~ safest rule is to drive within the speed limit.

_____ 13. Rex Bishop plays real^[ly] good golf.

__C__ 14. Dr. Hellwig looks different today.

_____ 15. Smoking is the ~~least~~ [lesser] of the two evils.

_____ 16. The accountant's report is ~~more~~ better than the manager's.

_____ 17. Ms. Pearts prepared the invoices as careful^[ly] as possible under the circumstances.

__C__ 18. The food tasted really greasy and smelled even worse than at the old restaurant.

_____ 19. The mashed potatoes tastes[~] badly[~] too.

__C__ 20. He writes more clearly than his supervisor.

_____ 21. Those kind^s of apples ~~is~~ [are] sweet.

_____ 22. The institute was ~~busyier~~ [busier] than anyone could have expected.

_____ 23. Which is the ~~most~~ [more] beautiful city, Dallas or Miami?

_____ 24. When you ski faster, you negotiate turns ~~easier~~ [more easily].

_____ 25. She hasn't gone ~~nowhere~~ [anywhere] yet.

_____ 26. We don't want to do business with them anyways[~].

_____ 27. These kind^s of letterheads are out of date.

Name: _____ Date: _____ Number right: _____ out of 27

POP QUIZ FOR READS AND REPLAYS 45–47 (CHAPTER 8)

A. Underline the adjectives and circle the adverbs in the following sentences. Draw an arrow to show which word or words each one modifies. See No. 1 as an example.

1. Early radio focused mostly on music and news.
2. Adventure-mystery dramas targeted adolescents chiefly.
3. Television was commercially introduced to the United States in the early 1940s.
4. The Clampetts of the 1960s television show lived—uneasily at times—in a Beverly Hills mansion.
5. What old TV shows did your parents see regularly when they were young children?

B. Most of the following sentences have adjective or adverb errors. Keep your dictionary handy. Show corrections in the blanks, or write **C** beside correct sentences. Complete 16 and 17.

quietly	6. The children sat as quiet as possible through the movie.
casually	7. She dresses too casual for this office.
Those	8. Them people are always chosen for these jobs.
prettier	9. Natasha is the prettiest of my sister's two daughters.
_C_	10. She is unhappy about Andy's behavior.
anywhere	11. She hasn't gone nowhere yet.
more easily	12. When you ski faster, you negotiate turns easier.
really	13. We are real unhappy about his behavior.
_C_	14. These mashed potatoes taste good.
_C_	15. He must learn to speak and write more clearly than he does now.
_____	16. The four kinds of adjectives are _pointers, articles, describing, limiting_ .
_____	17. Adverbs modify _verbs, adjectives, and other adverbs_ .
_C_	18. Many adverbs are formed by simply adding *ly* to the end of an adjective.

C. Turn these adjectives into adverbs.

19. happy _happily_
20. joyful _joyfully_
21. busy _busily_
22. generous _generously_
23. real _really_
24. smart _smartly_
25. most _mostly_

Name: _____ Date: _____ Number right: _____ out of 25

POP QUIZ FOR READS AND REPLAYS 48–50 (CHAPTER 9)

A. Fill in the blanks with the correct forms of the nouns or pronouns in parentheses. Some nouns in parentheses are already correct. Write **C** in the blank if no correction is required.

1. From Mr. (Saunders) letter we learned that prices have advanced 11 percent since Ms. (James) article was published. _____Saunders'_____ _____James's_____

2. If (Schultz) gets the layout into Mr. (Austerlitz) hands today, the advertisement will be run on time. _____C_____ _____Austerlitz' or Austerlitz's_____

3. The signature on the contract is my (brother-in-law). _____brother-in-law's_____

4. Has (Mr. Fingles) committee reached (its) quota of contributions in the (PTA) annual fund-raising campaign? _____Fingles'_____ _____C_____ _____PTA's_____

5. A successful business considers all the (customers) (needs). _____customers'_____ _____C_____

6. (Girls) (jeans) are selling fast at (Macys). _____Girls'_____ _____C_____ _____Macy's_____

7. My (mother-in-law) will was probated by (Atlanta) most famous (attorneys). _____mother-in-law's_____ _____Atlanta's_____ _____C_____

8. This is a (weeks) supply of (groceries). _____week's_____ _____C_____

9. (Mens) clothing is on the first floor, (womens) on the second, and (girls) on the third. _____Men's_____ _____women's_____ _____girls'_____

10. Both his (daughters-in-laws) (offices) are on Fifth Avenue in Manhattan. _____daughters-in-law's_____ _____C_____

B. Insert apostrophes and s where needed. If no apostrophe is needed, write **C** to the left of the number. Circle incorrect apostrophes.

_____ 11. Why isnt there ever enough time to get everything done? _____isn't_____

_____ 12. The casting agent wouldnt choose him for that part. _____wouldn't_____

_____ 13. I was surprised to see five As on my transcript. _____A's_____

_____ 14. Bostons average winter temperature is in the 40s. _____Boston's_____ _____40's_____

_____ 15. They told me the company already had too many MBAs from Harvards class of '01. _____Harvard's_____

_____ 16. The childrens teacher gave them each two Bs on their report cards. _____children's_____ _____B's_____

_____ 17. Mr. Perkins girlfriend wore the bosss hat; thats Matt Perkins were writing about. _____Perkins'_____ _____boss's_____ _____that's_____ _____we're_____

_____ 18. We received two c.o.d.s in todays mail. _____c.o.d.'s_____ _____today's_____

_____ 19. General Motors president read all of Aristophanes plays. _____Motors'_____ _____Aristophanes'_____

_C_ 20. Moses' speech problem is discussed in *Exodus*, one of the books of the *Bible*. _____

_____ 21. The president was unhappy with Congress lack of action regarding Medicare. _____Congress's_____

_____ 22. We think Vitex sales force was very effective during last years heat wave. _____Vitex's_____ _____year's_____

_____ 23. Garlic and onion plants make Ms. Forbis farm very smelly. _____Forbis's_____

_C_ 24. Do yours, hers, and ours require any apostrophes? _____

_C_ 25. He won the sweepstakes and scored 25 points two years ago. _____

_C_ 26. The students are taking final exams two weeks later. _____

Name: _____ Date: _____ Number right: _____ out of 26

A. Write **T** (true) or **F** (false).

_____T_____ **1.** To be a sentence, a group of words requires identity, action, and independence.

_____F_____ **2.** The identity requirement of a sentence means every sentence must have a verb.

_____T_____ **3.** An independent clause can stand alone as a complete sentence.

_____F_____ **4.** If a dependent clause is joined to an independent clause, the result is a comma splice.

_____T_____ **5.** A clause is a word group with a subject and a verb.

B. Fill in the blanks.

6. A noun or pronoun that tells who or what a sentence is about is called the _____subject_____.

7. A word that tells what the subject does, is, or has is what part of speech? _____verb_____

8. A word group beginning with a preposition that has no subject or verb is called a _____prepositional phrase_____.

9. A clause beginning with a dependent conjunction is called a _____dependent clause_____.

10. A comma between the independent clauses of a run-on creates a _____comma splice_____.

11. Two independent clauses joined with a semicolon can create a _____sentence_____.

C. Write **F** for fragment, **S** for sentence, and **CS** for comma splice. Insert one apostrophe in a possessive noun.

_____F_____ **12.** A boxer, which is a very friendly dog.

_____S_____ **13.** A boxer is a very friendly dog.

_____F_____ **14.** Because of the establishment of this law.

_____S_____ **15.** The hanging of greens, such as holly and ivy, is a British winter tradition that originated before the Christian era.

_____F_____ **16.** Since Austria's contribution to Christmas is the beloved carol, "Silent Night."

_____S_____ **17.** During the weeks preceding Christmas, the windows of Parisian department stores contain fabulous displays of animated figures.

_____F_____ **18.** Although in Mexico the home is decorated and ready to receive guests by December 16, the beginning of the Mexican posadas.

_____S_____ **19.** Japanese businesspeople who understand English say it is often difficult for them when Americans speak too fast.

_____F_____ **20.** Thanking you for your attention to this problem.

_____F_____ **21.** While she is the purchasing manager and he is the assistant manager.

_____F_____ **22.** Hoping to hear from you soon.

_____CS_____ **23.** She is the purchasing manager, he is the assistant manager.

D. Replace incorrectly used commas with periods, and capitalize the first word of each sentence. If a comma does not create a comma splice, leave it alone.

24–30. When working at the computer, take frequent breaks before you get too tired. Do desk work for no more than 30 minutes at a time, then take a mini holiday of one to two minutes, use that time for physical activity like getting up to look out the window, juggling two or more balls, running up a flight of steps, or organizing items on a high shelf.

31–35. In the international marketplace, wearing certain clothing or colors may offend the businesspeople of another country, americans tend to let others know who they are by the way they dress, some styles are inappropriate when you are with people of another country, it is wise to wear conservative clothes in quiet colors for your business meetings.

Name: _____ Date: _____ Number right: _____ out of 35

A. Correct the comma splices with a semicolon and transition expression or with a coordinate conjunction after the comma. Possible solutions

1. When appointments are made in China, a 30-minute "courtesy time" is often understood; however, businesspeople are usually punctual.

2. Mike likes Lotus for spreadsheets, but I prefer Excel.

3. Katie and Art arrive today; however, Ron won't arrive until Friday.

4. The 90-minute lecture starts at noon, but we can't meet you until 1:30.

5. The computers are down this morning; however I'll do some filing.

B. In the blanks, write **CS** for a comma splice, **R** for a run-on, or **C** for a correct sentence; make needed corrections.

_____R_____ 6. Many U.S. firms trade with the EU*; therefore, they have large phone bills.

_____CS_____ 7. Many U.S. firms trade with the EU; consequently, they have large phone bills.

_____C_____ 8. Many U.S. firms trade with the EU. They have large phone bills.

_____C_____ 9. Since many U.S. firms trade with the EU, they have large phone bills.

_____CS_____ 10. Many U.S. firms trade with the EU; they have large phone bills.

C. Correct the five comma splices in Part A above by making one clause dependent. Use a different dependent conjunction for each one. For No. 15 use the informal, conversational, two-letter coordinate conjunction.

11. When appointments are made in China, a 30-minute "courtesy time" is often understood, although Chinese businesspeople are usually punctual.

12. While I prefer Excel, Mike likes Lotus for spreadsheets.

13. Even though Katie and Art arrive today, Ron won't arrive until Friday.

14. Because the lecture starts at noon, we can't meet you until 1:30.

15. The computers are down this morning, so I'll do some filing.

Name: _____ Date: _____ Number right: _____ out of 15

*European Union

POP QUIZ FOR READS AND REPLAYS 57–60 (CHAPTER 11)

Write **C** beside the correct sentences. Insert commas as needed. Three of the sentences also need an apostrophe.

C   **1.** A letter of transmittal for a report may include background information, supplementary information, or confidential information.

_____   **2.** The title page includes the report's title, its author, the date, and the name of the person for whom the report was prepared.

C   **3.** Personality conflicts are upsetting and cause serious drops in productivity.

_____   **4.** According to a *Wall Street Journal* article, misplaced commas are the most common error in business writing.

_____   **5.** Some important morale factors are satisfaction with the job itself, opportunity to learn, and compatibility with co-workers.

_____   **6.** In a corporation having several thousand employees, top-level managers seldom know more than a small fraction of those on the payroll.

_____   **7.** A human resources department is responsible for activities concerning employees, and it advises other departments and management regarding personnel matters.

_____   **8.** If these ideas are interesting to you, perhaps you could take courses in management, human relations, or industrial psychology.

_____   **9.** You might discover that you have a career interest in such a position as human resources director, industrial psychologist, or human relations consultant.

_____   **10.** Formal business management courses alone will not make you effective, but they will enable you to learn about supervision and management.

_____   **11.** If becoming a supervisor sounds interesting to you, you should start to prepare now.

_____   **12.** Some decisions may seem trivial to you, but the quality of each decision is important to a department's success.

_____   **13.** Yes, you will need to talk with your most intelligent and loyal employees on many subjects.

_____   **14.** A supervisor must learn how to conduct group meetings, to discipline and terminate employees, and to be a good member of the management team.

_____   **15.** My sister's friend manages a small, elegant specialty shop in Hartford, Connecticut.

C   **16.** Six months ago Michael and Rachel were assigned stores to manage.

_____   **17.** The four words in the English language that end with *dous* are tremendous, horrendous, stupendous, and hazardous.

C   **18.** Yesterday Austin Silver decided he wanted to get into the management field.

_____   **19.** He may not know everything about motivating others, but he is willing to learn.

_____   **20.** Should you move into a supervisory job from a technical one, you must quickly become a people-oriented individual.

_____   **21.** Whenever you can take some management courses, do so.

C   **22.** Managers must learn early in their careers that what others do under their guidance is more important than the actual work they do themselves.

C   **23.** Some top executives ask their assistants to be alert to employee gripes and try to correct them if possible.

C   **24.** Only those who risk going too far can possibly find how far they can go.

C   **25.** Pack a few small gifts for emergencies when traveling abroad for business.

Name: _____ Date: _____ Number right: _____ out of 25

POP QUIZ FOR READS AND REPLAYS 61–65 (CHAPTER 11)

Apply Chapter 11 comma and end-of-sentence punctuation rules. Also be alert for missing apostrophes; insert them where they belong. Add a capital letter if it's essential to begin a new sentence. Write **C** to the left of items that are already correct.

_____ C _____ 1. The name of the book is *The Encyclopedic Dictionary of Business Terms.*

_____ 2. The president of a major corporation said,"To get to the top, it takes a sense of urgency about getting things done."

_____ C _____ 3. Will you please send us your check today.

_____ 4. Would you be able to send the scripts to my nephew Alex,who moved to Toronto, Canada,last year?

_____ C _____ 5. We recommend that business travelers ask the concierge at their hotel for advice on tipping in that country.

_____ 6. First he keyboarded the messages to be faxed,and then he replied to the e-mails.

_____ C _____ 7. The average wage earner in 1914 worked 12 hours a day for about $800 a year.

_____ 8. They lived at 40536 Picket Fence Road, Levittown,Pennsylvania,in an older home that was in very good condition.

_____ 9. Pierre,visitors to the United States need to be aware that dinner,which is usually between 7 and 9 p.m.,is the main meal of the day.

_____ 10. It was the expense of the work,not its difficulty,that caused us to refuse to begin the project.

_____ 11. She wants to learn aerobics,not modern dance.

_____ 12. Allen Yoshimoto,Ph.D.,will conduct the research for a fee of $52,000.

_____ 13. The professor asked,"Are these the boys who stole the bicycles?"

_____ 14. The boys said to the police,"We didn't do it."

_____ C _____ 15. May I hear from you by return mail.

_____ 16. Sun World,growing and shipping millions of boxes of fruit every year,is one of this state's largest fruit marketers.

_____ 17. It is,I am pleased to report,an idea that the CEO finds promising.

_____ C _____ 18. Tigers have striped skin, not just striped fur.

_____ 19. "Nonessential words,"however,should be enclosed in commas.

_____ C _____ 20. Please visit us in August if you can't make it in July.

_____ 21. My aunt said,"If you can't make it in July, please visit us in August."

_____ 22. Is George pleased with his Alaskan transfer and Twileen with her management job?

_____ 23. The young woman exclaimed excitedly,"I want a diamond,not a rose!"

_____ 24. Brad Pitt won the title of the "Sexiest Movie Star."

_____ 25. Did you know that Hofstra University had a 1.3 million volume library, which was larger than the libraries of 95 percent of the nation's universities.?

_____ 26. "No,"he shouted angrily,"I didn't!"

_____ 27. The manager said,"You won the million dollar lottery!"

_____ 28. Is it true that Jell-O can be used to make hair mousse,hair dye,or wrestling mats?

_____ 29. That's amazing!

_____ C _____ 30. Microsoft Encarta College Dictionary has more than 320,000 entries and definitions.

Name: _____ Date: _____ Number right: _____ out of 30

POP QUIZ FOR READS AND REPLAYS 66–68 (CHAPTER 12)

Insert all needed punctuation and one capital letter.

1. During the tour we shall visit four points of interest in Washington:
   Library of Congress
   White House
   Capitol
   Smithsonian Institute

2. "Do you really believe that?" she asked.

3. A Confederate soldier wrote the bugle melody we call "taps."

4. Our campaign to increase sales is ready: The salespeople are trained, their morale is good, our product is excellent, and we have an ample supply.

5. Campers must take these items: tents, pillows, sleeping bags, cooking utensils, bottled water, and food.

6. Our prices are competitive; for example, $3.99 for a ream of laser paper, $3.42 for 500 envelopes, and $1.59 for an Elmers Glue Stick.

7. Get a good education; then you can compete for better jobs and earn more money.

8. We have four requirements for the job: reliability, common sense, courtesy, and accuracy.

9. "Do not use hyperbole; not one writer in a million can use it effectively," wrote Dr. Alan Dundes, with tongue in cheek. [Hi! Look up *hyperbole*—meaning and pronunciation.—Leila]

10. Each king in a deck of playing cards represents a great king from history: spades, King David; clubs, Alexander the Great; hearts, Charlemagne; and diamonds, Julius Caesar.

11. Originally there were four flavors of Jell-O: strawberry, raspberry, orange, and lemon.

12. Consumer demand can be changed; you can be influenced to select one item over another.

13. The judges at the canine show rejected all but two of the dogs: a cocker spaniel and a bulldog.

14. His phenomenal success is due to one thing: hard work.

15. Twileen will call on customers in Oakland, California; The Bronx, New York; and Tampa, Florida.

16. The following hotels have rooms for conference attendees: Hyatt Regency, 208 Barton Road; Embassy Suites, 300 South Street; and Four Seasons, 98 Rodeo Drive.

17. Some people get a great deal of work done on rainy days; however, when it's sunny, they take the day off.

18. "A semicolon," said the anatomy expert, "is not part of the small intestines; it is the mark that may be placed between independent clauses instead of a period."

19. Michael has an overpowering ambition; namely, to propose to Kim atop the Empire State Building.

20. English for Careers is more fun to learn from than most other textbooks.

Name: _____ Date: _____ Number right: _____ out of 20

Insert all needed punctuation. DO NOT CHANGE CAPITALIZATION; IT IS ALREADY CORRECT IF YOU PUNCTUATE APPROPRIATELY. Write "C" beside the correct sentence.

_____ 1. Eric told us, "Stockbrokers must make decisions on an hour-to-hour basis."

___C___ 2. To be a good salesperson, you must be self-motivated.

_____ 3. He likes to eat in first-class restaurants in far-off places.

_____ 4. This (my little business) is just a small-scale operation.

_____ 5. He shouted, "Five hundred movies have been overrated!"

_____ 6. Please realign the columns of figures.

_____ 7. The president's wife said, "This comment is off-the-record."

_____ 8. Why should parents buy school-age kids the latest styles in school clothes when uniforms are less expensive and easier to manage? or !

_____ 9. Mr. Perkins' office is close to Ms. Adams' store.

_____ 10. Place A's beside the alumni's names on the contributions list.

_____ 11. "It's easy to see," wrote Mr. Potter, "how developing a tractor to pull two plows instead of one doubles the number of acres a farmer can plow."

_____ 12. Then he added, "The difficulty is visualizing a tool or technique to double the productivity of a chef, an artist, or a teacher."

_____ 13. Today, as indicated above, most people are employed in service or information industries, thus making production increases more difficult to measure.

_____ 14. During the 1990s, I saw ex-president Ford in the children's department of Macy's at a pre-Christmas sale.

_____ 15. Excerpts from the article "How to Be a Perfect Speller" are in the chapter called "Writing on the Job."

With a line (I) show where these words may be divided at the end of a line. If a word shouldn't be divided at all, place the line at the end. Use your dictionary.

16. re|ferred   **17.** pre|ferred   **18.** per|mit|ting   **19.** catches| **20.** con|junc|tion

Use dashes for emphasis of a nonessential expression ordinarily enclosed with commas.

21. Sales last year—the best year in our history—were well over 10 million dollars.

22. The president—who was once penniless—was always generous to her staff.

Insert parentheses in the next sentence that will de-emphasize the nonessential expression.

23. American men own an average of three coats (an overcoat, a topcoat, and a raincoat) and five pairs of trousers.

Insert a comma, dash, or parentheses where required.

24. A CPU, keyboard, monitor, and printer—these were all essential parts of a computer system.

25. Our profits (see chart, page 3) were the highest in the history of this company.

Name: _____ Date: _____ Number right: _____ out of 25

POP QUIZ FOR READS AND REPLAYS 72–76 (CHAPTER 13)

A. Review Chapter 10, "Secret Life of a Sentence Revealed," if necessary, so that you readily identify the following word groups as *fragment, comma splice, run-on,* or *correct*. Write **F** for fragment, **R** for run-on, **CS** for comma splice, or **C** for correct.

_____F_____ 1. If it looks like a duck, talks like a duck, and quacks like a duck.

_____C_____ 2. Ace in the hole, blue chip, feed the kitty, and follow suit are examples of gambling slang.

_____F_____ 3. Noah Webster, who is still considered the dean of American dictionary makers.

_____C_____ 4. In British English a period at the end of a sentence is called a *full stop*.

_____CS_____ 5. Avoid word mix-ups, use words in proper context.

_____R_____ 6. Your decision will not be easy several capable applicants want the job.

_____F_____ 7. If Professor Lundin will attend the convention.

_____CS_____ 8. Professor Lundin will present a paper at the convention, she will also conduct a workshop.

_____C_____ 9. We often identify people by their work; for example, we might introduce someone as a lawyer.

_____CS_____ 10. Some of our business students are majoring in management, others are more interested in accounting.

_____C_____ 11. Fear is that little darkroom where negatives are developed.—Michael Pritchard

_____C_____ 12. We can do no great things, only small things with great love.—Mother Theresa

_____CS_____ 13. Shared joy is double joy, shared sorrow is half sorrow.—Swedish proverb

_____CS_____ 14. Love is not blind, it sees less.

_____CS_____ 15. Life is either a daring adventure or nothing, security is mostly a superstition.—Helen Keller

_____C_____ 16. A complete sentence has a subject, a verb, and independence.

_____CS_____ 17. He thought carefully about the move, then he discussed it with his family.

_____R_____ 18. This session the best one will begin at 2 p.m. however, our guest speaker may be late.

_____CS_____ 19. Experience is not what happens to you, it is what you do with what happens to you.—Aldus Huxley

_____C_____ 20. He thought carefully about the move; then he discussed it with his family.

Name: _____ Date: _____ Number right: _____ out of 20

B. On a separate sheet of paper, rewrite the **F, R,** and **CS** items so that they are correct sentences.

C. Identify sentences with the following faults, and write the appropriate letters in the blanks: (CH) "choppy," (LPC) lacks parallel construction, (G) gobbledygook, (M) misplaced words, (VP) vague pronoun usage, (F) fragment, (OK) correct sentence, (CS) comma splice.

LPC 1. Christine Lundin is attending the convention for the purposes of presenting a paper and to hear the keynote address.

_OK_ 2. The agenda includes introducing new officers and discussing the prospectus.

_M_ 3. Our bank approves loans to reliable people of any size.

_CH_ 4. Professional people must keep their skills up-to-date. They must take classes. They need to network.

_M_ 5. In the parade will be 300 children carrying flags and the governors of several states.

LPC 6. Sometimes we must choose between giving up our ideals or to remain faithful to them.

_G_ 7. Not long after the beginning of the 17th century, which means the early 1600s, a game of long times past with the common appellation of ninepins commenced to be an activity engaged in for recreation.

_VP_ 8. Charlene told Ms. Seefer that she needs to go to Concord.

LPC 9. The plan for the next meeting includes:

(a) introduce new officers

(b) reviewing current budget

(c) to decide agenda for annual meeting

LPC 10. Carlos is not only excellent in keyboarding but is also an excellent cook.

_F_ 11. Because when you write, you should avoid using long words as much as possible.

_OK_ 12. If you can think of a short word to express what you want to write, choose it over a long one.

_F_ 13. Enclosed are several carpeting samples. Each of which will be on sale next week.

_CS_ 14. Our new catalog is in the mail, it should reach you by the end of the week.

D. On a separate sheet of paper, rewrite the sentences, correcting the sentence faults.

E. Change these passive voice sentences to active voice. In the blank, rewrite the sentence adding any missing information needed for the sentence to make sense.

15. The contracts were signed by all the college presidents. _All the college presidents signed the contracts._

16. The depositions were taken yesterday. _Tricia took the depositions yesterday._

17. The package was mailed yesterday. _My assistant mailed the package yesterday._

18. Marcia Stranix's name was added to each document. _My secretary added Marcia Stranix's name to each document._

19. New multinational markets will be developed by many companies within the next year.
Many companies will develop new multinational markets within the next year.

F. Change these active voice sentences to passive voice.

20. The king purchased the emerald necklace for $100,000. <u>The emerald necklace was purchased by the king for $100,000.</u>

21. Lorraine visited several other Ohio campuses last year. <u>Several other Ohio campuses were visited last year by Lorraine.</u>

22. Several managers expressed approval of the new software. <u>Approval of the new software was expressed by several managers.</u>

23. One member of our sales staff has sold 1,000 old pocket calculators. <u>About 1,000 old pocket calculators have been sold by one member of our sales staff.</u>

24. Schroeder predicted a daily attendance of about 500 at the Houston meeting. <u>A daily attendance of about 500 was predicted by Schroeder at the Houston meeting.</u>

G. Write **D** in the blank if the sentence has a dangler. Otherwise, write **C** for correct.

<u>D</u>   **25.** Walking down Marietta Street in Atlanta, a statue of the famous editor Henry Grady caught the tourists' attention.—Error example from *SPELL Handbook*

<u>D</u>   **26.** Having been on the phone for three hours, Dad took it away from her.

<u>D</u>   **27.** Her car was stolen while watching the beautiful sunset.

<u>D</u>   **28.** While watching the beautiful sunset, her car was stolen.

<u>C</u>   **29.** While they were watching the beautiful sunset, her car was stolen.

<u>C</u>   **30.** While watching the beautiful sunset, they suddenly discovered her car had been stolen.

<u>C</u>   **31.** Since we were young and foolish, the partying went on all night for us.

<u>D</u>   **32.** Being young and foolish, the party went on all night for us.

<u>D</u>   **33.** To be well baked, you should leave the potatoes in the oven for an hour.

<u>C</u>   **34.** If the potatoes are to be well baked, leave them in the oven for an hour.

<u>C</u>   **35.** If you can think of a short word to express what you want to write, choose it over a long one.

Name: _____ Date: _____ Number right: _____ out of 35

Write **T** (true) or **F** (false) in the blank.

___F___ **1.** One correct way to express the date in a business letter is June 21st, 2006.

___T___ **2.** All the following are correct as salutations:

Dear Billy,   Dear Mr. Crystal: Dear Mr. Crystal

___F___ **3.** All the following are correct as complimentary clauses:

Sincerely   Sincerely,   Sincerely;

___T___ **4.** The open punctuation style means you eliminate punctuation after the salutation and complimentary close.

___F___ **5.** A modified block letter means you begin the date and closing at the left margin, and you indent the first line of each paragraph.

___T___ **6.** Attention lines should rarely be used in business letters.

___F___ **7.** It's advisable to keyboard e-mails in all capital letters.

___T___ **8.** For two-page letters, key the addressee's name and address at the top of the second page.

___F___ **9.** Avoid using a conversational tone in important business letters.

___F___ **10.** Try to keep all paragraphs in business letters about the same length.

___F___ **11.** Use "To Whom It May Concern" as a salutation if you don't know the name of the person to whom you must write.

___F___ **12.** "We are herewith enclosing the price list" is an example of good language style for a business letter.

___T___ **13.** After keyboarding a business letter, proofread it first on the __screen/monitor__ before you print. After printing a copy, proofread again from the printed copy.

___T___ **14.** European business stationery is called A4 paper and it is 8 1/4 by 11 1/2 inches.

___F___ **15.** Avoid using a conversational tone when writing a business letter.

___T___ **16.** A concisely written letter might be very long or very short.

___T___ **17.** Always include a meaningful subject line in an e-mail message.

___F___ **18.** Most e-mails should be written in all capital letters.

___F___ **19.** Even if you know the person well, avoid using informal salutations like "Hi, Joe."

Write the answer in the blanks for each of the following questions.

20. Name in order at least 10 parts that might be included in a typical business letter. __letterhead, date,__
__special mailing instructions, name and address of addressee, subject line, salutation, body, complimentary close,__
__handwritten signature, typed signature and signer's title or department, typist's initials, postscript, copy notation__

21. What is the principal disadvantage of faxing a letter or a long message rather than mailing it?
__Confidentiality may be lacking since you can't always be sure of who will retrieve the message.__

22. After keyboarding a letter, proofread it first on the ___screen___ and then on the __paper copy__.

Name: _____ Date: _____ Number right: _____ out of 22

POP QUIZ FOR READS AND REPLAYS 81 AND 82 (CHAPTER 14)

A. Write your answers in the blanks.

1. What are some differences between traditional and multimedia writing? <u>(answers will vary but may</u> <u>include: Traditional writing is read by a reader; letters, memos, and the like are written on paper and held in the</u> <u>reader's hands. The recipient can read the document at his or her own pace from beginning to end. Long, detailed</u> <u>sentences, paragraphs, and columns of text are the norm. **Multimedia** writing, however, is **viewed** by the **viewer.**</u> <u>Text is presented on a screen. It isn't held, but rather reviewed. The recipient has less time to digest the text and may</u> <u>not read it from beginning to end. For that reason, briefer sentences, paragraphs, and columns are preferred.</u>

2. Name five ways to emphasize text in a slide show. <u>placement (or "negative space"), color, size, font style,</u> <u>movement or animation, bullets, text boxes, boldface, italics, underlining, spacing</u>

3. Copying text from a Web site without saying where you obtained it is <u>illegal—also called plagiarism</u> .

B. Answer **T** (true) or **F** (false).

_____F_____ 4. When you prepare to make an oral presentation to your co-workers and/or superiors, the first step should be to practice your presentation.

_____T_____ 5. The content of your presentation should consist of an introduction, a body, and a conclusion.

_____F_____ 6. When you make an oral presentation, avoid specific recommendations.

_____T_____ 7. When making an oral presentation on the job, it's a good idea to begin with a story or joke.

_____F_____ 8. It's advisable to introduce a few new ideas in the conclusion of an oral presentation to co-workers and supervisors.

_____T_____ 9. Everything you plan to do or say during your presentation should be expressed in a way that helps you meet your objective.

_____T_____ 10. Make the concluding portion of your presentation interesting by urging the audience members to take action on your recommendations.

Name: _____ Date: _____ Number right: _____ out of 10

B Spelling and Vocabulary for Careers

With your computer's spelling checker, you decrease the likelihood of spelling errors; however, the spelling checker is not always available. For example, when filling out a job application form in a personnel office, your opportunity for a good job may be ruined by a single spelling error. Surveys and interviews of managers reveal that in many occupations accurate spelling is essential to get the job, succeed at it, and maintain eligibility for promotion. Not only are electronic spelling checkers often unavailable, but they are not always used correctly. To use the checker correctly:

- Remember that it is not a college dictionary; it recognizes a limited number of words and, of course, doesn't include all proper nouns. A spelling checker will highlight as incorrect a perfectly spelled word that doesn't happen to be in its list.
- If you can add words, add your work-related terms, including proper nouns not already listed.
- Carefully proofread each document both before and after activating the spelling checker. The checker does not eliminate the need to proofread. In fact, it may give a false sense of security resulting in uncorrected errors. If you use the spelling checker without proofreading, a document might be left with errors that form different words such as *a line* for *align*, *affect* for *effect*, or *thorough* for *through*.

1. BE AN EXPERT SPELLER

The **1–3–2–1 Plan** is an efficient method for mastering correct spelling of an entire list of words. Here's how:

1 Ask someone to dictate the words in one of the lists. After writing or typing them **once,** note those you misspelled or were unsure of. Follow the same procedure with the other lists.

3 Correctly say and spell aloud each misspelled word. Then write or type the misspelled words **three** times each. Use the dictionary if in doubt about pronunciation.

2 Next, write correctly each previously misspelled word **twice.**

1 Now use an audiotape or ask someone to dictate the originally misspelled words to you, and write each one **once.** After checking the spelling of each word, list any words you misspelled or felt unsure of. Practice these, using the **1–3–2–1 Plan** again.

2. BUILD YOUR VOCABULARY

- Work with a small group of people interested in vocabulary growth and expertise.
- When necessary, verify pronunciation with your instructor or with a dictionary.
- Divide the list of words among the group members, who will write sentences showing they understand the meaning of the assigned words.

You'll probably find words that are new to you. You'll increase your vocabulary, and you'll have fun with some new words.

- Finally, check one another's sentences for spelling and correct usage.

Below are 14 groups of 25 commonly misspelled words. Many of these words are listed in the *Microsoft Encarta College Dictionary* under the heading "Commonly Misspelled Words." Perhaps you can master one list along with each of the 14 chapters of the text. Be sure you know the meaning of each word as well as the pronunciation and spelling. I guarantee that you will increase your vocabulary as well as improve your spelling. ☺

Word Power

Shanty Hogan, former football coach at Phoenix Community College, says this is a true story: He asked his freshman players to fill out a card to be used in case of serious injury. The card lists whom to notify and other such information. In the blank for religion, one player wrote "Bhaptizz." Amused, Hogan asked, "Now, son, what religion are you?" The young man answered, "Presbyterian." But you wrote "Baptist," the coach said. "I know," said the player, "but I can't spell *Presbyterian*."

CHAPTER 1	CHAPTER 2	CHAPTER 3	CHAPTER 4
abbreviate	bulletin	carriage	debatable
absence	buoy	commission	disappointed
absurd	bureaucracy	chronic	deductible
acquaintance	blizzard	concede	deferred
ambiguous	bureau	ceiling	deluge
announcement	benefited	customary	definitely
ambassador	beginning	commencement	dependent
acquire	battalion	confidently	deterrent
appearance	belligerent	confidentially	develop
admissible	belief	chargeable	diligent
aggressive	bankruptcy	candidate	dimension
accommodate	balloon	calendar	disastrous
acceptance	bargain	camaraderie	discrepancy
apparatus	basically	chronic	disappearance
approximately	bazaar	condemn	disapprove
adjournment	believe	connoisseur	disguise
align	bookkeeping	conscience	dissatisfied
allotment	boundary	camouflage	desperate
allotted	bouquet	campaign	deficient
altogether	brilliant	career	default
amateur	bachelor	cemetery	discretion
analysis	beneficial	collaborate	deceive
amendment	broccoli	curriculum	dilemma
accessible	buoyant	copyright	defendant
advantageous	broach	compelled	dragster

CHAPTER 5	CHAPTER 6	CHAPTER 7	CHAPTER 8
especially	forfeit	generalize	harass
eerie	fiscal	government	hazardous
economics	fluorescent	grandeur	hesitant
effervescent	forth (forward)	grateful	hindrance
efficiency	fourth (after third)	grievance	hypocrite
eighth	fundamentally	guarantee	hierarchy
eligible	foreseeable	genealogy	handful
eliminate	foreign	government	hygiene
emphasize	flexible	governor	habeas corpus
encouragement	feasible	graffiti	handkerchief
endorsement	fictitious	grammar	happiness
enforceable	fortunately	granddaughter	hoping
enormous	freight	grieve	humorous
en route	fulfill	guardian	hemorrhage
enthusiastically	February	guidance	handicapped
enumerate	forty	guitar	haphazard
environment	frieze	gnash	Hawaii
equipped	feint	gigantic	headhunter (business)
erroneous	forehead	grievous	healthful
especially	fascism	guaranty	hence
espionage	fraudulent	guesstimate	hesitant
embarrass	furor		hindsight
	FAQ		hiring hall
			hoax

CHAPTER 9	CHAPTER 10	CHAPTER 11
impromptu	jeopardize	khaki
improvement	jewelry	kaleidoscope
incidentally	judgment	knowledgeable
indispensable	justifiable	kumquat
initiative	jealous	kung fu
insistence	journal	kook (slang)
intangible	jack-of-all-trades	kopek
interpretation	jackpot	Kitty Hawk
intolerable	jagged	kleptomaniac
irrelevant	JAMA	knead
itemize	jargon	kinesiology
itinerary	java	knuckleball
idiosyncrasy	Jaycee	kibitz
illegal	jaywalk	laboratory
imaginary	jeep	launch
immediately	jeer	legitimate
inaccurate	jell	lucrative
incredible	jester	likelihood
innuendo	jetliner	library
inoculation	jettison	liaison
interference	jicama	liqueur
irrational	jillion	luscious
	jitney	leisure
	job lot	larceny
	joystick	length

CHAPTER 12	CHAPTER 13	CHAPTER 14
manageable	occasionally	relevant
management	occupation	reference
media	odyssey	recurrence
maneuver	occurrence	referred
miscellaneous	omission	relegate
marriage	occurred	subsidize
maintenance	observant	syllabus
mediator	omitted	silhouette
messenger	opponent	tariff
memorize	optimistic	tedious
mortgage	outrageous	tournament
mischievous	override	ultimately
misspelling	ozone	vacillate
Mediterranean	oppression	vacuum
medieval	pamphlet	weird
millennium	participant	wholly
naïve	perceive	workable
necessary	per annum	write-off
ninety	precedent	X-rated
noticeable	plausible	Xerox
nuclear	perceptible	yo-yo
neighbor	prominent	yield
ninth	quota	yottabyte*
notable	questionnaire	zealous
necessitate	queue	zillionaire

*__Yes,__ it's really a word—a new one; so is *zettabyte!* If you don't believe me, look them up in a new dictionary. ☺

C Final Rehearsal

Before rehearsing for your final exam, review the Practice Quiz at the end of each chapter. Then look carefully at each of the following sentences. Look for errors based on what you have studied in the 14 chapters of this text. Correct these errors, but write **C** beside the sentences that are already correct.

Please don't change words because you think other words might "sound better"; correct specific errors only. Many sentences have more than one error; careful proofreading is part of the "final rehearsal." For sentences that may be corrected in more than one way; choose the simplest way.

_____ 1. The sidewalks of Paris are an art display in ~~theirself~~ themselves.

_____ 2. The passer_s_by throw coins to the musicians who especially pleases them.

_____ 3. One of the most spectacular sights ~~are~~ is the light demonstration at Notre Dame.

_____ 4. _Per annum_ is Latin for _by the ~~day~~ year_.

___C___ 5. An overdue amount that is unpaid is _in arrears_.

_____ 6. Twileen_s_ grandmother in addition to her cousins, like_s_ George better than Jesse.

_____ 7. Watterson's student_s_ wrote good stor_ie_s about Twileen and George.

_____ 8. ~~Although~~ many students already have good English skills, and do not require this practice.

_____ 9. For some students, however, learning to use verbs according to Standard English can make a big differen_ce_ts in career potential.

_____ 10. Neither Ben nor he know_s_ the names of the witnesses we subpoenaed.

_____ 11. I saw the man today whom they say will be the new CEO.

___C___ 12. The smallest unit of space in a computer image is called a voxel.

_____ 13. I told Seinfeld that you ~~done~~ did the work ~~good~~ well and had come to work on time every_day.

_____ 14. Most businesspeople want their business to be in_solvent.

_____ 15. If you ~~was him~~ were he, you would have ~~axed~~ asked her for a date.

_____ 16. He ~~don't~~ doesn't plan to give you _ac_excess to your text book during the final exam.

_____ 17. Mathematics is important to every one, but each department m~~o~~ust choose ~~their~~ its own ~~coarses~~ courses.

_____ 18. A chart and a photograph ~~was~~ were in the file.

_____ 19. "A_n_ adverb describes a verb, a_n_ adjective, or another adverb," said the instructor.

_____ 20. Each one of us ~~have~~ has succeeded.

_____ 21. The defendant has no liabilities; consequently, you have no chance of recovering any of your losses.

_____ 22. The title page of a report include_s_ the report_s_ title, it_s_ author, the name of the person or organization for who_m_ the report is prepared, and the date it_s_ to be submitted.

_____ 23. Pronounce the second syllable of the word _preface_ like ~~the word face~~ _fis._

_____ **24.** Some reports also have a preface;^it comes before the introduction.

_____ **25.** A report's introduction sets the stage;^that is, it tells readers what the report contains○

_____ **26.** A^n upper berth on a ship is less comfortable than a lower berth.

_____ **27.** Do you know whose^is in charge of this company^?

_____ **28.** He was using a common idiot^m when ⚡he said,^"I wouldn't touch this job with a 10-foot pole."

_____ **29.** *Mischievous* is correctly pronounced with ~~four~~ three syllables.

_____ **30.** ~~Their~~ They're planning to expand the Teaneck,^New Jersey,^plant.

_____ **31.** Of the two applicants we're considering, the first one seems the ~~best~~ better qualified.

__C___ **32.** He dug and hoed his little field and planted sweet potatoes.

_____ **33.** ^"Good morning, Little Red Riding Hood,^" said the big bad wolf.

_____ **34.** Please destroy ~~them~~ those letters before April 1,^which is her birthday.

_____ **35.** Twileen replied by just saying,^"Happy Birthday.^"

__C___ **36.** He sells his invention to YMCAs all over the country.

__C___ **37.** *Lien* rhymes with *seen.*

_____ **38.** Thank you for sending us the ~~free~~ gift. [A gift is always free; otherwise it's not a gift.]

_____ **39.** Judging by his ~~passed~~ past performance, he is a "loser○^"

_____ **40.** Each one of us have^s succeeded.

_____ **41.** My brakes suddenly stopped working while^I was driving to the conference.

_____ **42.** The YWCA's campaign has elicited the support of women's organizations.

_____ **43.** Just to stand up in the face of life's problems. That^takes courage.

_____ **44.** While opening a can of juice,^his hand was cut real^ly bad^ly.

_____ **45.** My family goes on separate vacations each August.

_____ **46.** Many alumnus^i attended the home coming game and dance.

_____ **47.** Three items^—the bookcase, cash register and display rack—are not for sale.

__C___ **48.** Because I failed to punch the time card, the manager docked my pay.

__C___ **49.** While going over the work carefully, she found a major error.

_____ **50.** The native food tasted strangely to the American tourists.

_____ **51.** Some reports have a preface;^it precedes the introduction.

_____ **52.** COLA ~~is the favorite drink at the NYSE~~ means cost of living adjustment.^

__C___ **53.** The term *arrears* applies to an unpaid bill.

__C___ **54.** Time between starting a task and completing it is called "turnaround time."

__F___ **55.** (True or False) Since Alaska has many polar bears, it is called a *bear market.*

__C___ **56.** A pledge of property as security for a loan is a mortgage.

_____ **57.** Pronounce *affluent* with the accent on the ~~second~~ first syllable.

__C___ **58.** The correct pronunciation of *versatile* is with the accent on the first syllable.

_____ **59.** Use a ~~thesaurus~~ dictionary to find definitions when you're not sure of the meaning of a word.

_____ **60.** *Etymology* is the ~~history of computer~~ origin of words.

_____ **61.** Diacritical marks in the dictionary show how a word is ~~spelled~~ pronounced.

_____ **62.** Because of a misunderstanding,^we lost the customer and discharged one of our best sales reps.

_____ 63. If you don't know where you're going, you'll probably end up somewhere̶s̶ else.

_____ 64. The general costs of running a business, such as taxes, rent, and heating, are called the ~~markup~~ *overhead.*

_____ 65. ~~A analasis of the discrepencys prooved disasterous.~~ *An analysis of the discrepancies proved disastrous.*

_____ 66. The children were told to sit quiet^(ly) and read their books.

_____ 67. After adjusting the carburetor, the car ran smooth^(ly).

_____ 68. Are all employees required to have lunch in the employee's cafeteria?

_____ 69. Punctuation marks act like road signs; they tell us when to slow down, and when to stop.

_____ 70. Antonyms are words with ~~similar~~ *opposite* meanings.

___C___ 71. *We, us, I, he, she, you* are examples of personal pronouns.

_____ 72. The state and federal governments i̶s̶ *are* helping people get off o̶f̶ welfare.

___C___ 73. Your speaking style is clear, forceful, and pleasant.

_____ 74. "Why did the peanut cross the road?" asked Professor Johnson.

___C___ 75. "To get to the Shell Station," replied the Delta students.

_____ 76. We visited a factory. I̶t̶ *that* makes computer parts.

_____ 77. Seattle is the close̶s̶t̶ *closer* of the two cities.

_____ 78. Upon what criteria̶s̶ did you base your opinions o̶n̶?

_____ 79. The congregation we̶r̶e̶ *was* quiet while the rabbi spoke.

_____ 80. The plural of *knife* is *knives*, and the plural of *handkerchief* is ~~handkerechiefs.~~ *handkerchiefs*

_____ 81. Mathematics a̶r̶e̶ *is* my favorite subject.

_____ 82. The plural of alumna is *alumnas̶.* *alumnae.*

_____ 83. Everyone must make t̶h̶e̶i̶r̶ *his or her* own decision.

_____ 84. Who̶'̶s̶ *Whose* laptop computer is missing?

_____ 85. Everybody̶s̶ *Everybody's* going to the office party.

_____ 86. Who̶m̶ *Who* did you say will handle the new account?

_____ 87. H̶i̶m̶ *He* and m̶e̶ *I* will report it to whom̶ever is in charge.

_____ 88. Ms. Hixon explained that many fine people live̶d̶ *live* in Torrance now.

_____ 89. Accuracy in figures mark̶ *marks* the expert accountant.

_____ 90. You have broke̶ *broken* one of the rules.

_____ 91. I politely said, "Go home now."

_____ 92. Gretta writes ~~good.~~ *well.*

_____ 93. I feel badl̶y̶ *bad* about losing my motorcycle.

_____ 94. These kind̶ *kinds* of games are too much like gambling.

_____ 95. We would sure̶ *surely* like to meet him.

_____ 96. The butler stood at the doorway and called the guests' names.

_____ 97. Five minutes' planning can sometimes save an hour's work.

_____ 98. "He was tall, blonde, and had brown eyes," she said.

_____ 99. Business is part of our society; there is no escape.

_____ 100. Thomas Carlyle referred to music as the "speech of angels."

True/False

_F_   **101.** Typical business stationery in the United States is 8 1/4 inches by 11 1/2 inches.

_F_   **102.** A Web page that looks gray is desirable.

_F_   **103.** "You'll lose your good credit rating if you don't send your check today." is good wording to persuade your customer to pay.

_F_   **104.** "Can you meet me at 10 a.m. tomorrow?" is an example of redundancy.

_T_   **105.** Even though a letter writer's name is typed at the end of the letter, it's necessary to include his or her handwritten signature.

_T_   **106.** A good formula for an oral presentation is:

     (a) introduction: tell them what you're going to tell them

     (b) body: tell them

     (c) conclusion: tell them what you've told them

_F_   **107.** In the body of your presentation, avoid details about how your plan would work and how it will save the company money.

_F_   **108.** To make a good impression, use as many big words and job-related technical vocabulary as you can during your presentation.

D Mini Reference Manual Read and Replay

READ ABOUT NUMBERS IN BUSINESS

Numbers are important to business, professional, and technical writing. Write them in figures on invoices, orders, requisitions, statistical documents, memos, and most e-mails. The following information—a consensus of the style used by better business writers—enables you to decide whether to spell out a number or use figures in letters, reports, and other workplace documents.

1. General

a. Spell out numbers up to ten; use figures for specific numbers over ten. If numbers under ten and over ten are used in a related way, use figures for all.

> We need **five** computer engineers in our Akron office.
>
> We need **5** computer engineers, **25** clerks, and **30** assemblers in our Cleveland office.

b. Spell out approximate numbers that can be written in one or two words.

> Nearly **five thousand** employees were laid off last year due to downsizing.
>
> We have hired over **two hundred** new employees this year and have developed **104** new products.

c. When a number begins a sentence, spell it out if you can do so in one or two words. If more than two words are required, rephrase the sentence.

> **Six hundred** crates were shipped to you yesterday.
>
> Yesterday **642** crates were shipped to you. [four words required for spelling out]

d. To express millions or billions, combine figures with words to make reading easier.

> We produced **1 1/2 million** electric fans last year. [or 1.5 million]
>
> Our gross profit last year was **66 million dollars.** [or $66 million]

e. When two numbers appear together, spell out the number that can be written with fewer letters.

> Lloyd **designed twenty 16**-unit apartment buildings.
>
> Frank designed **110 sixteen**-unit apartment buildings.

2. Time

a. Use figures with a.m. and p.m. Use figures or words before *o'clock* or similar words. Type a.m. and p.m. in lowercase letters with no space after the first period. Avoid the colon and zeros for "on the hour" time.

OK   The teleconference will take place from **9 a.m.** through **5:30 p.m.** We should be there at **9 o'clock** in the evening.

NO   nine p.m. 9:00 p.m.

b. Use just one way to express time; avoid redundancy.

OK   9 a.m. or 9 in the morning or nine o'clock in the morning

NO   9 a.m. in the morning

3. Dates

a. Use figures if the date follows the name of the month; do not use *th, d,* or *st* after the figure.

> The American Bankers Association will meet here again on **May 7, 2008.**

b. In military, international, and some government correspondence, write the date before the month; do not use a comma in this style date.

> The International Bankers Association will meet in Oslo on **7 May 2004.**

c. If you write *of* between the date and the name of the month, use *th, nd, rd,* or *st** after the figure or spell out the number. Sometimes the date is given without the name of the month. The same rule applies.

> The World Bankers Association will meet on the **7th of May in 2008.**
>
> The World Bankers Association will meet on the **seventh of May in 2008.**
>
> [or . . . on the seventh . . . or . . . on the 7th . . . or . . . on the 7th of this month.]

d. Spell out or use figures for centuries and decades.

> This book is about nineteenth-century poets. [or 19th-century poets]

4. Money

a. Use figures for amounts of money. The decimal point is unnecessary with even dollar amounts (no cents).

> Twileen paid $250 to register and $15.50 a month to exercise at the gym.
>
> They need about $150,000 to remodel the gym.

b. Use a dollar sign and decimal point for cents to be consistent with other amounts used in the same context. If other amounts are not involved, use figures for the number but spell out the word *cents.*

> We sold **1,000** cookies at **$.25** each, **42** pies at **$7.80** each, and two big cakes at **$50** each. The plugs cost **8 cents** each.

c. In legal documents amounts are often spelled (notice capital letters) and then written in figures enclosed in parentheses. In ordinary correspondence, do not repeat numbers in this legal style.

> The fee for use of said property is to be **Two Hundred and Fifty Dollars ($250.00)** a month. BUT We paid **$250** for those tickets.

5. Addresses

a. Spell out names of streets under 11th.

> The store is on **Sixth Street.**
>
> Professor Maxey of Oroville moved to **11th Avenue** in Newport.

b. Use figures for all house numbers except One.

> Their new suite of offices is at **One Abercrombie Street.**
>
> The Atlanta factory is at **8 Leland Avenue.**

---

*These are *cardinal* numbers; *ordinal* numbers don't have the two-letter ending—e.g., ORDINAL 1, one; 2, two; 3, three, etc., vs. CARDINAL 1ST *first, second, third,* etc.

6. Percentages and Fractions

a. Spell out *percent*, but use figures for the number (use the % sign in statistical and technical forms).

> The unemployment rate was **5 percent** that year.

b. Spell out a common fraction when it is the only one in the sentence. Use figures for less common fractions or when several fractions are in a sentence.

> We have received only **one-fourth** of our order.

> The specifications for the blue widgets are **3/8** of an inch and for the purple widgets, **4/5** of an inch.

c. Use figures for a mixed number (fraction and whole number).

> Our profits are **4 1/2** times those of last year.

7. Measurements

> In business letters and reports, spell out measuring words, such as *feet, pounds,* and *inches;* however, use figures for the numbers.

> The boards are **5 by 6 by 2 inches.** [Use *by*, not ×.]

> Each one weighs **6 pounds 4 ounces** and is **8 feet** long.

8. Age

> Usually spell out an age expressed in years only—unless it immediately follows a person's name. Notice that no commas are needed in the third example, which illustrates age expressed in years as well as months and days.

> Mr. Weber will be **sixty-three** on the day his medal is presented.

> Carl Weber, **63,** is the new supervisor.

> The records show her age at death was **24 years 5 months and 6 days.**

9. Books

> Use figures for numbers of pages, chapters, volumes, etc.

> The information you need is on **page 46** in **Chapter 5.**

REPLAY NUMBERS IN BUSINESS

Change the number style where necesssary to make the following sentences correct in the paragraphs of a business letter or formal document.

1. At 9 a.m. ~~in the morning~~ on June 4th, we will have checked ~~forty-two~~ [42] leases.

2. I know about ~~50~~ [fifty] ways to make about 5 million dollars.

3. [On 6 June 2004,] 210 boxes were shipped to your London office by Acme Air Freight ~~on 6 June 2004~~.

4. Please send $15.00 for the book and ~~fifty cents~~ [$.50] for postage.

5. On page ~~six~~ [6] his age is given as ~~40~~ [forty].

6. Almost ~~5,000~~ [five thousand] people attended the Alliance for Survival rally.

7. Our new office is at 62 ~~4th~~ [Fourth] Street.

8. The prime interest rate went to ~~eleven %~~ [11 percent] today.

9. We need ~~12~~ [twelve] 8 ~~x~~ [by] 10 offices in the new building.

10. Tracy's opened its doors for business at ~~ten~~ [10] a.m. on June 30.

READ ABOUT CAPITALIZATION

Capitalization gives a word special importance or emphasis. Capitalize the first word of a sentence to give it the emphasis a word in this position requires. In addition, many specific things, people, and places have two names: the classification name, such as *girl,* and the official name, which is capitalized, such as *Cindy.* Many capitalization principles are also shown in Read 26.

1. General

a. Capitalize official names of specific people, animals, places, days, months, holidays, religions, gods, documents, and historical events.

Joseph	United Airlines	Uganda	Veteran's Day
Declaration of Independence	Fabulous Forties	Wednesday	Atomic Age

b. Do not capitalize seasons.

fall	spring	winter	summer

c. Capitalize titles and headings but only the first word of an item in an outline or list. Use lowercase for short prepositions, articles, *to* in infinitives, and the conjunctions *and* or *or* unless one of these words begins the line.

Titles or Headings: How to Cook with Electricity or

HOW TO COOK WITH ELECTRICITY

Outlines or Lists: I. How to cook with electricity [Only the first word is capitalized]

2. Titles of People

a. Capitalize a title that directly precedes a person's name.

Professor Washington   Reverend Juan Perez   Madame Curie

b. Do not capitalize the title if a comma separates it from the name.

Our English professor, Dolores Denova, is . . . .

The captain, Patrick O'Connor, seemed . . . .

c. Do capitalize the title in an address or typed name at the end of a letter.

Address: Professor Carol Baker
Computer Studies Department

End of Letter: Carol Baker
Professor of Computer Studies

d. Do not capitalize occupations, such as manager, lawyer, accountant, engineer.

My accountant is also an attorney.

3. Titles of Publications and Art Works

a. Capitalize the first word and all principal words of titles of books, films, plays, songs, and so on. Do not capitalize articles (*a, an, the*), prepositions of three or fewer letters, or *to* in an infinitive—unless one of these is the first word. Italicize titles of full-length published works, or if italics are not available, underline them.

I saw the movie *Gone With the Wind.* (or Gone With the Wind)

The best-selling book that year was To Kill a Mockingbird. (or *To Kill a Mockingbird*)

b. Use quotation marks for titles of portions of full-length works such as newspaper or magazine articles and chapters of books, as well as poems, songs, and other short literary or musical works.

Chapter 8 is entitled "The Taming of the Apostrophe."

"The Star Spangled Banner" is the national anthem of the United States.

4. Names of Organizations

a. Capitalize names of organizations and specific government groups.

> Supreme Court, Fullerton College, Palm Springs Tennis Club, Department of Motor Vehicles, Royal Canadian Air Force, Arkansas Paper Mill

b. Except in legal documents and formal communications, do not capitalize words like *company, department, college* when used without the name.

> Melissa and Steven attend Seattle court reporting **s**chools.
>
> This **c**ompany will not issue common stock this year, but Knott **C**ompany will.
>
> Give the papers to the **c**ommittee at the **c**ollege.

c. Capitalize names of departments within your own organization but not departments within other organizations.

> Our **S**hipping **D**epartment packed the order and shipped it via Federal Express.
>
> Does your **p**urchasing **d**epartment have our newest catalog?

5. Names of Places

Capitalize official names of specific places.

Charleston, Victory Boulevard, Atlantic and Pacific Oceans, Yosemite National Park, Mississippi River BUT Let's swim in the **o**cean and then go to the **p**ark.

6. Compass Points

a. Capitalize compass points that name areas that are geographical, cultural, or political units.

> Far East, West Coast, Midwest

b. Do not capitalize compass points indicating direction or general areas.

> The sun sets in the west.
>
> Drive south along Main Street.
>
> He would like to settle in northern Massachusetts, not Southern California. (names specific area)

c. Capitalize derivatives of compass points referring to people.

> I believe Northerners usually appreciate Southern hospitality.

7. Trade Names

Capitalize the trade name of a product but not the product word itself unless you work for the company or it is part of the product name appearing on the container.

Do you like Chef Boyardee Pizza? [you work for the producer]

Do you like Nescafe coffee? [you don't work for the producer]

8. Business Letters

a. Capitalize all courtesy or professional titles used in the inside address. Do not use a courtesy title—such as Dr., Mr., or Ms.—with a signature or with the writer's typed name at the end of the letter. Do use a courtesy or a professional title in an address when you know what the title is.

Addresses:	Professor Tony Carter	Lt. Wendy Gelberg	Mr. Robert Reiss
			General Manager
End of Letter:	Arthur Khaw	John Link, M.D.	Sean Chandra, J.D.
	Program Coordinator		

b. Capitalize the first word of a complimentary close.

> Sincerely yours

c. Capitalize the first word of a salutation and any noun or title in the salutation.

> Dear Friends   My dear Mr. President

9. Family Relationships

a. Capitalize a family relationship title when used as part of the name or instead of the name.

> Do you think Uncle George will retire soon?
>
> Do you think Uncle will retire soon?

b. Do not capitalize a family relationship title when a possessive noun or pronoun precedes it.

> Steve and Sue's uncle was the principal of Saturn Street School.
>
> I believe my cousin Esther should apply for the job.

10. School Subjects and Degrees

a. Capitalize official names of courses. Do not capitalize the name of a subject or course that isn't the official name—except languages, which are always capitalized.

> Professors Denova, Hixon, and Shay will teach Business 31 this year.
>
> Professor Dennison was asked to teach a business English course next year.
>
> Do you plan to take a Spanish or business law class next year?

b. Capitalize the name of a spelled-out degree directly after a person's name—or anywhere in a sentence if abbreviated. Do not capitalize a spelled-out degree unless it immediately follows the person's name. Notice the comma before and after the degree following a name.

> Sarala Chandra's MBA is specialized in management of nonprofit organizations.
>
> However, Jonathan Waller has a master of arts degree in speech pathology.
>
> With an associate's degree such as AA or AS, you can qualify for many good jobs.
>
> Sean S. Chandra, Doctor of Jurisprudence, is the newest Supreme Court judge.
>
> Nina Chandra's BS is in animal husbandry.

11. Government Terms

> **Don't capitalize geographical terms used instead of the full name.**

> This state has the highest income tax rate in the nation.
>
> We do not wish to violate city ordinances.

12. Ethnic Terms

   a. Capitalize names of nationalities and religions.

        Dutch, Korean, British, Catholic, Hindu, Jewish

   b. Capitalize names of races except for races named by a color.

        Asian   African American   Caucasian   BUT   black   white

REPLAY CAPITALIZATION

Capitalize where needed in the following sentences.

1. the atomic age began on August 5, 1945, when the atomic bomb was dropped on the city of hiroshima.

2. we flew on american airlines last summer with the president of israel and the senator from maine.

3. eric a. smith, cfp, manages investment portfolios of $100,000 or more in seattle.

4. winston churchill wrote *triumph and tragedy,* an important book about world war II.

5. the judge said that the supreme court decision was favorable to my company.

6. i will know chinese well enough by september to take a university course in chinese literature.

7. the atlantic and pacific oceans are natural borders of the united states.

8. use hunt's tomato sauce for making lasagne on tuesdays in march.

9. lalitha, a hindu woman from the south of india, has a bachelor's degree in education and is married to an American md.

10. the salutation of the letter is "dear customer," and the complimentary close is "sincerely yours."

READ ABOUT WORD DIVISION

Do not use right-margin justification for business letters. For neat right margins, however, you may choose to divide words. Because a divided word may distract readers, certain conventions have been established for workplace documents. The word processing hyphenation feature on your computer (though usually appropriate for newpapers, books, etc.) may not be acceptable for business letters and reports. If so, change the division place of the word, or don't divide at all.

If your organization has a manual showing preferences for its business documents, see whether word division is permitted. If it is permitted, activate your computer's hyphenation feature, but proofread to see that word divisions conform to the principles below.

1. Never divide:
   a. the last word on a page
   b. a word containing an apostrophe
   c. a number expressed in figures
   d. an abbreviation

    e. on more than two consecutive lines

    f. a word with fewer than five letters

    g. a word of only one syllable

    h. a proper noun

    i. between the number and a.m., p.m., noon, midnight, or percent

    j. unless at least three letters can be carried to the next line

2. It's all right to divide:

    a. between syllables. When consulting the dictionary, refer to syllables in the entry word, not in the pronunciation. Dots, spaces, or accent marks—depending on the dictionary—show syllables.

       fol.low.ing [may be divided between the *l*'s or after the *w*]

       stopped [may not be divided because it is a one-syllable word]

    b. when at least two letters—preferably three—can be typed on the line before the hyphen.

OK   rec-ognition (3 letters)

OK   re-veal (2 letters)

NO   a-gainst (1 letter)

    c. when at least three letters can be typed on the next line.

OK   compil-ing

NO   compa-ny

       If syllables don't permit minimums of rules b and c, do not divide the word.

    d. after a one-letter syllable unless the vowel is part of a suffix such as *ible* or *able.*

OK   cata-log

OK   credit-able

NO   credita-ble

    e. between double consonants unless the division breaks the spelling of a root word.

OK   begin-ning

OK   spell-ing

NO   spel-ling

       [If the second of the double consonants forms a suffix, place the hyphen between the double letters.]

OK   posses-sion

NO   possess-ion

    f. at the hyphen if it is part of the spelling of the word.

OK   self-confidence

NO   self-confi-dence

    g. nonhyphenated compound words between the main parts of the word.

OK   under-developed

NO   underdevel-oped

3. It's all right to separate a date between the day and the year.

       We were married on September 5,
       1954, in New York City.

4. It's all right to carry the last name to the next line. A short or abbreviated title and a middle initial must stay with the first name.

OK   The most capable auditor we have had is Dr. Martin L.
Simon. OR The most capable auditor we have had is Dr. Martin
Simon.

NO   The most capable auditor we have had is Dr. Martin
L. Simon. NOR The most capable auditor we have had is Dr.
Martin L. Simon.

5. It's all right to separate a spelled-out title (but not an abbreviated or short title) from the name.

OK   Our most outstanding English teacher is Professor
Isha Chandler.

NO   The dermatologist doing Jean's new face is Dr.
Barry Chang.

6. It's all right to separate an address between the street name and the city or between the city and the state.

We have moved our office to 990 Kennedy Street,
Minneapolis, Minnesota 55416.

We have moved our office to 990 Kennedy Street, Minneapolis,
Minnesota 55416.

7. If it's essential to divide an e-mail address or a URL,
    a. divide the e-mail address before @ or the dot
      ITALBCJones   or   ITALBCJones@QRS
      @QRS.com   .com
    b. divide the URL after the double back slashes
      http://
      orgmoveup.com

REPLAY WORD DIVISION

Write **T** (true) or **F** (false) in the blank.

_F_   **1.** It's all right to divide a four-letter word if it has two syllables.

_T_   **2.** Never divide the last word on a page.

_T_   **3.** It's all right to divide the following e-mail address before the dot: somlovr@grr.com.

_F_   **4.** It's all right to divide a one-syllable word if it has more than five letters.

_T_   **5.** Never divide a word containing an apostrophe.

_T_   **6.** If you can't carry at least three letters to the next line, don't divide the word.

_F_   **7.** If you must divide a date, type the month on one line and the day and year on the next.

_F_   **8.** To divide an address, separate the house number from the street name.

_T_   **9.** A word spelled with a hyphen may be divided only at that hyphen.

_T_   **10.** Divide between double consonants unless the division breaks the spelling of a root word; for example, *press-ing* and *win-ning*.

READ ABOUT ABBREVIATIONS

Spelling words out instead of abbreviating them helps create an image of thoroughness, carefulness, and accuracy; **when in doubt, spell it out.** In workplace letters and reports, avoid abbreviations except as shown below. Abbreviations are appropriate, however, in e-mail, interoffice memorandums, telephone messages, catalogs, statistical reports, and business forms such as invoices and orders.

1. Abbreviations in Business Letters and Reports

a. Except for commonly used abbreviations such as *USA* and acronyms such as *laser,* spell out an abbreviation or acronym the first time it is used and follow it with the abbreviation in parentheses. Thereafter in the same document, use just the abbreviation or acronym.

> Wide Area Information Systems (WAIS)

b. Always abbreviate the following titles: *Mr., Mrs., Dr., Jr., Sr., Esq.,* * *Ph.D., CPA,* and academic degrees following names (see item 1h regarding degrees). Type a period after *Ms.* even though it isn't an abbreviation. *Ms.* is the feminine form for *Mr.,* neither of which indicates marital status. The two feminine courtesy titles that do show marital status are *Miss* with no period after it and *Mrs.,* which ends with a period since it abbreviates *Mistress.*

> Ronald Rosenberg, Esq.   Ms. Sarala Waller   Dr. Spencer Blackman

c. Military and professional titles before a full name may be abbreviated, but if only the last name is given, spell out these titles. The abbreviation *Dr.,* however, is used with either the full name or last name only.

> The *Rev.* Jonathan Flaherty OR The *Reverend* Jonathan Flaherty OR The Reverend Flaherty; Dr. Marvin Belzer OR Dr. Belzer

d. Names of well-known organizations are usually abbreviated in all but the most formal documents. However, do not abbreviate unless you're sure the reader will understand. Spelling out these abbreviations (as in *a* above) may be required for international communications.

> AT&T   CIA   FBI   AFL-CIO   CBS (See your dictionary if in doubt about these.)

e. Time words may be abbreviated.

> a.m.   EST   BCE   AD (See your dictionary if in doubt about these.)

f. Abbreviate parts of names and addresses, such as *Inc., Co.,* or *Ltd.* (Limited) when they appear that way in the company's letterhead and *St.* (Saint) or *Mt.* (Mount) when this agrees with the dictionary spelling of the name of this place. Spell out words like *Street, Avenue,* or *Boulevard* and names of cities. State abbreviations are shown later in part *i.*

g. Here are common abbreviations and acronyms acceptable in most workplace documents—if the reader has good English skills. Many of these are used with or without periods. The trend is to omit periods in those most easily recognized. Some of the following are acceptable in lower- or uppercase; for example cod or COD (collect on delivery).

> CEO—chief executive officer
>
> CFO—chief financial officer

---

*In the United States, *Esq.* (Esquire) is used only after an attorney's name—either male or female. In Great Britain, however, *Esq.* is a courtesy title equivalent to *Mr.* but used *after* a man's name regardless of occupation.

CFP—certified financial planner

cm.—centimeter

cu.—cubic

COD—collect, or cash, on delivery

CPA—certified public accountant

EEO—equal employment opportunity

e.g.—for example

e-mail—electronic mail

EOM—end of month

etc.—and so on

Ext.—when followed by a number for a telephone extension (Ext. 32)

FAX—facsimile copy

FDIC—Federal Deposit Insurance Corporation

FOB—free on board (the point from which the customer pays shipping charges)

GE—General Electric

GM—General Motors

GDP—gross domestic product

HRD—human resources department

IBM—International Business Machines

ID—identification

Inc., Corp., or Ltd.—when part of a company name (incorporated, corporation, limited)

IRS—Internal Revenue Service

memo—memorandum

MIS—management information systems

NASA —National Aeronautics and Space Administration

OPEC—Organization of Petroleum Exporting Countries

OSHA—Occupational Safety and Health Administration

PC—personal computer (usually refers to an IBM or IBM clone)

P. O. Box—Post Office Box No. 0000 (as part of a mailing address)

PR—public relations

PS—postscript

RE—regarding or concerning

R&D—research and development

RSVP—Please respond (translated from French—Répondez vous s'il vous plaît)

SEC—Securities and Exchange Commission

VIP—very important person

h. Abbreviate an academic degree, usually with periods, when it follows a person's name. When it doesn't follow a name, the periods may be omitted; or "If in doubt, spell it out" (with all lowercase letters).

His AA degree qualifies him for the job. She has a bachelor of science degree.

Jose Hernandez, D.D.S., will speak on dental health.

A.A.—associate in arts (two-year degree)

A.S.—associate in science (two-year degree)

B.A. or A.B.—bachelor of arts

B.B.A.—bachelor of business administration

B.S.—bachelor of science

D.A.—doctor of arts

D.B.A.—doctor of business administration

D.D.—doctor of divinity

D.D.S.—doctor of dental surgery or of dental science

Ed.D.—doctor of education

Ed.M.—master of education

J.D.—doctor of jurisprudence

J.M.—master of jurisprudence

J.S.D.—doctor of the science of laws

LL.B.—bachelor of laws

LL.M.—master of laws

M.A.—master of arts

M.B.A.—master of business administration

M.D.—doctor of medicine

M.S.—master of science

Ph.D.—doctor of philosophy

Th.D.—doctor of theology

i. It is correct to abbreviate names of states, territories, and provinces (as in Canada) in addresses if the official abbreviation is used. The ZIP or other mailing code follows the abbreviation on the same line: Mobile, Alabama 36600 or Ontario, Canada LOS 1JO.

States

Alabama	AL
Alaska	AK
Arizona	AZ
Arkansas	AR
California	CA
Colorado	CO
Connecticut	CT
Delaware	DE
District of Columbia	DC
Florida	FL
Georgia	GA
Hawaii	HI
Idaho	ID
Illinois	IL
Indiana	IN
Iowa	IA
Kansas	KS
Kentucky	KY
Louisiana	LA

Maine	ME
Maryland	MD
Massachusetts	MA
Michigan	MI
Minnesota	MN
Mississippi	MS
Missouri	MO
Montana	MT
Nebraska	NE
Nevada	NV
New Hampshire	NH
New Jersey	NJ
New Mexico	NM
New York	NY
North Carolina	NC
North Dakota	ND
Ohio	OH
Oklahoma	OK
Oregon	OR
Pennsylvania	PA
Rhode Island	RI
South Carolina	SC
South Dakota	SD
Tennessee	TN
Texas	TX
Utah	UT
Vermont	VT
Virginia	VA
Washington	WA
West Virginia	WV
Wisconsin	WI
Wyoming	WY

Territories

Canal Zone	CZ
Puerto Rico	PR
Guam	GU
Virgin Islands	VI

Canadian Provinces

Alberta	AB
British Columbia	BC
Manitoba	MB
New Brunswick	NB

Newfoundland	NF
Northwest Territories	NT
Nova Scotia	NS
Ontario	ON
Prince Edward Island	PE
Quebec	PQ
Saskatchewan	SK
Yukon Territory	YT

2. Other Abbreviations and Acronyms in Workplace Communication

The following abbreviations are usually spelled out in the body of a business letter or in a formal report but are often abbreviated in less formal documents.

acct. or a/c	account
amt.	amount
anon.	anonymous, nameless
ARM	adjustable rate mortgage
ASAP	as soon as possible
assn. or assoc.	association
ATM	automatic teller machine
bal.	balance
b.l.,b/l, or B/L	bill of lading
BTW (e-mail only)	by the way
c/o	care of
ctn.	carton
cwt.	hundredweight
e.g.	for example
ESL	English as a second language
et al.	and others
FAQ (e-mail only)	frequently asked questions
frt.	freight
ft.	foot or feet
FYI (e-mail or memo)	for your information
g.	gram
gal.	gallon
GATT	General Agreement on Tariffs and Trade
GNP	gross national product
i.e.	that is
lb.	pound
LCL	less than a carload lot
mfr.	manufacturer
misc.	miscellaneous
NAFTA	North American Free Trade Agreement
p. pp.	page pages
pd.	paid
PLC	public limited company (British equivalent of U.S. corporation)
rec'd	received
viz.	namely
vs.	versus
yd.	yard

REPLAY ABBREVIATIONS

Answer **T** (true) or **F** (false) for Numbers 1–9.

_____F_____ **1.** To save time in preparing written business communications, abbreviate as much as possible.

_____F_____ **2.** It's inadvisable to abbreviate on invoices, receipts, or memos.

_____F_____ **3.** Periods are required in abbreviations of names of well-known organizations.

_____T_____ **4.** It's correct to abbreviate an academic degree that follows the person's name.

_____T_____ **5.** _Ave._ (for _Avenue_) should not be used in the address of a business letter.

_____F_____ **6.** If _Saint_ is the first part of a city's name, always spell it out.

_____F_____ **7.** Miss., Okla., Ca., and Mass. are examples of correct abbreviations for use in a business letter address.

_____T_____ **8.** A military title may be abbreviated if it precedes the full name of the person.

_____F_____ **9.** It's correct to abbreviate business terms like _acct._ for _account_ or _amt._ for _amount_ in business letters and reports.

10. What do the following abbreviations mean?

BTW	By the way	SEC	Securities and Exchange Commission
MIS	Management Information Systems	LL.B.	Bachelor of Laws
mph	miles per hour	CEO	Chief Executive Officer
CFP	Certified Financial Planner	CFO	Chief Financial Officer

Notes:

Recap and Replay Answers E

Chapter 1

Memo From the Wordsmith: Naval

Chapter 1

Replay 1
a. 18, b. 14, c. 15, d. 8, e. 11, f. 4, g. 7, h. 2, i. 10, j. 6, k. 9, l. 16, m. 12,
n. 13, o. 3, p. 19, q. 1, r. 20, s. 5, t. 17

Replay 2
a. 21, b. 22, c. 27, d. 30, e. 38, f. 23, g. 39, h. 34, i. 40, j. 31, k. 37, l. 36, m. 26,
n. 32, o. 29, p. 24, q. 28, r. 25, s. 33

Replay 3
a. 41, b. 44, c. 43, d. 56, e. 53, f. 57, g. 60, h. 52, i. 42, j. 59, k. 48, l. 50, m. 54,
n. 49, o. 45, p. 58, q. 51, r. 55, s. 47, t. 46

Replay 4
a. 61, b. 62, c. 73, d. 64, e. 63, f. 79, g. 70, h. 65, i. 67, j. 72, k. 74, l. 77, m. 76,
n. 71, o. 69, p. 66, q. 68, r. 75

Replay 5
a. 98, b. 88, c. 81, d. 87, e. 83, f. 86, g. 82, h. 99, i. 94, j. 101, k. 89, l. 100,
m. 92, n. 91, o. 95, p. 97, q. 93, r. 85, s. 84, t. 90

Replay 6

A.
1. amalgamation
2. depreciate
3. deficit
4. balance sheet
5. beneficiaries
6. euro
7. foreclosure
8. slander
9. libel
10. mortgage

B.
11. concierge
12. multinational
13. dividend
14. spyware
15. American Stock Exchange (or Amex)

Chapter 2

Replay 7
1. accept
2. except
3. add
4. ad
5. dying
6. dyeing
7. dissent
8. descent
9. bizarre
10. bazaar
11. coarse
12. course
13. site
14. cite
15. sight
16. counsel
17. Council
18. access
19. excess
20. affect
21. effect
22. capitol
23. capital
24. allot
25. a lot
26. principle
27. principal
28. elicit
29. illicit

Replay 8

1. imminent
2. eminent
3. regardless
4. eligible
5. illegible
6. devise
7. device
8. dessert
9. desert
10. desert
11. choose
12. chose
13. biographical
14. bibliography
15. defer
16. differ
17. conscience
18. conscious
19. besides
20. beside
21. envelop
22. envelope
23. fiscal
24. physical
25. complimented
26. complemented
27. guise
28. guys
29. reality
30. realty
31. whether
32. weather
33. rapt
34. wrapped

Replay 9

1. personal
2. lose
3. than
4. thorough, suite
5. Whether, suit
6. prosecute
7. that
8. whether, weather
9. proceeds, rye
10. We're, quite
11. proceed, morale, personnel
12. prospective, perks, wry, Personnel
13. morale, moral
14. Where, we're
15. perspective
16. respectfully
17. We're, whether, through
18. persecuted
19. Morale, personnel
20. proceed, suit
21. wry, lose
22. compliments
23. We're, where
24. Then, than
25. that, realty

Replay 10

1. apprise, appraise
2. bloc, block
3. canvass, canvas, block
4. everyday, every day
5. foreword, forward
6. have, halve, half, half
7. key, quay
8. lesson, lessen
9. Marquis, Marquise, marquee
10. navel, Naval
11. ode, owed
12. precede, proceed
13. rein, rain, reign
14. peak, pique, peekΩ
15. serge, surge
16. stationery, stationary
17. taught, taut
18. vise, vice
19. wary
20. wear and tear
21. waive, wave

Replay 11

1. eager
2. anxious
3. besides
4. beside
5. disinterested
6. uninterested
7. enthusiastic
8. enthusiasm
9. Explicit
10. implicit
11. per annumnon
12. per diem

13. inflammable	18. indigent	23. prerequisite perquisites	28. RSVP
14. nonflammable	19. emigrate	24. thought thorough	29. proceed
15. none	20. immigrate	25. through	30. proceeds
16. regardless	21. fewer	26. simple	31. ingenious
17. Indigenous	22. less	27. simplistic	32. ingenuous

Replay 12

A.

1. 2   **2.** 4   **3.** 3   **4.** 3   **5.** 2

B.

6. s and es   **7.** s   **8.** s   **9.** b e   **10.** h e

C.

11. T   **12.** F   **13.** T   **14.** F   **15.** F   **16.** F   **17.** F   **18.** T   **19.** F   **20.** F   **21.** T,
22. F   **23.** F   **24.** F   **25.** F

CHAPTER 3

Replay 13

1. Answers will vary.

2. Asia

3. Guide words show the first and last words of that page.

4. Answers will vary.

5. False

6. spelling, syllables, pronunciation, definitions, parts of speech

7. unabridged

8. pocket, college, encyclopedic, online

9. college, pocket or small electronic

10. thesaurus

11. etymology . . . . . brackets

12. appendixes . . . . . index . . . . . front matter

13. Answers will vary.

14. Answers will vary but should include everything after the main section Z words.

15. Longman's

Recap

1. F   2. F   3. F

Replay 14

A.

1. T   **2.** T   **3.** F   **4.** F   **5.** F   **6.** T   **7.** F   **8.** T   **9.** F

10. circled—but corrected here: pronunciation, separate, recommend, congratulate, pursue, bachelor

B.

1. b   **2.** c   **3.** a   **4.** a   **5.** c

C.

1. weird	6. recommend
2. accommodate	7. privilege
3. bachelor	8. persistent
4. congratulate	9. embarrassed
5. pronunciation	10. pursued

D.

1. accommodations	3. judgment	5. recurrence
2. indispensable	4. consensus	6. acknowledgment

Recap—Read 15

1. b<u>a</u>nan<u>a</u>   c<u>o</u>llect   eas<u>i</u>ly   gall<u>o</u>p   circ<u>u</u>s

Recap

1. electrical power generated in relatively small quantities
2. a protective cover for a CD
3. war in which computer systems damage or destroy enemy systems
4. obtaining legal evidence by examining computer networks and data
5. personal journal showing links to a Web site

Replay 15

1. x and s
2. three
3. first
4. 3rd
5. 2nd
6. verb
7. 2nd
8. matchless, unsurpassed, etc.
9. Answers vary depending on the dictionary.
10. false
11. Answers vary depending on the dictionary.
12. A secondary school that prepares students for attending a university
13. first
14. subt'le ra'tion al' in'fra.struc'ture
15. catalogue catalog noun and verb

Recap—Read 16   1, 2, and 3. Answers will vary. 4. Earl of Sandwich
Recap   Answers will vary.
Recap   Answers will vary for 1 and 2; 3 examples: lefty, dandy, stoned

Replay 16

1. inactive, lazy, useless, futile
2. slang
3. queer or eccentric person, practical joke, hoax, to make fun of
4. ap.pen.dage—adjunct, addition, accessory

5. Individual Retirement Account, also Irish Republican Army
6. 1822
7. 1606
8. hon 'r able
9. to be indecisive
10. Janus
11. the study of handwriting
12. proofreader's mark meaning "let it stay, don't change it"
13. 1959
14. Answers will vary
15. thyroid gland

Chapter 4

Recap—Read 17

1. porch
2. baby
3. day
4. interviewers, applicants, energy, competence, loyalty, skill, ambition, flexibility
5. Answers will vary; for example, window, sofa, carpet
6. Answers will vary; for example, Seymour—principal, pilot, pal

Replay 17

A.

1. smile, situation
2. correspondence, United States, Asia, South America, Europe, Africa
3. career, people, cultures, individuals
4. slang, co-workers, English, language
5. neighborhood, classroom, workplace, United States, Canada, Great Britain, parts, world

B.

1. It   2. He   3. Who   4. She, her   5. They, her, him

C.

1. His, hers
2. He, it, she
3. She, herself, these, them
4. Somebody, Everyone, who, anyone, it
5. We, nothing, our, we, everything

D.

1. I, me, my, mine, myself
2. we, us, our, ours, ourselves
3. you, your, yours, yourself, yourselves
4. she, her, herself, hers
5. he, him, his, himself
6. it, its, itself
7. they, their, them, theirs, themselves

Recap—Read 18

action verbs: wrote, proofread, received, looked, read, invited, has, helped   pronouns: it, it, she, it, she, she, she, who, her

Recap

is, appears, am, sounds, were, seems, was, smells, feel, remains

Replay 18

A.

being verbs: is, appears, am, are, were, was, sounds, feels, be   action verb: think

B.

helping and main verbs: do dance, haven't met, may read, are reading, has danced, will be going, will have been, might have danced, does sign

C.

1. seek, have   **2.** spend   **3.** are asking   **4.** is   **5.** spend

D.

Verbs

1. are judged	**11.** have been sent
2. is	**12.** know, will represent
3. has	**13.** rescheduled
4. do call, are, is	**14.** should deliver
5. was	**15.** is
6. were keyboarding	**16.** will mail
7. should have completed	**17.** received
8. should go	**18.** checked
9. is known	**19.** built
10. will ask	**20.** dressed

Nouns

1. Applicants, behavior, knowledge	**11.** Southeastern Community College
2. Etiquette, part, activities, job	**12.** lawyer, building
3. data, topic	**13.** Barbra Streisand, performances, Anaheim
4. clients, colleagues, names, organization	**14.** Larry, tickets
5. Sheela Danielle, winner, scholarship	**15.** drawer, computer
6. secretaries, answers, blanks	**16.** receptionist, Monday
7. job, Tuesday	**17.** Aldrich Company
8. conference, Las Vegas	**18.** auditors, books, accuracy
9. husband, manager, courtesy	**19.** Frankenstein, factory, city, moon
10. mail	**20.** children

Pronouns

1. their, their	**7.** Someone
3. one	**8.** Who
4. their, you, it, your	**9.** his

10. He

11. These

12. They, who, me

14. you

15. It

16. Their, these

17. everything

18. Our, everyone's

20. themselves

Recap—Read 19

1. idea, music, dresses, potatoes

2. You, Ethics, Ms. Parks

3. paper, blossom, bird, nun

Recap

ADVERBS: quietly, exceptionally, intelligently, attractively, finally, really

Replay 19

A.

NOUNS: rooms, air-conditioning, windows, office, homes, street, tri-level, attorneys, firm, year

LIMITING ADJECTIVES: any, several, some, Ten

POINTING ADJECTIVES: These, That, this, this, this

ARTICLE: the

B.

1. an

2. an exclusive resort

3. the beautiful

4. the crystal blue

5. the white sandy

6. exquisite guest

7. Mediterranean

8. luxurious

9. The smogless, sunny, mild

10. this

C. Answers will vary.

1. Some people in that room have a good attitude.

2. This morning I found two dimes near a red phone.

3. These companies bought a new computer for every manager.

D. Possible Answers

1. yesterday, today, immediately

2. well, poorly, quickly

3. carefully, recklessly, fast

4. correctly, accurately, slowly

5. always, never

6. home, away, there

7. more, less

8. rarely, seldom, never

9. really, carefully, never

10. elegantly, tastefully

11. extremely, very, such

12. most, least

13. not, very, sometimes

14. usually, often, always

15. much, even

Recap—Read 20   and, but, or, nor, for, yet, so

Recap   Dependent conjunctions—if, because; coordinating conjunctions—but, and, or; conjunctive adverb—therefore

Recap

1. you, plane, village, mountain
2. *Hid* is a verb and *newly* is an adverb. A prepositional phrase begins with a preposition, ends with a noun or pronoun, and never includes a verb.

Replay 20

A.

1. nor   2. but   3. and, when   4. although   5. Since, while
6. and   7. until   8. If   9. because   10. or

B.

1. into   2. by   3. to   4. through
5. over   6. across   7. below   8. under

C.

1. **in** the workplace, **of** kindness and consideration, **toward** other employees and customers
2. **on** the behavior, **of** people, **on** their lifestyles
3. **between** a dream and a goal
4. **from** your office, **to** my home
5. **at** producing spot announcements **for** their customers

Replay 21

A.

1. verb   2. adjective   3. noun

B.

1. noun   2. verb   3. adjective

C. Possible Answers

1. Ozzie is reading a report from the president.
2. Reading is my favorite activity.
3. She needs new reading glasses.

D. Possible Answers

1. Let's plant roses and gardenias this year.
2. In our Seattle plant, we hire many environmentalists.

E.

1. adjectives
2. Young children are probably better off in the country than in the city.

Replay 22

A.

1. pronoun   2. verb   3. preposition   4. adjective (article)   5. conjunction
6. adjective   7. noun   8. conjunction   9. verb

B.

1. believe   2. conduct   3 paid   4. had driven   5. were   6. existed

7. drove   8. We   9. your   10. we   11. we   12. we

13. it   14. We   15. your

C.

1. for this adjustment   5. at the time

2. of the exclusion provisions   6. of purchase

3. on page 3   7. without charge

4. of the warranty

CHAPTER 5

Recap   crops, Nassau, diary, customs

Recap—Read 23   Globetrotters, Magicians, skills, Trotters, years

Replay 23

A.

1. allies   6. injuries

2. accessories   7. moneys or monies

3. itineraries   8. authorities

4. proxies   9. ferries

5. facilities   10. surveys

B.

1. tattoos   6. altos

2. dominos, dominoes   7. potatoes

3. cargoes, cargos   8. zeros, zeroes

4. pianos   9. portfolios

5. heroes   10. mementos, mementoes

C.

1. thieves   6. halves

2. handkerchiefs   7. safes

3. knives   8. wolves

4. tariffs   9. plaintiffs

5. wives   10. chiefs

D.

tours, States, Globetrotters, fans, games, players, audiences, routines

Replay 24

A.

1. corps   3. deer   5. series

2. economics   4. Larrys   6. Chinese

7. Joneses   **9.** fish (or fishes)   **11.** feet

8. aircraft   **10.** stepchildren   **12.** Floreses

B.

1. P   **2.** S/P   **3.** S/P   **4.** S/P   **5.** S   **6.** S/P

C.

1. have   **6.** are

2. is   **7.** is

3. are   **8.** was

4. were   **9.** is

5. are   **10.** were

Replay 25

A.

1. formulas, formulae   **8.** crocuses, croci

2. alumni   **9.** appendixes, appendices

3. bases   **10.** concertos, concerti

4. censuses   **11.** indexes, indices

5. criteria, criterions   **12.** analyses

6. axes   **13.** media, mediums

7. parentheses   **14.** diagnoses

B.

1. S   **2.** P   **3.** S   **4.** P or S   **5.** P   **6.** P

C.

1. medium   **2.** vertebrae   **3.** criteria   **4.** parenthesis   **5.** alumni

Recap—Read 26   brother-in-law, stock car, trade-ins, price tags, letterhead

Replay 26

A.

1. follow-up, follow-ups   **6.** spaceflight, spaceflights

2. textbook, textbooks   **7.** headhunter, headhunters

3. trade-in, trade-ins   **8.** bush league, bush leagues

4. editor in chief, editors in chief   **9.** chief of staff, chiefs of staff

5. runner-up, runners-up   **10.** volleyball, volleyballs

B.

See Appendix D for Read and Replay Capitalization

C.

1. general manager, Telephone, tostadas, tacos   **5.** President

2. business teachers   **6.** Catholic

3. C   **7.** C

4. Spanish   **8.** yeast, summer, City

9. city, state

10. C

11. Anthropology, colleges

12. compan

13. C

14. C

15. Academy Award, Nobel Prize

Replay 27

1. Society or humanity instead of mankind

2. Flight attendant instead of stewardess. Delete second sentence of No. 2.

3. Change *girls* to *women*

4. Change *chairman* to *chairperson*

5. Change *wives* to *spouses*

6. C

7. Change *lady policemen* to *police officers*

8. End this sentence after *intelligent.*

9. Change *cameramen* to *photographers*

10. Change *girls* to *women*

11. C

12. Delete first part of sentence. Begin with *He.*

13. Change *man-sized* to *excellent*

14. Delete *male*

15. Delete entire sentence.

16. Delete entire sentence.

17. Change *manmade* to *synthetic*

18. Change *man* to *agent*

Replay 28

1. businesspersons

2. plaintiff

3. software

4. premises

5. editors in chief

6. cargo

7. corps

8. portfolio

9. proxy

10. proceeds

11. itinerary

12. memento

13. notaries public

14. chassis

15. write-offs

CHAPTER 6

Recap—Read 29 Possible Answers

1. I love you.   I am in love with him.

2. He loves her.   We are in love with it.

3. She loves him.   He is in love with her.

4. Who loves them?   They are in love with whom?

Replay 29

1. she and I

2. me

3. They

4. We

5. I

6. us

7. me

8. her

9. we	14. him	19. her	24. me
10. he	15. He and I	20. them	25. they
11. me	16. me	21. me	26. me
12. us	17. he	22. them	
13. them	18. We	23. me	

Recap—Read 30

1. she   **2.** he   **3.** do   **4.** he likes

Replay 30

A.

1. does   **2.** he loves   **3.** does   **4.** you know   **5.** can

B.

1. C	**2.** themselves	**3.** he can	**4.** he	**5.** me
6. we (do)	**7.** I leave (or I do)	**8.** C	**9.** themselves	**10.** you
11. I	**12.** C	**13.** C	**14.** C	**15.** C
16. themselves	**17.** C	**18.** himself	**19.** myself	**20.** C

Replay 31

1. Everybody's	**6.** Yours	**11.** Everybody's
2. Nobody's	**7.** hers	**12.** No one's
3. its, it's	**8.** mine, theirs	**13.** anyone's
4. C	**9.** You're	**14.** Who's
5. ours, yours	**10.** you're, your	**15.** Whose

Recap—Read 32

1. whom	**3.** whomever	**5.** whoever	**7.** whoever	**9.** whom
2. who	**4.** whoever	**6.** whom	**8.** Whom	**10.** whom

Recap   Who

Replay 32

1. whoever	**6.** Whom	**11.** whom	**16.** whomever	**21.** whomever
2. whoever	**7.** Who	**12.** who	**17.** who	**22.** whom
3. who	**8.** Whom	**13.** whom	**18.** who	**23.** who
4. who	**9.** Whoever	**14.** whom	**19.** whom	**24.** whoever
5. Whom	**10.** whomever	**15.** who	**20.** whoever	**25.** who

Recap—Read 33

1. No one   **2.** Every one   **3.** *his* instead of *their*   **4.** someone

Replay 33 Corrections

A.

1. Members of this department (or All department members) should be sure their nouns and pronouns agree in number in their written communications.

2. All the mechanics finished their work quickly. OR Each mechanic finished the work quickly.

3. Every child in the class needs a book OR All the children in the class need their own book.

4. All applicants should write their name in the blank. OR Each applicant should write his or her name in the blank.

B. Correct Forms

1. Anybody	5. Everybody	9. their
2. Every one	6. someone	10. Every body, its
3. No one	7. Any one	11. its
4. anyone	8. no body	12. their

C.

1. c/b   2. a/d   3. b/d   4. b/c   5. d/b

Recap—Read 34   1. its   2. its   3. their

Replay 34

A. Possible Answers

1. group   2. class   3. jury   4. team   5. staff   6. committee

B.

1. plural	6. its	11. its	16. he
2. singular	7. its	12. its	17. I
3. subject	8. I	13. the janitorial service	18. they
4. its	9. its	14. trash shouldn't be thrown	19. I
5. it is	10. its	15. she	20. us clerks

CHAPTER 7

Replay 35

A.

1. works	6. are	11. wants	16. will (or should) consider
2. needs	7. looks	12. wanted	17. is considering
3. moved	8. climbed	13. will (or would) want	18. stay
4. are sailing	9. will/should find	14. influences	19. watched
5. waxed	10. want	15. selected	20. discussed

B.

1. is	3. need	5. is	7. flow	9. is
2. is	4. are	6. are	8. knows	10. being

Replay 36

1. broken, worn
2. began (or had just begun), rang
3. chooses
4. chosen
5. does, stands
6. risen
7. have eaten
8. saw, did

9. delete *had* or change *flew* to *flown*   15. taken
10. quit   16. wear
11. run, broken   17. saw
12. worn or omit *had*   18. doesn't
13. stayed   19. swung
14. saw, spoken   20. gave

Replay 37

A.

1. being, was, been
2. biting, bit, bitten
3. blowing, blew, blown
4. coming, came, come
5. costing, cost, cost
6. falling, fell, fallen
7. forgetting, forgot, forgotten
8. freezing, froze, frozen
9. hiding, hid, hidden
10. leading, led, led
11. paying, paid, paid
12. shaking, shook, shaken
13. sinking, sank, sunk or sunken
14. singing, sang, sung
15. throwing, threw, thrown

B.

1. beaten, winning   2. paid   3. broken, hidden   4. hung, forgotten   5. stayed, written

Recap—Read 38

1. you, (you "understood") *finish, leave*
2. Everyone . . . *was discharged.*
3. Lewis and Martin *told*

Replay 38

SUBJECTS	VERBS (helping and main verb):
1. you	do enjoy
2. analysts	are doing
3. she	will get
4. clothing, you	select, should be
5. you, appearance, that	ask, sends, will benefit
6. women	should limit
7. diner, he, she	puts, cuts
8. knife, you	remains, dine
9. advancement, who	is, get
10. work	turns
11. Playing	is
12. sales	have risen
13. Cleveland, women	said, do . . . want
14. Everyone	is working
15. Xerox	manufactured, sold
16. assistant, who	is, will help
17. you	would like
18. turnover	is
19. poverty, riches	are
20. grammar, spelling	are

Recap—Read 39   Change *are* to *is*.

Replay 39

1. Change *were* to *was*.   **2.** Change *seem* to *seems*.   **3.** Change *was* to *were*.   **4.** C
5. Change *were* to *was*.   **6.** C   **7.** C   **8.** Change *are* to *is*.   **9.** Change *are* to *is*.
10. Change *have* to *has*.   **11.** Change *is* to *are*.   **12.** Change *rides* to *ride*.   **13.** C
14. C   **15.** C   **16.** Change *have* to *has*.   **17.** Change *has* to *have*.   **18.** Change *are* to *is*.   **19.** Change *don't* to *doesn't*   **20.** Change *was* to *were*.   **21.** Change *is* to *are*.
22. Change *are* to *is*.   **23.** Change *have* to *has*.   **24.** Change *greets* to *greet*.   **25.** Change *have* to *has*.   **26.** Change *pack* to *packs*.   **27.** C   **28.** Change *We'd* to *We would*.
29. Change *leaves* to *leave*.   **30.** Change *do* to *does*.   **31.** Change *don't* to *doesn't*
32. Change *work* to *works*.   **33.** Change *are* to *is*.   **34.** Change *goes* to *go*.   **35.** Change *have* to *has*.

Replay 40

1. Change *was* to *were*.   **6.** C
2. C   **7.** Change *was* to *were*.
3. Change *was* to *were*.   **8.** Change *were* to *was* and *weren't* to *wasn't*.
4. C   **9.** Change *was* to *were*.
5. Change *was* to *were*.   **10.** Change *is* to *were*.

Replay 41

1. Change *takes* to *take*.   **6.** Change *was* to *were*.
2. C   **7.** Change *favor* to *favors*.
3. Change *have their* to *has its*   **8.** C
4. Change *need* to *needs*.   **9.** Change *are* to *is* and *their* to *its*.
5. Change *was* to *were*.   **10.** Change *work* to *works*.

CHAPTER 8

Replay 42   Possible Answers

1. that kind   **6.** delete *here*
2. Those or These   **7.** delete *there*
3. types   **8.** Ask *these* or *those* people . . . .
4. sorts   **9.** delete *there*
5. those types or that type of person   **10.** delete *an*

Recap—Read 43

1. an	**5.** a	**9.** a	**13.** an	**17.** an	**21.** a
2. a	**6.** an	**10.** a	**14.** an	**18.** a	**22.** a
3. an	**7.** a	**11.** a	**15.** an	**19.** an	**23.** an
4. an	**8.** an	**12.** an	**16.** an	**20.** a	**24.** an

Replay 43

1. a, an   **3.** an, a, an   **5.** A, an
2. A, an, an   **4.** a, an   **6.** An, an

7. A, a, a   **10.** a, a, an   **13.** An, an

8. A, a, an   **11.** An, a   **14.** an, a

9. a, a, an   **12.** an, an   **15.** A, a, an

Replay 44

1. rarely or hardly ever   **6.** doesn't, anything   **11.** isn't ever

2. anything, anybody   **7.** anywhere   **12.** could

3. any   **8.** any   **13.** any

4. any   **9.** C   **14.** Nobody dislikes or Everybody likes

5. C   **10.** anywhere   **15.** I will never use double negatives again.

Recap—Read 45   wisest, wiser

Replay 45

A.

COMPARATIVE	SUPERLATIVE
1. farther/further	farthest/furthest
2. worse	worst
3. littler or less	littlest or least
4. more	most
5. better	best

B.

1. delete *most*   **6.** delete *more*

2. change *worst* to *worse*; delete *the worst*   **7.** more recent

3. C   **8.** change *brightest* to *brighter*

4. change *older* to *oldest*   **9.** delete *more*

5. change *best* to *better*   **10.** change *biggest* to *bigger*

Recap—Read 46   smoothly

Recap

1. more clearly   **2.** writes   **3.** adverb

Replay 46

1. bad   **2.** clearly/correctly   **3.** sad   **4.** carefully   **5.** legibly   **6.** C   **7.** deeply
8. fairly   **9.** efficiently   **10.** gracefully   **11.** bad   **12.** C   **13.** sweet   **14.** quietly
15. more quietly   **16.** more smoothly   **17.** satisfactorily   **18.** calm   **19.** C
20. louder   **21.** worse   **22.** more quietly   **23.** most capable   **24.** more concisely
25. older

Recap—Read 47

1. well   **2.** well   **3.** either good or well   **4.** well   **5.** bad

Recap

1. very or extremely   **2.** really or especially

Replay 47

1. really, very   **3.** logically   **5.** better

2. surely, certainly   **4.** better   **6.** more widely

7. delete *more*	10. C	13. well, an
8. really well	11. differently	14. kinds, an
9. most poorly	12. bad	15. C

Checkpoint—Word to the Wise

1. It's darned well   2. The dog smells badly

CHAPTER 9

Word Power Answer: 1

Recap

1. minutes' planning, hour's work, New Year's Day, George's mother, Twileen's father
2. "Seward's Folly," foolishness, critics', Seward's foolishness

Word to the Wise

1. noun   2. verb

Replay 48

1. C	11. crew's strength
2. editor's *stories*	12. California's mines, groves
3. C	13. nation's wine, raisins
4. law's manager	14. world's manufacturers
5. C	15. Men's College
6. attorneys' offices	16. C
7. Dakota's resources	17. Tom's book
8. Men's women's clothes	18. hours' work
9. Ms. Lopez's orders	19. Penney's success
10. industry's directors	20. Barbie's fame Mattel's success

Recap—Read 49

1. Smith's Perkins' City's week's
2. law's son's

Recap

1. brothers' Martinezes   2. Women's women's men's   3. daughters-in-law's

Replay 49

A.

1. representative's	representatives	representatives'
2. week's	weeks	weeks'
3. witness's	witnesses	witnesses'
4. James's	Jameses	Jameses'
5. country's	countries	countries'
6. Filipino's	Filipinos	Filipinos'
7. man's	men	men's
8. Asian's	Asians	Asians'

9. wife's	wives	wives'
10. father-in-law's	fathers-in-law	fathers-in-law's
11. congresswoman's	congresswomen	congresswomen's
12. family's	families	families'
13. Webster's	Websters	Websters'
14. hour's	hours	hours'
15. Wolf's	Wolfs	Wolfs'
16. wolf's	wolves	wolves'
17. organization's	organizations	organizations'
18. boss's	bosses	bosses'
19. woman's	women	women's
20. child's	children	children's

B.

1. C	**6.** Childress's, days	**11.** coaches', players'
2. years'	**7.** ladies' coats	**12.** Keats's
3. person's	**8.** Goldsteins, days	**13.** Hendrix's
4. Men's fashions women's	**9.** Brunswick's, years	**14.** brothers' films
5. minute's	**10.** Jenkins'	**15.** guests' names

Replay 50

1. couldn't, o'clock	**6.** Don't	**11.** I's, they'll, t's
2. i's, t's	**7.** C	**12.** Won't
3. C	**8.** it's	**13.** A's
4. C	**9.** workers	**14.** women's
5. CFO's 2007	**10.** C	**15.** Couldn't

CHAPTER 10

Recap—Read 51

1. We	**5.** They
2. Ms. Hirsch	**6.** companies
3. You	**7.** (understood you) (understood you)
4. crime, punishment	**8.** he, Eric Smith

Recap   communication, is, to the nation's productivity

Replay 51

A.

2. from Ergonomic, Inc.   **3.** of way   **4.** NONE   **5.** at the annual meeting . . . . in Omaha
6. with honesty and social responsibility   **7.** NONE   **8.** NONE

B. Subjects/Verbs

1. Steam Gene	could have shot
2. You	must be

3. winner	sees
4. loser	sees
5. books	were put
6. you/you	earn/learn
7. members	have
8. visitors	arrived
9. (You)	give
10. Professor Friede	found

C. Independent

—Nos. 1, 4, 6, 8, 10, 11

Dependent Conjunction

2. —because   **3.** although   **5.** who   **7.** when   **9.** if   **12.** after

Sentences

1. She couldn't arrange the reports in chronological order.
2. Many organizations have similar problems with employees.
3. An accountant needs excellent communication skills.
4. In the workplace you may converse with many people.
5. They have limited English skill.
6. Don't laugh at someone's pronunciation or grammar error.
7. You transport something by car.
8. It's called a shipment.
9. However, you transport it by ship.
10. It's called cargo.
11. Isn't that strange?
12. You've learned a few phrases in other languages.

Recap—Read 52

1. She dines at the college cafe.
2. The beans taste like caviar.

Recap

1. the beans taste like caviar.
2. The beans taste like caviar when she dines at the college cafe.

Replay 52

A.

1. F	**3.** F	**5.** S	**7.** F	**9.** F
2. S	**4.** S	**6.** F	**8.** S	**10.** S

B.

1. If you do a good job on the Mendocino project, you will get a salary increase after the first of the month.
2. Some errors may never be found although most corporations use the services of an auditor to examine the books.

3. Don't ever think that you know it all.

4. Because sexual harassment is illegal and immoral, Twileen filed a complaint.

5. When Mr. Lopez became general manager, he asked the staff to greet Spanish-speaking customers in Spanish.

C.

1. F	**3.** F	**5.** S	**7.** S	**9.** S
2. F	**4.** S	**6.** F	**8.** F	**10.** S

D. Answers will vary.

1. The man whom we met yesterday in Fullerton is the treasurer of the Cambridge Corporation.

2. Our Human Resources Department is on the third floor.

3. Twileen, believing she was right, took the matter to the Human Resources Department.

6. George, having been on vacation last week, was shocked by the news.

8. The team that won all the games in Oklahoma City last year will be hard to beat.

Replay 53

1. R	**6.** R	**11.** C	**16.** C
2. C	**7.** C	**12.** CS	**17.** C
3. CS	**8.** CS	**13.** CS	**18.** C
4. C	**9.** C	**14.** CS	**19.** C
5. R	**10.** C	**15.** C	**20.** C

Recap—Read 54

1. Don't fill a business letter with long sentences or with words of many syllables, for it doesn't impress anyone.

2. Don't fill a business letter with long sentences or with words of many syllables; it doesn't impress anyone.

3. Don't fill a business letter with long sentences or with words of many syllables. It doesn't impress anyone.

Replay 54

A.

1. CS	**2.** CS	**3.** R	**4.** C	**5.** R	**6.** C	**7.** CS

B.

1. R—quickly since he	**5.** R—conference, but maybe
2. C	**6.** CS bones; words; or insert but
3. R—floor, but	**7.** C
4. CS once, and you	**8.** C

C.

1. however, therefore, for example, also, yet, then (answers vary)

2. a **3.** b

D.

1. C   **2.** today; then   **3.** communication; however,   **4.** rehearsals; that is,   **5.** C
6. sleeps; consequently,   **7.** C

Replay 55
A.

1. R when   **2.** C which, when   **3.** R work; they   **4.** C Although   **5.** C
although   **6.** CS than . . . person. They   **7.** R than   **8.** C because   **9.** C Because
10. C
11. CS performers;   **12.** C   **13.** R questions; also   **14.** CS interview; make
15. C that   **16.** C   **17.** CS tasks; begin   **18.** C yet   **19.** CS plans; it   **20.** C as

B.

1. C   **2.** since   **3.** Although   **4.** When   **5.** because

Replay 56

Dear Professor Head

Thank you for the time and courtesy you extended to our representative, Laura Mann, at your college last month. She enjoyed her visit with you.

At Laura's request we have sent you the new edition of *Mathematics of Business.* This was sent to you several weeks ago, and you should have it by now. We do hope you'll look it over carefully. In addition your name has been placed on our mailing list to receive an examination copy of *Business Math: Practical Applications.* A new edition of this book by Cleaves, Hobbs, and Dudenhef is expected off the press sometime next month.

We'll send your copy just as soon as it is available. If there is any way we can be of help to you, Professor Head, please let us know. Best wishes for a happy holiday season.

Cordially,

Hal Balmer
Vice President

lrs

PS Don't write comma splices or run-ons. They are hard to read.

CHAPTER 11

Replay 57

1. . . . videos, Web sites, books, and personal advisers . . . .

2. C

3. Buicks, Fords, and Hondas . . . Jaguars, Lincolns, or Audis

4. Taiwan, Hong Kong, Singapore, and Korea . . . .

5. . . . report, . . . accuracy, and . . . .

6. Major . . . language, religion, politics, and . . . .

7. . . . in, . . . at, and . . . .

8. When . . . lunch, . . . appropriately, . . . manners, . . . eat, . . . beverages.

9. The job . . . . company, ... them, and . . . . possibilities.

10. Executive . . . meet, greet, and eat . . . . religions

Recap—Read 58

1. . . . intelligent, loyal . . . .

2. C

Recap

1. C

2. . . . highly paid, famous . . . .

Replay 58

1. A dazed, demure . . . disheveled, double-dealing . . . .

2. . . . unique, exciting . . . .

3. . . . why you came, what you wanted, and what you expect to do.

4. Our showroom is in a small, elegant . . . .

5. C

6. . . . Texas City, Dallas, and Austin.

7. . . . profitable, highly . . . .

8. . . . bonus, . . . salary, and stock options.

9. Use *nd, rd, st, or th* . . . . itself; for example, the 4th of May, but May 4.

10. . . . settled, well educated . . . .

Recap—Read 59

1. . . . Pam's work, and we . . . .   **3.** . . . like sports, but she . . . .

2. . . . Rosalyn's work . . . .   **4.** C

Replay 59

1. C   **3.** C   **5.** today, for   **7.** C   **9.** rich, but

2. C   **4.** dance, but   **6.** job, nor   **8.** C   **10.** available, and

Recap—Read 60

1. discrete, she . . . Hideaway, Rosie's retreat, or Larry's Lair

2. C

3. When George went to his car, he. . . .

4. . . . to meet her, he didn't . . . .

5. C

6. . . . note, . . . Hernando's Hideaway.

7. Twileen, we . . . .

Replay 60

1. . . . Flintridge Inn, he spotted . . . .

2. As he approached her, he . . . .

3. In this attractive, . . . room, ...

4. As . . . know, real . . . .

5. . . . igloo, kayak, moccasin, skunk, and persimmon.

6. C

7. . . . Twileen, ...

8. . . . world's . . . .

9. Being . . . salesperson, he . . . . prospect's

10. No, ...

11. C

12. If it is to be, it is up to me.

13. C

14. . . . the aviary, but . . . Cucamonga.

15. C

Replay 61

1. On March 9, 2004, ...   **2.** C   **3.** . . . Smith, PhD., ...   **4.** . . . Long Beach, New York, ...   **5.** . . . Friday, January 4, 8:30 a.m., in my office.   **6.** . . . new shop, and . . . .
7. Little Rock, Arkansas, was. . . .   **8.** . . . Stores, Inc., . . . Yonkers, New York, store . . . .
9. Edinburgh, Scotland, ...   **10.** C

Replay 62

A.

1. Jeff told the students, "Use last names in the workplace unless you're sure first names are appropriate."

2. "I've e-mailed the price list he requested," said Ms. Kato.

3. The instructor said, "Beginning some sentences with 'I' is all right," and then gave item 2 as an example.

4. "Love," Jesse said, "is often combined with heartache."

5. "I'm married!" shouted Jeff.

B.

1. Use commas for direct address. Use commas to separate the parts of an address that are in sentence form. Do not use a comma between state and ZIP.

2. Use a comma before a coordinate conjunction that joins independent clauses.

3. (1) Enclose elements of dates within commas. (2) In the U.S. style date, enclose the year in commas when it follows the day. (3) A comma may be used in four-digit figures if the number refers to quantity.

4. (1) Use a comma before and after direct address. (2) Use a comma after an introductory expression that includes a verb.

5. (1) When needed for clearness, use a comma after an introductory phrase. (2) Insert commas between items of a series.

Recap—Read 63

1. Home, where I learned to walk and talk, holds . . . .

2. . . . parts, "Job Content" and "Work Environment."

3. . . . want, therefore, may . . . .

4. . . . system, which . . . Association, ...

5. . . . courthouse, . . . wife's injury.

Word to the Wise

Brad thought Ron was very generous.

Recap

1. c   **2.** delete comma   **3.** delete all commas

Replay 63

1. stockholders, . . . family,   **2.** C   **3.** C   **4.** C   **5.** C   **6.** C   **7.** . . . Internet, . . . communications, ...   **8.** loudest, however, ...   **9.** costs, . . . know,
10. instructor, John Spinell, who teaches English, ...   **11.** . . . office, Room No. 103, ...
12. Office, which is on the third floor, ...   **13.** . . . Simon, the auditor, . . . $100,000 error.
14. Endive, a plant . . . leaves, is. . . .   **15.** C   **16.** C   **17.** . . . worth, ...   **18.** C
19. booklet, which . . . information,   **20.** C   **21.** . . . idea, it has been said, ...
22. . . . bought, *Complete Business Etiquette Handbook*, ...   **23.** . . . Royce, . . . secret, . . . long.   **24.** C   **25.** . . . printer, ... at Dewey, Cheatem, & Howe, doesn't work.

Replay 64

1. 10 p.m.,   **2.** George,   **3.** brother,   **4.** C   **5.** system, . . . Inc.,   **6.** C
7. morning,   **8.** October, . . . clerk; . . . November, a Web site manager.   **9.** C   **10.** C
11. C

Word Power 1

Replay 65

1. . . . software.   **2.** good, . . . where's . . . pepperoni?   **3.** us.   **4.** . . . fell. . . . him.
5. . . . decisions. We. . . .   **6.** is?   **7.** you?   **8.** now!   **9.** . . . lunch.
10. Wonderful!

Word Power

Woman! Without her, man would be uncivilized.

Checkpoint—Word Power

misogynist

philogynist

CHAPTER 12

Recap—Read 66

1. more; now

2. high, your

3. money; fold . . . half,

4. C

5. employer, Anton, was . . . criticism; and

6. Banks, he explained, . . . accounts; but

7. speed; however, the

8. C

9. mousetrap; then

10. limited, their

Replay 66

A.

1. acronym; that is, it's . . . Powers, Europe.   **2.** England's . . . hero, Lord Nelson,
3. . . . show, Mr. Ho, . . . you;   **4.** you;   **5.** . . . you've . . . someone's   **6.** Fuller,
president; Hall, vice president; Hecht,   **7.** properly; it's isn't   **8.** ambition; namely, Willy's

9. don't . . . about; for example, **10.** University's . . . department, **11.** C **12.** understand; **13.** wealth, **14.** applicants, . . . good; however, **15.** Track, Inc., . . . dividend; . . . is, . . . company's **16.** . . . history, **17.** C **18.** harder, **19.** believe, therefore, . . . you're **20.** isn't . . . kills, it's

B.

Charles the First walked and talked; half an hour after, his head was cut off.

Replay 67

A.

1. following: two
2. C
3. life: revenge
4. life: It goes on.—Robert Frost
5. cities: Winston Salem, Bisbee, Honolulu, and

6. shoplifting: Place
7. community: Nearby
8. about: A
9. duty: they
10. C

B.

interested. degree, bachelor's. qualifications: . . . Word, Excel, communication, relations. The . . . applicant's . . . experience.

Replay 68

A.

1. He shouted, "Your house is on fire!"
2. "Your house is on fire!" he shouted.
3. Twileen whispered, "Are you sure you love me?"
4. "Are you positive you love me?" she whispered again.
5. Do you know whether Jesse said, "I love you"?
6. Germany's Helmut Kohl has been a strong force behind the euro, the universal currency for many European countries.
7. Jim Rogers wrote, however, that many European politicians had been eager to establish the new currency.
8. What does the phrase negotiable instrument mean? (or italics)
9. He faxed us as follows: "We depart from O'Hare on Unix Flight 23 at 8 a.m. and arrive at Kennedy at 10:30 p.m."
10. Was it Mr. Higgins, the character in *My Fair Lady*, who said, "Results are what count"?
11. "You need more tacos for the company party!" exclaimed Jim angrily.
12. "This shipment of boys' wear," the manager said, "will arrive in time for your January sale."
13. "Are you all right?" asked Ann. "Yes," groaned Dad as he lifted *Merriam Webster's Unabridged Dictionary.*
14. "Bach's Suite No. 2 in B Minor" is first on the program at Laredo.
15. University of South Carolina Professor Benjamin Franklin is tired of people asking him, "Why aren't you out flying your kite?"

B.

When the New York YWCA offered typing instruction for women in the late 1800s, the school's managers were called "well-meaning but misguided ladies." The female mind and constitution were considered "too frail" to survive a six months' typing course. Upon completion of the course, the girls still faced strong opposition. The business world was a man's world: Men spoke "male" language, smoked strong "male cigars," and ignored the niceties.

Recap—Read 69

1. brother-in-law	5. off-limits
2. vice president	6. de-escalate
3. semisweet	7. ex-boyfriend
4. up-to-date	8. rediscover

Recap

1. back-to-school, Windows-based software
2. C
3. part-time job, never-to-be-forgotten
4. seven-foot-tall basketball player, problem-solving

Replay 69

A.

1. I work in a 100-story building.
2. Do you need a 10- or a 20-foot ladder to do the repairs at Golden West College?
3. My father is a hard-working man.
4. The case against the Dallas-based company was handled in Seattle.

B.

5. re-creation   6. C   7. C   8. C   9. father-in-law, commander in chief   10. self-made
11. self-sufficient   12. C   13. twenty-one   14. Johnny-come-lately   15. anti-intellectual

C.

func/tion be/lieve hori/zontal wouldn't/ thou/sands punctu/ation aligned/ impos/sible
inter/rupt syl/lables stopped/ guess/work

Replay 70

1. representative's, assistant's	6. Germany's, America's
2. Hill's	7. weekend's, Teachers'
3. women's, girls'	8. wife's, New York's
4. Here's, Gary's, money's, businessperson's	9. Lopezes
5. Watkins', year's	10. Dr. Lopez's husband

Replay 71

1. Union's aim—to countries—is
2. grammar—it
3. following:
4. (one blank line).
5. Company (I'll check the address)
6. *Roget's Thesaurus*—a treasury . . . words—was first . . . Roget (look up pronunciation of *Roget*)
7. The 7-11 chain—which . . . -added straws . . .

8. charm—Stella

9. C

10. attributes: money,

11. once—not

12. $152 (not $156) was

13. to live—a crime-free, auto-free, dog-free new island—right

14. companies (fewer than . . . .).

15. government—the executive, the legislative, and the judicial—derive

CHAPTER 13

Recap—Read 72

1. S   **2.** F   **3.** F   **4.** S   **5.** S

Recap

1. R (explanation. You)   **5.** C (service, it)

2. CS (service. Therefore,)   **6.** R (marketing. It)

3. R (service; therefore,)   **7.** CS (marketing; it)

4. C (service. Therefore,)   **8.** C

Replay 72

A.

Mr. Bruce Finlay, President . . . Dear Mr. Finlay: requested, I . . . preferences. Since. . . newsletter, I'm . . . valid. article "Five . . . Heart" Day. Use . . . articles, however, fit. subjects. logo. Then . . . available. Call Teesdale, our . . . artists, . . . need. Sincerely,

B.

1. downtown;   **6.** C

2. it;   **7.** agree,

3. map;   **8.** eat; example, lobster, pasta, . . . sandwiches.

4. C   **9.** first; hungry,

5. working; flow."   **10.** C

C.

1. CS   **2.** C   **3.** CS   **4.** C   **5.** F

Recap—Read 73

1. He finished dinner and returned to work. OR After finishing dinner, he returned to work.

2. If you go to China on a business trip, avoid giving your hosts expensive gifts; they might be embarrassed or even refuse them. (or replace ; with . and capitalize They)

3. Possible solution: When you dine in China, hold the rice bowl close to your mouth and sample every dish offered.

Replay 73 Possible Solutions

1. The beautiful movie star smashed a bottle of champagne over the stern as the ship slid gracefully into the sea.

2. We got up at 6 o'clock and had a quick breakfast.

3. A college pennant and a Monet painting are on the wall of the student's room.

4. A house in need of painting sits far back among the trees.

5. Dennis and Mike spent all afternoon walking to the telephone, gas, and electric company offices to pay their bills.

6. An animal that appeared to be a hyena paced restlessly back and forth in the cage.

7. We realize travelers choose an airline based on the quality of service, and we're sorry we let you down.

8. Enclosed is our check for $635.23 in payment of your May statement.

9. I missed the final examination because I had been out late the night before. Nevertheless, the instructor and the dean refused to change my F grade to Incomplete or Withdrew.

10. Respectable lady seeks comfortable room where she can cook on an electric stove.

11. FOR SALE—Discarded clothing collected by women of the First Presbyterian Church. The clothing may be seen in the church basement any day after six p.m.

12. For those of you who didn't know it, we have a nursery downstairs for small children.

13. This is Professor Loup's friend, who lives in Binghamton. (or rephrase if it's Professor Loup who lives in Binghamton.)

14. Because Mr. Anderson is so unkind, we don't recommend that Mr. Sedirko go to his office. (or rephrase if it is Mr. Sedirko who is unkind?)

15. Despite our repeated reminders about nonpayment of your long overdue account, we have still not received your check.

Recap—Read 74 Possible Answers

1. We sell gas in a glass container to anyone.

2. We sat there in awed silence listening to his singing.

3. Ms. Grigg worked for IBM's Information Systems Department during her vacation.

4. Genevive Astor died at the age of 96 in the home in which she had been born.

5. The fire department brought the fire under control before much damage was done.

Recap—Read 74 Possible Answers

A.

1. With the new software we hope to improve response time, reduce input errors, and identify systems problems more readily.

2. Keyboarding accurately can be more important than keyboarding fast.

3. Linda is a full-time securities analyst, and her husband is a part-time insurance agent.

4. We would appreciate learning your views on how to introduce change, control quality, and motivate employees.

5. Please bring to the meeting our new production schedule, the names of all new employees, and the vacation dates for the administrative assistants.

6. Opthalmologists and optometrists may examine eyes and may also issue prescriptions for glasses or contact lenses.

Replay 74

1. M	3. P	5. P	7. P	9. P	11. M	13. P	15. P	17. P	19. P
2. M	4. M	6. P	8. P	10. C	12. C	14. M	16. C	18. C	20. M

Replay 75

1. P	**3.** P	**5.** A	**7.** A	**9.** A	**11.** A	**13.** A	**15.** P	**17.** P	**19.** A
2. A	**4.** A	**6.** P	**8.** A	**10.** P	**12.** P	**14.** P	**16.** A	**18.** P	**20.** A

Recap—Read 76

While strolling along the beach, you can find unusual shells and pebbles.

Recap

1. Since you are one of our most discriminating customers, we invite you to attend this private showing.
2. When he turned the corner, the new building was right in front of him.
3. While I was using the computer, the cursor became stuck in the middle of the screen.
4. Unlike many others who became millionaires in the '90s, she became financially successful through her talents in fine art and classical music.
5. Before he has dinner with the woman he met through Ultimate Encounters Dating Agency, he needs to improve his table manners.

Replay 76

1. D *Having keyboarded just half the report,*
2. D *On examining the goods,*
3. D *Like many people living in Alaska,*
4. D *Having recovered from his illness,*
5. C
6. D *Before going to lunch,*
7. D *before going to lunch.*
8. D *Walking quickly down the aisle,*
9. D *While doing the daily chores,*
10. D *Handing me the $50,000 order,*
11. D *Having produced a printout,*
12. D *After looking the cars over for a while,*
13. D *Looking marvelously glamorous in a midnight green gown adorned with baguette rhinestones,*
14. D *Upon landing in Dallas,*
15. C
16. C
17. D *Being in dilapidated condition,*
18. D *If invited to dinner at a British home,*
19. C
20. C
21. C
22. D *After standing and repeating the pledge,*
23. D *While walking home,*
24. C *While I was walking home,*
25. D *after being on vacation for two hours,*

CHAPTER 14

Replay 77

1. 8 1/2 by 11 inches	6. proofread, hard copy
2. No. 6 3/4	7. 4
3. 9 or a 10	8. false
4. signature	9. letterfoot
5. 2 . . . 3	10. upper left corner

Replay 78

A.

1. F   2. F   3. F   4. T   5. T   6. T   7. T   8. T   9. F   10. F

B.

1. Date	6. Writer's name
2. Inside Address	7. Writer's title
3. Salutation	8. Typist's initials
4. Body	9. Enclosure notation
5. Complimentary close	10. Blind copy notation

Replay 79

1. T   2. F   3. F   4. F   5. T   6. F   7. F   8. F   9. F   10. T
11. F   12. T   13. F   14. F   15. F   16. T   17. F   18. F   19. T   20. T

Recap—Read 80 Possible Answers

1. Thank you for sending *Writing Better Letters*.
2. Enclosed is the price list you requested.

Recap Possible Answers

1. We hope you will be pleased.
2. We are open from 10 a.m. until 8 p.m. on weekdays and until 5 p.m. on weekends.
3. We'll be glad to give you a credit slip that you can use for any other item in the store.

Recap Possible Answers

1. We suggest you pack the essentials first.
2. The investigation revealed a small, triangular purple UFO.
3. I believe the event is usually in December.
4. Please repeat the instructions.

Replay 80

A.

1. F   2. F   3. F   4. T   5. F

B. Possible Solutions

1. About half the flowers in my garden are blooming.
2. The widget parts you ordered were shipped today via Airborne Express.

3. Our newly designed widgets have been carefully tested for accuracy.

4. We faxed the price list to you yesterday. Let us know if you have any questions.

5. To maintain your good credit rating, please send your check for $6,453 in the enclosed envelope today.

Replay 81

A.

1. T   **2.** T   **3.** F   **4.** T   **5.** F   **6.** F   **7.** T

B.

1. underline, negative space   **4.** bold, text boxes

2. italics, animation   **5.** large font size

3. color, bullets   **6.** font style, sound

Recap—Read 82

1. T   **2.** F   **3.** T

Replay 82

1. F   **2.** F   **3.** F   **4.** F   **5.** F   **6.** F   **7.** F   **8.** F   **9.** T   **10.** F

Index

V

W

Y

Z